AF443201

PHOTON CREATION-ANNIHILATION

Continuum Electromagnetic Theory

PHOTON CREATION-ANNIHILATION

Continuum Electromagnetic Theory

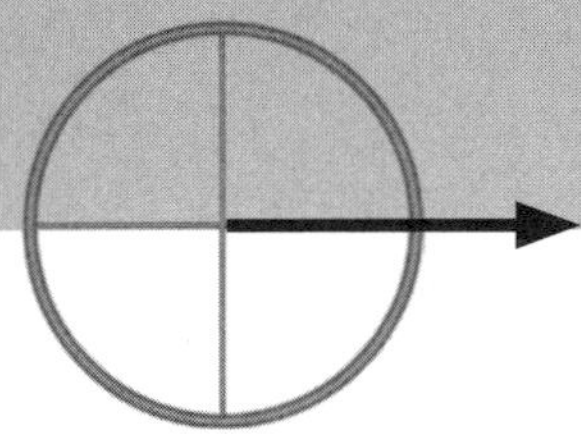

Dale M Grimes

formerly

The University of Michigan, USA

The Pennsylvania State University, USA

Craig A Grimes

Nanjing University of Technology, China

NEW JERSEY · LONDON · SINGAPORE · BEIJING · SHANGHAI · HONG KONG · TAIPEI · CHENNAI

Published by

World Scientific Publishing Co. Pte. Ltd.

5 Toh Tuck Link, Singapore 596224

USA office: 27 Warren Street, Suite 401-402, Hackensack, NJ 07601

UK office: 57 Shelton Street, Covent Garden, London WC2H 9HE

British Library Cataloguing-in-Publication Data
A catalogue record for this book is available from the British Library.

PHOTON CREATION–ANNIHILATION
Continuum Electromagnetic Theory

ISBN-13 978-981-4383-36-3
ISBN-10 981-4383-36-8

Printed in Singapore by World Scientific Printers.

Prelude

Many fundamental questions about light remain unanswered, including: How can a short electromagnetic pulse be monochromatic, long-lived, and fully directed? What is an appropriate photon model? Why does a transitioning electron radiate a monochromatic frequency? Why does a photon diffract and reflect light as a wave but act as a particle in creation and annihilation? What is an appropriate model for a photon-emitting electron? Herein, we address and propose answers to these questions. We find and describe a new method for the emission of an electromagnetic field, a method in stark contrast with those of antennas. It is based upon Maxwell's equations, it produces a unique fully z-directed field set, and it supports a zero-valued radiation Q: there is no reactive- or stored, energy associated with the emitted radiation. Zero Q permits field generation by an electrically small source and it permits the field to exit the source region. Each differential length of that radiation possesses both the electromagnetic and the kinematic properties of a photon and the structures necessary for its creation teaches us how to form an electron model and it, in turn, offers insight to many properties of quantum physics.

Generation of this unique field set requires matched electric and magnetic multipolar modes that increase in magnitude with increasing modal number. Within electrically small sources this, in turn, requires a superregenerative drive; we find the active mechanism is the googol-sized ratio of reactive-to-real power flow in electrically small, high-order multipolar sources. The result is a continuum field theory description of a pulse of fully directed monochromatic energy: a photon. The electron model predicts both a probabilistic time and direction of spontaneous emission and equal probabilities of induced photon creation and annihilation. In a manner foreseen by Einstein, the model places quantum theory within classical theory in a manner analogous with the placement of statistical mechanics within classical mechanics.

Acknowledgments

We are pleased to acknowledge assistance provided by Professor Emeritus J. Samuel Bitler and the ongoing support of Janet L. Grimes.

Contents

Introduction

$\mathcal{M}$any fundamental questions about light remain unanswered after several centuries of carefully crafted scientific theories and experiments. For example: What is an appropriate model for a photon? What is an appropriate model for an active electron? How can a short electromagnetic pulse, *i.e.* a photon, be both monochromatic and long-lived? How can it be fully directed? Derivations, discussion, and suggested answers are the primary subjects of this book.

Comparisons of the corpuscular and wave theories of light have been ongoing since the seventeenth century days of Huygens and Newton. Acceptance of Newton's preferred view of light as corpuscles dominated until the experiments of Young and Fresnel early in the nineteenth century. Those works coupled with Maxwell's 19th mid-century treatise on electromagnetism again seemed to settle the issue, but this time the preferred view was electromagnetic waves. That conclusion seemed even more secure after Hertz used radio waves to communicate between separated objects, objects designed on the basis of Maxwell's equations.

But in 1900 Planck published his derivation of the law that describes electromagnetic radiation within an enclosure and, to do so, found it necessary to add the *ad hoc* requirement that light is exchanged in granules of energy proportional to frequency. Five years later Einstein published an explanation of the newly discovered phenomenon of photoelectricity and that also required energy to be packaged in quantized energy units proportional to the frequency. More than a decade later, 1917, Einstein provided a succinct, energy-based argument for Planck's radiation law and then noted that momentum, like energy, must be conserved in radiation exchanges. He re-derived the radiation law on that basis and showed the two derivations yield the same result if and only if the exchanged energy-to-momentum ratio is the same for emission and absorption. On the basis of the early 20th century understanding of electromagnetic radiation he concluded two things: (1) a quantum theory of radiation is almost unavoidable, and (2) the weakness of the theory is that it does nothing to reconcile the conflicting requirements of continuum and quantum field theories. The Compton effect was reported soon afterward, 1923, and provided confirming and convincing experimental evidence for the corpuscular nature of light.

The dilemma about radiation for those who were trying to reconcile newly discovered facts with continuum electromagnetics is described in Moore's biography of Schrödinger. He cites Heisenberg's book, which, in turn, cites an August 1926 conversation between Schrödinger and Bohr. Schrödinger's points included: {1} if an electron revolves periodically it must radiate, {2} if it changes energy gradually then it must change its rotation frequency gradually, and {3} if it changes suddenly, it should emit a continuous spectrum. Bohr argued that the obvious difference between theory and observation only means we cannot visualize what actually happens.

Eventually attempts at satisfactory explanations ceased and, instead, the radiative pulses were named and their properties recorded. The issue, although of primary importance to Planck, Einstein, and many others of the time, was set-aside. Subsequent work ultimately led to the development of the magnificent structure that today is quantum electrodynamics, see Weinberg for example, but the dichotomy remains as described by Einstein in 1917 and by Schrödinger in 1926. Nonetheless we believe an analytical coupling of continuum and quantum electrodynamics is possible, significant, and timely. In this book we construct and analyze an electron model that satisfies all physical laws and conservation theorems. On the basis of that model we conclude Schrödinger's equation applies only during equilibrium and near-equilibrium conditions; although equilibrium conditions prevail both before and after a quantum jump, the jump itself is not an equilibrium event and therefore his equation does not apply. The same electron model, using continuum fields, predicts creation and absorption of a short electromagnetic pulse that is both monochromatic and fully directed, *i.e.*, a photon.

$\mathcal{T}$o analyze an eigenstate electron as a photon-generating source we first circumscribe a virtual sphere, of radius a, snuggly about the host atom. Next, and without loss of generality, we make a mathematical expansion of the exterior field using a superposition of collocated TE and TM field sources centered on the atom. This model shows radiation cannot be created in a way similar to the way an antenna creates a radiation field. To understand the difficulty with antenna-like radiators, consider that an atomic diameter is on the order of 0.1 nm and the wavelength at mid-optical range is 500 nm. On the circumscribing surface, and with $k = 2\pi/\lambda$, the factor ka is on the order of 0.001. Since the radiation Q of physically small dipoles is $1/(ka)^3$ the real-to-reactive power ratio is on the order 10^9. That is, only one of every 10^9 units of power exchanged between the atom and the field travels into the far field. If the process extends to quadrupoles Q is on the order of 10^{19}, and it continues to increase by about a factor of 10^{10} for every increase in modal order, the large reactive field creates new and dominatingly large forces inside the emitter that squelches the radiation and prevents emission. In other words, Q is an emission turnoff-switch. Electromagnetic pulses obey a parabolic relationship between pulse width and bandwidth, and radiative Q equals the wavetrain-to-wavelength ratio. For radiation with Q on the order of 10^9 and λ on the order of 500 nm the wave train is on the order of 500 m. That is, applying the usual rules governing such pulses shows the wavetrain of a mid-optical photon is about 500 m; for quadrupolar radiation it is 10^{12} meters, a thousand astronomical units! Radiation is Q-trapped.

Our analysis begins by imposing the kinematic properties of a photon as a boundary condition on the exterior fields; this permits us to evaluate, within an arbitrary constant, each of the infinite number of constant multiplying coefficients in the field expansion. We note from the uniqueness and completeness theorems that the resulting expression is both complete and unique: there can be no other independent analytical description of radiation with the kinematic properties of a photon. It has the form of an infinite sum over multipolar sources collocated on the surface of the atom-encompassing virtual sphere. Each modal term describes a matched TE and TM field and each is proportional to a spherical Hankel function and hence to $\exp[i(\omega t - kr)]$; analysis shows that the sum, quite differently from individual terms, is z-directed and proportional to $\exp[i(\omega t\text{-}kz)]$. Most significantly, the sum is not subject to the Q-trap. Field generation requires that (1) all electric moments are parallel, (2) all magnetic moments are parallel, (3) the two sets of moments are spatially orthogonal, (4) all moments obey a unique recursion relationship, and (5) the entire system of moments rotates about an axis perpendicular to the moment directions at radian frequency ω. Under these very special conditions the radiation Q of fields emanating from an electrically small source is zero; unlike antennas, both near and far fields contribute to the output pulse.

$\mathcal{T}$he general structure of this book is as follows: Beginning from first principles, the cosmological theorem, **Chapter 1** shows that the science of electromagnetism is a fundamental property of the universe and hence an analysis of them is an appropriate basis for understanding atomic to galactic sized events. **Chapter 2** introduces the fundamental techniques and tools necessary for a critical examination of radiation properties, including fields, power, field stresses, radiation reaction forces, and both linear and angular momenta. The chapter includes a discussion of stresses on the surface of a source of multipolar radiation.

The analysis techniques and tools introduced in **Chapter 2** are put to use in **Chapter 3** and **Chapter 4.** They provide mathematically exact solutions of both transmitting and receiving biconical antennas. Biconical antennas are the only antenna embodiment that approximates linear dipole antennas at one extreme and spherical antennas at the other, and for which exact field expansions are available. They arguably serve as a quintessential example of the power of electromagnetic analysis techniques. Expressions are derived for the input impedance and for the full set of electric and magnetic fields at all points about a biconical antenna with perfectly conducting arms and caps. Surface currents, pattern directivity, field power density, *etc.* follow from the fields. The mathematical techniques inherent to the analysis of biconical antennas are critical to the rest of the book. Taken together, the two chapters demonstrate the full power of electromagnetic analysis techniques and serve as a foundation for the analysis of general radiation sources.

Now, nearly a century after quantum mechanics became a fundamental branch of physics, many basic questions remain unanswered. For example, is quantum mechanics a separate base upon which classical physics is built or is it a derivable result of classical physics on the atomic level of dimensions? In **Chapter 5** our electron model provides a statistical basis for quantum theory that, in turn, leads to a modified interpretation of Schrödinger's equation. The resultant interpretation of nonrelativistic quantum mechanics follows and is discussed in Sec. 5.1 through 5.7. The remainder of the chapter is a brief review of subjects important to our later arguments. The first four sections of **Chapter 6** are devoted to Planck's black body radiation law, its implications, and the background electromagnetic radiation field; later sections are dedicated to understanding radiation properties associated with atoms. We next consider many-electron systems and the exclusion principle, first formulated by Pauli, stating that only antisymmetric wave functions exist. From this we find that, on some dimensional scale, an electron's internal charge distribution is granular.

Both the properties of matched pairs of multipolar radiation sources and sums over them are considered in **Chapter 7**. The radiation and kinematic properties of matched pairs are quite different from those of individual ones. This is a key result since the lack of appreciation of that difference played a significant role in the early history of the quantum theory of radiation. In his 1917 paper Einstein first showed that Planck's equation for the thermal radiation field requires energy exchanges to be fully directed and then stated that Maxwell's equations show the radiation has spherical (*sic*) symmetry. On this basis he concluded quantum and classical results are inconsistent and a separate quantum theory of radiation is required to explain observed phenomena. It was 43 years later when Harrington, in quite a different context, showed that matched multipolar radiators could create directed radiation. In **Chapter 7** we detail a continuum radiation field that creates the kinematic pulse properties Einstein showed were necessary to explain quantum properties. In the first five sections of **Chapter 7** we analyze increasingly inclusive, but always continuum, radiation fields. Our calculations show there is porthole on the z-axis that permits zero-Q radiation to emanate. That is, independently of the Q per individual multipolar pair, the sums of Eq. 7.4.1 describe fully directed radiation and that radiation supports neither reactive power nor standing energy. The angular distribution of angular fields forms a Dirac delta function on the z-axis, $\delta(\theta,0)$, and dramatically different from antennas the entire electromagnetic radiation field, all fields near and far, contribute to the output. The event is uniquely described by the sums of Eq. 7.4.1.

For an electrically small source to radiate it is necessary to somehow create the field coefficients of Eq. 7.4.1. We detail a method for doing so in **Chapter 8**. In Sec. 8.1 we also show that the Manley Rowe equations satisfactorily explain the observed power-frequency relationships, but if and only if internal electronic frequencies interact nonlinearly. In Sec. 8.2 we show that intermodal and modal energies of Eq. 7.4.1 are nearly equal and that forces resulting from intermodal energies can drive the system to ever-higher orders. In Sec. 8.3 and 8.4 we discuss positive feedback and its dependence upon ordered internal field structures; Sec. 8.6–8.10 show that the fields of Eq. 7.4.1 uniquely describe ordered structures. The combination provides a superregenerative buildup of modal orders that result in the fields of Eq. 7.4.1. That equation, in turn and as described in Sec. 7.11, describes the ejection of a zero-Q, fully directed pulse radiation; there is no lower limit on the size-to-wavelength ratio of such radiation.

In **Chapter 9** we conclude it is not feasible for the emitted radiation, *per se*, to support observed photon directivity; there must be something else. Therefore in Sec. 9.1 we create a photon model that supports the appropriate directivity. In Sec. 9.2 we give an example of a static binding force created by standing waves, and in Sec. 9.3 we use the photon model and the attractive photon-to-eigenstate electron force to examine photon creation and annihilation. In Sec. 9.4 we discuss how these results all follow from the appropriate electron model.

Our results are set within an historical perspective and summarized in **Chapter 10**.

References

A.H. Compton, *A Quantum Theory of the Scattering of X-Rays by Light Elements*, Phys. Rev., vol. 21, pp. 483–522 (1923)

A. Einstein, "The Quantum Theory of Radiation," *Phys. Z.*, vol. 18, 121–128 (1917); also in *The World of the Atom*, H. A. Boorse and L. Motz, eds., Basic Books (1966) p. 888–901

R.F. Harrington, "Effect of Antenna Size on Gain, Bandwidth, and Efficiency," J. Research, Nat. Bureau of Standards, vol. 64D, pp. 1–12 (1960)

J.C. Maxwell, *A Treatise on Electricity and Magnetism*, Dover Publicaitons, (1954); an unaltered replication of the third edition (1891)

W. Heisenberg, *Der Teil und das Ganze*, Munich, Piper Verlag (1969)

W. Moore, *Schrödinger, Life and Thought*, Cambridge University (1989)

M. Planck, "The Origin and Development of the Quantum Theory," Nobel Prize in Physics Address, 1919, in *The World of the Atom*, H. A. Boorse and L. Motz, eds., Basic Books (1966) pp. 491–501

S. Weinberg, *The Quantum Theory of Fields*, Cambridge University (1995)

Chapter 1

Classical Electrodynamics

1.1 Introductory Comments

We begin with, but *a posteriori* discard, the cosmological theorem as a basis for the special theory of relativity and then use the special theory as the basis for the science of electromagnetism. After deriving the underlying laws we use them to show that electromagnetic mass obeys Newton force laws; in this sense and for electromagnetic mass the laws of mechanics are a derived result of electromagnetism. Relativity shows that all energies obey the same force laws. Although our approach is conventional the order of presentation is not. Typically discussions follow the chronological order of development: first classical mechanics and then electromagnetics.

Our approach to electromagnetism is deductive; we begin with a single relativistic source potential and deduce from it the full slate of classical equations of electromagnetism. In 1959 Aharonov and Bohm, using the premise that potential has a special significance, predicted an effect that was confirmed in 1960. The Aharonov-Bohm effect is that a static magnetic potential is effective even in a region void of force fields. Therefore the potential has special significance in its own right and since that is true in the deductive approach, but not the inductive one, the deductive approach has special significance.

To begin, imagine a universe void of condensed matter but containing light. What is the speed of the light? Since there is no reference frame by which to measure it, the question is moot. Consider, if we may, an isolated asteroid just large enough to support a fully equipped observer, who uses it to determine the speed of light as v_A. Since there is nothing else in the universe, a question about the speed of the asteroid is moot. Next introduce a second asteroid that is identical to the first but far

1

enough away from it so the two are independent by any means of which we are currently aware. An observer on the second asteroid determines the speed of light as v_B. How will the measured values compare? By the Cosmological Principle, an experiment run in one local four-space yields the same results as an identical experiment run in a different local four-space. Therefore, on the basis of the Cosmological Principle we conclude that $v_A = v_B = c$.

Next, let the asteroids be in the same local region. Either the speeds depend upon the magnitude of the local masses or they do not, and if they do not there is no change in speed. However, in the local region, a relative speed between identical asteroids A and B may be determined. Since one asteroid cannot be preferred over the other in an otherwise empty universe, the two observers continue to measure the same speed. This condition requires the speed of light to be independent of the relative speed of the system on which it is measured. Next, bring in other material, bit by bit, until the universe is in its present form, and the conclusion remains the same. The speed of light is independent of the speed of the object on which it is measured, independently of the speed of other objects.

1.2 Space and Time Dependence upon Speed

Let a pulse of light be emitted from an origin in reference frame F and observed in reference frame F'. If the speed of light is the same in all reference frames, if the two frames are in relative motion, and if the origins coincide at the time the light is emitted, the light positions as measured in the two frames are:

$$x^2 + y^2 + z^2 - c^2t^2 = x'^2 + y'^2 + z'^2 - c'^2 t'^2$$

$$(1.2.1)$$

If the relative speed is such that F' is moving at speed v in the z-direction with respect to F, then at low speeds:

$$x' = x; \qquad y' = y; \qquad z' = (z - vt); \qquad t' = t \qquad (1.2.2)$$

Since Eq. 1.2.1 isn't satisfied by Eq. 1.2.2, it follows that Eq. 1.2.2 doesn't extend to speeds that are a significantly large fraction of c. To obtain a transition that is linear in the independent variables, and that goes to Eq. 1.2.2 in the low speed limit, consider the linear transformation form:

$$x' = x; \quad y' = y; \quad z' = \gamma(z - vt); \quad t' = At + Bz \qquad (1.2.3)$$

Parameters γ, A, and B are undetermined but independent of both position and time. Since Eq. 1.2.3 approaches Eq. 1.2.2 in the limit of velocity v much less than c, in that limit:

$$\gamma = 1; \quad A = 1; \quad B = 0 \qquad (1.2.4)$$

Since the coordinates are independent variables, combining Eq. 1.2.1 and 1.2.3 and solving shows that:

$$z^2\left(\gamma^2 - 1 - c^2 B^2\right) = 0; \quad t^2\left(c^2 + \gamma^2 v^2 - c^2 A^2\right) = 0$$
$$zt\left(\gamma^2 v + ABc^2\right) = 0 \qquad (1.2.5)$$

Solving Eq. 1.2.5 yields:

$$A = \gamma = \left(1 - v^2/c^2\right)^{-1/2}; \quad B = -\frac{\gamma v}{c^2} \qquad (1.2.6)$$

Combining yields the Lorentz transformation equations:

$$x' = x; \quad y' = y; \quad z' = \gamma(z - vt); \quad t' = \gamma\left(t - \frac{vz}{c^2}\right) \qquad (1.2.7)$$

This transformation preserves the speed of light in inertial frames.

Equation 1.2.7 is a sufficient basis upon which to determine results if events in one frame of reference are observed in another one. Let the observer be in the unprimed frame. A stick of length L_0 as determined in the moving frame, in which it is stationary, lies along the z-axis. It moves at speed v past the observer in the z-direction. A flash of light illuminates the region, during which time the observer determines the positions of the ends of the moving stick as z_1 and z_2. It follows from Eq. 1.2.7 that the measured positions are:

$$z'_1 = \gamma(z_1 - vt_0) \quad \text{and} \quad z'_2 = \gamma(z_2 - vt_0) \tag{1.2.8}$$

The length as measured in the stationary frame is:

$$L = (z_2 - z_1) = \left(z'_2 - z'_1\right)/\gamma = L_0/\gamma \tag{1.2.9}$$

It follows that:

$$L = L_0\left(1 - v^2/c^2\right)^{1/2} \leq L_0 \tag{1.2.10}$$

The observed length of the stick is less than is measured in the rest frame; this fractional contraction is the Lorentz contraction.

Next pulses of light are issued at times t'_2 and t'_1, again in the moving frame. When does a stationary observer see them, and what is the time interval between them? Using Eq. 1.2.7:

$$t'_2 = \gamma\left(t_2 - vz_2/c^2\right) \quad \text{and} \quad t'_1 = \gamma\left(t_1 - vz_1/c^2\right) \tag{1.2.11}$$

From Eq. 1.2.11 the time difference in the frame at which the two sources are stationary is:

$$T_0 = t'_2 - t'_1 = \gamma\left[(t_2 - t_1) - v(z_2 - z_1)/c^2\right] = \gamma T\left(1 - v^2/c^2\right) \tag{1.2.12}$$

T is the time measured in the stationary frame. Solving for T gives:

$$T = \gamma T_0 = \frac{T_0}{\left(1 - v^2/c^2\right)^{1/2}} \geq T_0 \tag{1.2.13}$$

The observer measures a greater time duration between pulses than is measured in the rest frame; this time expansion is time dilatation.

1.3 Four-Dimensional Space-Time

The equality of the speed of light in all inertial frames is the basis for a system of 4-vectors. Let x_1, x_2, x_3 represent the three spatial axes x, y, z of three dimensions and $x_4 = ict$ where $i = \sqrt{-1}$. The four space-time dimensions are:

$$(x_1, x_2, x_3, x_4) \tag{1.3.1}$$

Since three of the axes determine lengths and one determines time, a three-dimensional rotation represents a change in spatial orientation and a four-dimensional rotation includes a change in time. Such four-dimensional rotations are Lorentz transformations. These transformations are usually simple and contain a high degree of symmetry. Such transformations are covariant with respect to changes in coordinate systems: that is, an equation that represents reality in one reference frame has the same form in all other inertial frames.

The imaginary character of the fourth dimension represents an essential difference from spatial ones: the squares of the space coefficients and time coefficients have different signs. For notational purposes we use Roman or Greek subscripts to indicate, respectively, three- or four-dimensional tensors. For example, the rotation matrix element in four dimensions is $c_{\mu\nu}$ where, for velocities v directed along the x_1-axis:

$$c_{\mu\nu} = \begin{pmatrix} \gamma & 0 & 0 & i\gamma v/c \\ 0 & 1 & 0 & 0 \\ 0 & 0 & 1 & 0 \\ -i\gamma v/c & 0 & 0 & \gamma \end{pmatrix} \tag{1.3.2}$$

Four-dimensional and three-dimensional direction cosines follow similar laws:

$$c_{\mu\nu} = c'_{\nu\mu} \ ; \ c_{\mu\nu}c_{\mu\rho} = \delta_{\nu\rho} \ ; \ \det|c_{\mu\nu}| = 1 \tag{1.3.3}$$

The Lorentz direction cosines $c_{\mu\nu}$ are:

$$x'_{\mu} = c_{\mu\nu}x_{\nu} \tag{1.3.4}$$

The proper time interval, $\Delta\tau$, between two events with space-time coordinates spaced Δx_α apart is defined to be:

$$\left(\Delta\tau\right)^2 = -\frac{1}{c^2}\Delta x_\alpha \Delta x_\alpha \tag{1.3.5}$$

Using three-dimensional notation, the proper time difference is

$$\left(\Delta\tau\right)^2 = \left(\Delta t\right)^2 - \frac{\left(\Delta x\right)^2 - \left(\Delta y\right)^2 - \left(\Delta z\right)^2}{c^2} = \left(\Delta t\right)^2 - \frac{\left(\Delta r\right)^2}{c^2} \tag{1.3.6}$$

$$\left(\Delta r\right)^2 = \left(\Delta x\right)^2 + \left(\Delta y\right)^2 + \left(\Delta z\right)^2$$

Since $\left(\Delta\tau\right)^2$ can be zero, positive, or negative, $\Delta\tau$ may be zero, real, or imaginary. Since the speed of light is the same in all reference frames, by Eq. 1.2.1 the proper time is also the same in all reference frames. If it is real, it is "time-like" and if imaginary, it is "space-like". If time-like, the proper time is the time separation of the two events in the same frame. If space-like, there is a frame in which c times the proper time is the spatial separation of the two events that are simultaneous in that frame.

With τ as proper time, consider the 4-vector defined by the expression:

$$U_\mu = \frac{dx_\mu}{d\tau} \tag{1.3.7}$$

Since both x_μ and τ are independent of details of a specific inertial frame, so is U_μ and hence U_μ is a 4-vector with the four components:

$$U_1 = \frac{dx}{d\tau} = \frac{dx}{dt}\frac{dt}{d\tau} = \gamma v_x \quad ; \quad U_2 = \frac{dy}{d\tau} = \frac{dy}{dt}\frac{dt}{d\tau} = \gamma v_y$$

$$U_3 = \frac{dz}{d\tau} = \frac{dz}{dt}\frac{dt}{d\tau} = \gamma v_z \quad ; \quad U_4 = \frac{d\left(ict\right)}{d\tau} = \gamma ic \tag{1.3.8}$$

The three-dimensional velocity components are v_i and the 4-velocity components are U_μ.

A particle of mass m_0 with 4-velocity U_μ has 4-momentum given by:

$$P_\mu = m_0 U_\mu \qquad (1.3.9)$$

Combining shows the momentum components to be:

$$\boldsymbol{p} = \gamma m_0 \boldsymbol{v} \quad ; \quad p_4 = \gamma m_0 i c = iW/c \quad ; \quad W = \gamma m_0 c^2 \qquad (1.3.10)$$

The quantity W, defined by Eq. 1.3.10, is the energy associated with the moving mass.

The binomial expansion is:

$$(1 \pm a)^n = 1 \pm na + \frac{n}{2!}(n-1)a^2 \pm \dots \qquad (1.3.11)$$

This equation combines with the definition of γ, see Eq. 1.2.6, to show that:

$$\gamma = 1 + \frac{v^2}{2c^2} + \frac{3v^4}{8c^4} + \dots \qquad (1.3.12)$$

Combining Eq. 1.3.10 and 1.3.12 shows the total energy of the particle:

$$W = m_0 c^2 \left[1 + \frac{v^2}{2c^2} + \frac{3v^4}{8c^4} + \dots \right] \qquad (1.3.13)$$

The first term of Eq. 1.3.13 is the rest mass, the second term is the familiar kinetic energy at low speeds, and the remaining higher-order terms describe additions to high-speed kinetic energy.

The equation, first published by Einstein, shows that a particle's rest mass is a form of energy. An additional implication is that all forms of energy have a mass equivalence.

A note about notation: A 4-dimensional spatial vector, such as shown in Eq. 1.3.1, is denoted by the symbol x_α, where α represents an integer from 1 to 4. Such vectors are covariant in all inertial frames; with this choice of notation the time coordinate is imaginary. An alternate way of describing the same physical phenomena uses notation symbolized by x^α and called a contravariant vector. With the aid of metric $\eta_{\alpha\beta} = \eta^{\alpha\beta} = \delta_{\alpha\beta}$, where $\delta_{\alpha\beta}$ is the 4-dimensional unit matrix, the covariant and contravariant vectors are related to distance formula Eq. 1.2.1 as $x_\alpha \, \eta_{\alpha\beta} \, x^\alpha = x_\alpha \, x_\alpha$. Generally speaking vector calculus using covariant vectors is simpler to use in inertial reference frames but its usefulness does not extend to accelerating frames or to the curved space-times of gravitational fields. In this work we restrict ourselves to inertial frames and hence to the notation of Eq. 1.3.1.

1.4 Newton's Laws

The Minkowski force is defined to be:

$$F_\mu = \frac{d}{d\tau} P_\mu \qquad (1.4.1)$$

This force is a 4-vector with the x-directed component:

$$F_1 = \frac{d}{d\tau}\left(m_0 U_1\right) = \gamma \frac{\partial}{\partial t}\left(\gamma m_0 v_x\right) \qquad (1.4.2)$$

The corresponding three-dimensional force component is:

$$F_x = \frac{\partial}{\partial t}\left(\gamma m_0 v_x\right) \qquad (1.4.3)$$

This equation is a statement Newton's law. The factor γ in Eq. 1.4.3 was known before the full relativistic effect was understood. Although relativity makes it abundantly clear that the result is a space-time effect, it was historically interpreted as an increase in mass whereby the effective mass m is a function of speed:

$$m = \gamma m_0 \qquad (1.4.4)$$

The nomenclature remains and by definition the effective mass of a moving particle is that of Eq. 1.4.4.

Since the 4-momentum is a 4-vector, it is conserved between Lorentz frames. That is,

$$W_0^2 = W^2 - p^2 c^2 \tag{1.4.5}$$

The energy of a free particle is related to momentum. In any specific frame:

$$W^2 = m_0^2 c^4 + p^2 c^2 \tag{1.4.6}$$

Since W is second order in v/c, three-momentum is constant in low speed inertial frames. Energy is also nearly conserved. However, in high-energy systems only the combination of energy and momentum is conserved. This example illustrates the general characteristic of 4-tensors that at low speeds the real and imaginary parts are separately conserved but at high speeds it is the combination of the two quantities that is conserved.

Since, in this development, we seek only positive energies the energy-momentum dispersion relationship may be written:

$$W = \sqrt{m_0^2 c^4 + p^2 c^2} \tag{1.4.7}$$

For particles with zero rest mass and for those with very large kinetic energies the first term is negligibly small and the particle's energy is linearly proportional to its momentum:

$$\Delta W = c \Delta p \tag{1.4.8}$$

For particles moving slowly enough so the normalized kinetic energy is much less than the rest mass, keeping only the first term a similar expansion of Eq. 1.4.7 shows the energy-momentum relationship is quadratic:

$$\Delta W \equiv W - W_0 = \frac{1}{2m_0} \Delta\left(p^2\right) \tag{1.4.9}$$

This is the energy-momentum relationship of a free particle and with it Newtonian mechanics obeys the Lorentz transformation laws.

A clear implication of Eq. 1.3.13 is that rest mass is a form of energy and that, in turn, implies that all forms of energy have a mass equivalence. An equally clear implication of Eq. 1.4.8 is that electromagnetic fields have momentum. These conclusions blur pre-relativistic ideas about particles and waves: particles are not small and rigid objects and fields are not simply a means of explaining action at a distance.

1.5 Electrodynamics

The three defined scalars are: speed c, time interval τ between events in a rest frame, and mass m_0. A fourth is electric charge q; electric charge can have either sign. An intrinsic property of mass is the associated gravitational field, G, and similarly an intrinsic property of charge is the associated 4-vector potential field A_μ. If individual charges are much smaller than other dimensions and if there are many of them consider a differential volume, with dimensions (x_1, x_2, x_3), in which each dimension is much less than any macroscopic dimension of interest yet the volume contains large numbers of charges. If all conditions are met the tools of calculus apply. Charge density ρ is defined to be the charge per unit volume at a point. Charge density ρ_0 is defined in a frame in which the time-average position is at rest. Observers in fixed and moving frames see the same total charge but, because of the Lorentz contraction, the moving observer determines the volume containing it to be smaller by a factor of γ. Therefore, the charge density in a moving frame is increased by the factor:

$$\rho = \gamma \rho_0 \tag{1.5.1}$$

If the charge density moves with 4-velocity U_μ, define the 4-current density in a way similar to three dimensions:

$$J_\mu = \rho_0 U_\mu = \{\gamma \rho_0 \mathbf{v}, \gamma i c \rho_0\} = \{\mathbf{J}, i c \rho\} \tag{1.5.2}$$

The vector terms within the curly brackets, identified by bold font, indicate the first three dimensions and the scalar term represents the fourth dimension. The 4-divergence of the current density is:

$$\frac{\partial J_\mu}{\partial X_\mu} = \nabla \cdot \boldsymbol{J} + \frac{\partial \rho}{\partial t} = 0 \tag{1.5.3}$$

The first equality of Eq. 1.5.3 follows from definition of terms and the second is true if and only if net charge is neither created nor destroyed; pair production or annihilation may occur but there is no change in the total charge. The zero 4-divergence shows that the net change in the four-current is always equal to zero, there is no change in total amount of charge, and hence charges are created and destroyed only in canceling pairs. It is, therefore, defined to be the continuity equation.

The 4-vector potential field $A_v(X_\beta)$ is defined as the potential that satisfies the differential equation:

$$\frac{\partial^2 A_v}{\partial X_\beta \partial X_\beta} = -\mu J_v \tag{1.5.4}$$

By definition constant μ is a property of empty space. It is also a dimension determining constant and defined as $4\pi/10^7$ henrys/meter. Although the recommended symbol is μ_0, in this work we deal dominantly with fields in free space and hence for simplicity we use simply μ.

Taking the 4-divergence of Eq. 1.5.4 and then combining with Eq. 1.5.3 gives:

$$\frac{\partial}{\partial X_v}\frac{\partial^2}{\partial X_\beta \partial X_\beta} A_v = \frac{\partial^2}{\partial X_\beta \partial X_\beta}\frac{\partial A_v}{\partial X_v} = -\mu\frac{\partial J_v}{\partial X_v} = 0$$

Combining shows:

$$\frac{\partial A_\mu}{\partial X_\mu} = 0 \tag{1.5.5}$$

Equation 1.5.5 shows that the divergence of A_v is zero, from which it follows that, like charge, the total amount of 4-potential doesn't change.

If transitions are made between different reference frames changes occur in the components of the potential but not in the sum over all four components.

The four-dimensional Laplacian of Eq. 1.5.4 may be integrated over all space to obtain an expression for the potential itself. By Eq. A.6.2 the potentials due to a moving set of charges are:

$$A_1, A_2, A_3 \Rightarrow \boldsymbol{A}(\boldsymbol{r},t) = \frac{\mu}{4\pi} \int \frac{\boldsymbol{J}'}{\left[R - \boldsymbol{R} \cdot \boldsymbol{v}/c\right]} dV'$$

$$-icA_4(\boldsymbol{r},t) \Rightarrow \Phi(\boldsymbol{r},t) = \frac{\mu c^2}{4\pi} \int \frac{\rho'}{\left[R - \boldsymbol{R} \cdot \boldsymbol{v}/c\right]} dV'$$

$$(1.5.6)$$

The 3-dimensional integrals are taken over all source-bearing regions, dV' is differential volume at source point $\boldsymbol{r}'$, position and time at the field point is $\boldsymbol{r},t$ and retarded time, $t_R = t - R/c$ where R is the source-to-field point distance and R/c is the transit time for the radiation.

It is convenient to introduce constant ε, the permittivity of free space, defined as $\varepsilon = 1/(\mu c^2)$ farads/meter. Although, as with μ, the recommended symbol is ε_0 since in this work we deal dominantly with fields in free space for simplicity we use simply ε.

For a single point charge in free space the integral equations are:

$$A(\boldsymbol{r},t) = \frac{\mu}{4\pi} \frac{q\boldsymbol{v}'}{\left[R - \boldsymbol{R} \cdot \boldsymbol{v}'/c\right]}; \quad \Phi(\boldsymbol{r},t) = \frac{1}{4\pi\varepsilon} \frac{q}{\left[R - \boldsymbol{R} \cdot \boldsymbol{v}'/c\right]} \qquad (1.5.7)$$

These are the three-dimensional vector and scalar potential fields of individual charges.

It is apparent from Eq. 1.5.7 that a charge moving towards or away from a field point generates potentials with magnitudes respectively larger or smaller than the low speed value.

1.6 The Field Equations

If ρ_0 is the charge density in an inertial reference frame in which the average speed of the charges is zero, then $\rho = \gamma \rho_0$ is the charge density in a moving frame. The charge density and three-dimensional current density J_i were extended to form the 4-current density, as shown by Eq. 1.5.2, from which the Laplacian of the 4-potential was defined by Eq. 1.5.4. Other useful 4-tensors follow from 4-dimensional operations on the 4-potential A_α (X_γ), and significant ones follow.

A second rank antisymmetric tensor of physical interest is defined by the operation on the potential:

$$f_{\alpha\beta} = \frac{\partial A_\beta}{\partial X_\alpha} - \frac{\partial A_\alpha}{\partial X_\beta} \tag{1.6.1}$$

This 4-dimensional curl operation is the field strength tensor; it is a measure of the curvature of the 4-potential. Antisymmetric 4-tensors are spatial arrays of six numbers and, in common with all antisymmetric tensors, the trace is zero:

$$f_{\alpha\alpha} = 0 \tag{1.6.2}$$

Writing out the six components that appear in the upper right portion of Eq. 1.6.1 and using Eq. 1.5.6 to define function Φ gives:

$$f_{12} = \frac{\partial A_2}{\partial X_1} - \frac{\partial A_1}{\partial X_2} = \frac{\partial A_y}{\partial x} - \frac{\partial A_x}{\partial y} = B_z$$

$$f_{23} = \frac{\partial A_3}{\partial X_2} - \frac{\partial A_2}{\partial X_3} = \frac{\partial A_y}{\partial z} - \frac{\partial A_z}{\partial y} = B_x$$

$$f_{31} = \frac{\partial A_1}{\partial X_3} - \frac{\partial A_3}{\partial X_1} = \frac{\partial A_x}{\partial z} - \frac{\partial A_z}{\partial x} = B_y$$

$$\tag{1.6.3}$$

$$f_{14} = \frac{\partial A_4}{\partial X_1} - \frac{\partial A_1}{\partial X_4} = \frac{i}{c}\frac{\partial \Phi}{\partial x} + \frac{i}{c}\frac{\partial A_x}{\partial t} = -\frac{i}{c}E_x$$

$$f_{24} = \frac{i}{c}\frac{\partial \Phi}{\partial y} + \frac{i}{c}\frac{\partial A_y}{\partial t} = -\frac{i}{c}E_y$$

$$f_{34} = \frac{i}{c}\frac{\partial \Phi}{\partial z} + \frac{i}{c}\frac{\partial A_z}{\partial t} = -\frac{i}{c}E_z$$

With the deductive approach to electromagnetism Eq. 1.6.3 is the definition of field vectors $\boldsymbol{E}$ and $\boldsymbol{B}$. By inspection the 4-tensor of Eq. 1.6.1 using 3-dimensional quantities is:

$$\left(f_{\alpha\beta}\right) = \begin{pmatrix} 0 & B_z & -B_y & -iE_x/c \\ -B_z & 0 & B_x & -iE_y/c \\ B_y & -B_x & 0 & -iE_z/c \\ iE_x/c & iE_y/c & iE_z/c & 0 \end{pmatrix} \tag{1.6.4}$$

Differentiating $f_{\alpha\beta}$ with respect to X_β results in the equality chain:

$$\frac{\partial f_{\alpha\beta}}{\partial X_\beta} = \frac{\partial}{\partial X_\beta}\left(\frac{\partial A_\beta}{\partial X_\alpha} - \frac{\partial A_\alpha}{\partial X_\beta}\right) = \frac{\partial^2 A_\beta}{\partial X_\beta \partial X_\alpha} - \frac{\partial^2 A_\alpha}{\partial X_\beta \partial X_\beta} = \mu J_\alpha \tag{1.6.5}$$

Combining terms gives, after some algebra:

$$\frac{\partial f_{\alpha\beta}}{\partial X_\beta} = \mu J_\alpha \tag{1.6.6}$$

Evaluation of Eq. 1.6.6 in 3-dimensional terms gives:

$$\frac{\partial f_{1\beta}}{\partial X_\beta} = \frac{\partial B_z}{\partial y} - \frac{\partial B_y}{\partial z} - \frac{1}{c^2}\frac{\partial E_x}{\partial t} = \mu J_x$$

$$\frac{\partial f_{2\beta}}{\partial X_\beta} = \frac{\partial B_x}{\partial z} - \frac{\partial B_z}{\partial x} - \frac{1}{c^2}\frac{\partial E_y}{\partial t} = \mu J_y$$

$$\tag{1.6.7}$$

$$\frac{\partial f_{3\beta}}{\partial X_\beta} = \frac{\partial B_y}{\partial x} - \frac{\partial B_x}{\partial y} - \frac{1}{c^2}\frac{\partial E_z}{\partial t} = \mu J_z$$

$$\frac{c}{i}\frac{\partial f_{4\beta}}{\partial X_\beta} = \frac{\partial E_x}{\partial x} + \frac{\partial E_y}{\partial y} + \frac{\partial E_z}{\partial z} = \frac{\rho}{\varepsilon}$$

These are the nonhomogeneous Maxwell equations which, written in three-dimensional notation, are:

$$\nabla \times \boldsymbol{B} - \frac{1}{c^2}\frac{\partial \boldsymbol{E}}{\partial t} = \mu \boldsymbol{J}; \quad \varepsilon \nabla \cdot \boldsymbol{E} = \rho \tag{1.6.8}$$

These source equations relate electric field vector $\boldsymbol{E}$ to source ρ and magnetic field vector $\boldsymbol{B}$ to sources $\boldsymbol{J}$ and the time partial of $\boldsymbol{E}$.

Another coordinate-invariant law for electromagnetic fields follows using from the definition of $f_{\alpha\beta}$:

$$\frac{\partial f_{\nu\sigma}}{\partial X_\alpha} + \frac{\partial f_{\sigma\alpha}}{\partial X_\nu} + \frac{\partial f_{\alpha\nu}}{\partial X_\sigma} = 0 \tag{1.6.9}$$

Evaluation of Eq. 1.6.9 for each tensor component gives:

$$\frac{\partial f_{12}}{\partial X_3} + \frac{\partial f_{23}}{\partial X_1} + \frac{\partial f_{31}}{\partial X_2} = \frac{\partial B_z}{\partial z} + \frac{\partial B_x}{\partial x} + \frac{\partial B_y}{\partial y} = 0$$

$$\frac{\partial f_{24}}{\partial X_1} + \frac{\partial f_{41}}{\partial X_2} + \frac{\partial f_{12}}{\partial X_4} = \frac{1}{ic}\left(\frac{\partial E_y}{\partial x} - \frac{\partial E_x}{\partial y} + \frac{\partial B_z}{\partial t} \right) = 0$$

$$\frac{\partial f_{34}}{\partial X_2} + \frac{\partial f_{42}}{\partial X_3} + \frac{\partial f_{23}}{\partial X_4} = \frac{1}{ic}\left(\frac{\partial E_z}{\partial y} - \frac{\partial E_y}{\partial z} + \frac{\partial B_x}{\partial t} \right) = 0$$

$$\frac{\partial f_{14}}{\partial X_3} + \frac{\partial f_{43}}{\partial X_1} + \frac{\partial f_{31}}{\partial X_4} = \frac{1}{ic}\left(\frac{\partial E_x}{\partial z} - \frac{\partial E_z}{\partial x} + \frac{\partial B_y}{\partial t} \right) = 0$$

$$\tag{1.6.10}$$

These are the homogeneous Maxwell equations which, written in three-dimensional notation, are:

$$\nabla \times \boldsymbol{E} + \frac{\partial \boldsymbol{B}}{\partial t} = 0; \quad \nabla \cdot \boldsymbol{B} = 0 \tag{1.6.11}$$

The equations show that $\boldsymbol{B}$ has only the solenoidal source of Eq. 1.6.8 and electric field $\boldsymbol{E}$ has as a solenoidal source the time partial of magnetic field $\boldsymbol{B}$.

Another useful 4-vector is the volume force density as defined by the equation:

$$F_\alpha^\nu = f_{\alpha\beta} J_\beta \tag{1.6.12}$$

Superscript 'v' indicates volume density and is not a tensor index. This equation, Eq. 1.6.12, relates volume force density to the interaction of fields with charges. Term-by-term evaluation of each component of gives:

$$F_1^v = F_x^v = J_y B_z - J_z B_y + \rho E_x$$
$$F_2^v = F_y^v = J_z B_x - J_x B_z + \rho E_y$$
$$F_3^v = F_z^v = J_x B_y - J_y B_x + \rho E_z \tag{1.6.13}$$
$$F_4^v = i\left(E_x J_x + E_y J_y + E_z J_z\right)/c$$

In three-dimensional notation:

$$\boldsymbol{F}^v = \rho \boldsymbol{E} + \boldsymbol{J} \times \boldsymbol{B}; \quad -ic F_4^v = \boldsymbol{E} \cdot \boldsymbol{J} \tag{1.6.14}$$

To assist the interpretation of Eq. 1.6.12, consider the 4-scalar formed by taking the scalar product:

$$F_\alpha^v J_\alpha = f_{\alpha\beta} J_\alpha J_\beta = 0 \tag{1.6.15}$$

The second equality of Eq. 1.6.15 follows from the antisymmetric character of $f_{\alpha\beta}$ and shows that the 4-vector F_α^v is perpendicular to the 4-current density. Since the 4-current density is proportional to the 4-velocity, F_α^v is also perpendicular to the 4-velocity. Consider the differential with respect to proper time of the square of the 4-velocity:

$$\frac{d}{d\tau}\left(U_\alpha U_\alpha\right) = 2U_\alpha \frac{dU_\alpha}{d\tau} = \frac{d}{d\tau}\left(-c^2\right) = 0 \tag{1.6.16}$$

Both the 4-acceleration and F_α^v are perpendicular to the 4-velocity. This is a necessary but insufficient requirement for F_α^v to be the force density.

This approach to the Maxwell equations is based upon the original axiom relating a charge to its accompanying potential. The form of the source shows that only charges produce a 4-curvature of the 4-potential field. The technique is a way both to package the electromagnetic equations and to show they have the same form in all inertial coordinate systems.

The relationship between fields E and B and the potentials follows from Eq. 1.6.3. By direct comparison:

$$\frac{\partial A_j}{\partial x_i} - \frac{\partial A_i}{\partial x_j} = B_k \Rightarrow \nabla \times A = B$$

$$-\frac{\partial \Phi}{\partial x_i} - \frac{\partial A_i}{\partial t} = E_i \Rightarrow -\left(\nabla\Phi + \frac{\partial A}{\partial t}\right) = E \tag{1.6.17}$$

The above are the forces due to electromagnetic fields. There are, of course, other forces that may or may not be interdependent: gravitational, and weak and strong nuclear forces. In free space the electromagnetic and gravitational forces decrease as the inverse square of the distance of separation, but the nuclear forces decrease much more rapidly with distance

1.7 Accelerating Charges

Potentials due to electric charges in uniform motion are given by Eq. 1.5.6 and the force fields are related to the potential by Eq. 1.6.3. The partial derivative operations of Eq. 1.6.3 take place at the field position and time, (r,t) and the position and time at the source, (r',t') do not enter into the operations. To carry out required operations it is convenient to define S by the equation:

$$S = \left(R - \frac{R \cdot v}{c}\right) \tag{1.7.1}$$

Operating upon the potential while keeping terms involving charge accelerations gives:

$$E = \frac{q}{4\pi\varepsilon}\left\{\frac{1}{\gamma^2 S^3}\left(R - R\frac{v}{c}\right) + \frac{1}{c^2 S^3}R \times \left[\left(R - R\frac{v}{c}\right) \times \frac{\partial}{\partial t}v\right]\right\} \tag{1.7.2}$$

$$B = \frac{1}{Rc}R \times E$$

Keeping only first order terms in powers of v/c gives:

$$E = \frac{q}{4\pi\varepsilon R^3}\left\{\left(R - R\frac{v}{c}\right) + \frac{1}{c^2}R\times\left(R\times\frac{\partial v}{\partial t}\right)\right\}$$

$$B = -\frac{\mu q}{4\pi R^3}R\times\left(v + \frac{R}{c}\frac{\partial v}{\partial t}\right)$$

(1.7.3)

Both electric and magnetic field intensities exist about moving charges; each contains a term proportional to the speed of the charge that varies as the inverse square of the radius. If the charge is accelerating, both field intensities contain a term proportional to the acceleration and that varies as the inverse of the radius. Evaluation of the fields of Eq. 1.7.3 about distributed charge densities requires spatial integration over all sources. With no acceleration and low speeds the first order field terms are:

$$E = \frac{qR}{4\pi\varepsilon R^3}; \quad B = \frac{\mu q}{4\pi R^3}v\times R$$

(1.7.4)

Both fields vary as the inverse square of distance from the charge. The electric field exists for stationary charges but the magnetic field exists only if the charge is in motion.

1.8 The Electromagnetic Stress Tensor

Another useful four-dimensional array of coefficients is the electromagnetic stress tensor. By definition it is the symmetric second rank 4-tensor $T_{\alpha\beta}$ where:

$$\mu T_{\alpha\beta} = f_{\alpha\kappa}f_{\kappa\beta} + \frac{1}{4}\delta_{\alpha\beta}f_{v\sigma}f_{v\sigma}$$

(1.8.1)

A symmetric 4-tensor contains ten independent numbers. It may be shown that the force volume density 4-vector of Eq. 1.6.12 is related to the electromagnetic stress tensor as:

$$F_\alpha^v = \partial T_{\alpha\beta}/\partial X_\beta$$

(1.8.2)

The independent components of the Maxwell stress tensor $T_{\alpha\beta}$ follow from Eq. 1.6.7 and 1.8.1, and are:

$$T_{11} = \frac{\varepsilon}{2}\left(E_x^2 - E_y^2 - E_z^2\right) + \frac{1}{2\mu}\left(B_x^2 - B_y^2 - B_z^2\right);\ T_{12} = \varepsilon E_x E_y + \frac{1}{\mu}B_x B_y$$

$$T_{22} = \frac{\varepsilon}{2}\left(E_y^2 - E_z^2 - E_x^2\right) + \frac{1}{2\mu}\left(B_y^2 - B_z^2 - B_x^2\right);\ T_{23} = \varepsilon E_y E_z + \frac{1}{\mu}B_y B_z$$

$$T_{33} = \frac{\varepsilon}{2}\left(E_z^2 - E_x^2 - E_y^2\right) + \frac{1}{2\mu}\left(B_z^2 - B_x^2 - B_y^2\right);\ T_{31} = \varepsilon E_z E_x + \frac{1}{\mu}B_z B_x \qquad (1.8.3)$$

$$T_{44} = \frac{\varepsilon}{2}\left(E_x^2 + E_y^2 + E_z^2\right) + \frac{1}{2\mu}\left(B_x^2 + B_y^2 + B_z^2\right);\ T_{14} = \frac{1}{ic\mu}\left(E_y B_z - E_z B_y\right)$$

$$T_{24} = \frac{1}{ic\mu}\left(E_z B_x - E_x B_z\right);\quad T_{34} = \frac{1}{ic\mu}\left(E_x B_y - E_y B_x\right)$$

It is convenient for what lies ahead to group the terms into three-dimensional forms. The Poynting vector, N, is:

$$N = (E \times B)/\mu;\ N_j = T_{j4} = T_{4j} \qquad (1.8.4)$$

The energy density, w, is:

$$w = T_{44} = \frac{\varepsilon}{2}E^2 + \frac{1}{2\mu}B^2 \qquad (1.8.5)$$

The T_{44} component is also known as the Hamiltonian density for electromagnetic fields; conversely the stress tensor may be viewed as a covariant generalization of the Hamiltonian density.

The three dimensional portion of Maxwell's stress tensor, T_{ij}, is:

$$(T_{ij}) = \begin{pmatrix} \frac{\varepsilon}{2}\left[E_x^2 - E_y^2 - E_x^2\right] + \frac{1}{2\mu}\left[B_x^2 - B_y^2 - B_x^2\right] & \varepsilon E_x E_y + \frac{1}{\mu}B_x B_y & \varepsilon E_x E_z + \frac{1}{\mu}B_x B_z \\[2ex] \varepsilon E_y E_x + \frac{1}{\mu}B_y B_x & \frac{\varepsilon}{2}\left[E_y^2 - E_z^2 - E_x^2\right] + \frac{1}{2\mu}\left[B_y^2 - B_z^2 - B_x^2\right] & \varepsilon E_y E_z + \frac{1}{\mu}B_y B_z \\[2ex] \varepsilon E_y E_z + \frac{1}{\mu}B_y B_z & \varepsilon E_z E_y + \frac{1}{\mu}B_z B_y & \frac{\varepsilon}{2}\left[E_z^2 - E_x^2 - E_y^2\right] + \frac{1}{2\mu}\left[B_z^2 - B_x^2 - B_y^2\right] \end{pmatrix} \qquad (1.8.6)$$

In these terms, four-dimensional stress tensor Eq. 1.8.3 may be written:

$$\left(T_{\alpha\beta}\right) = \begin{pmatrix} T_{ij} & -\dfrac{i}{c}N \\[2ex] -\dfrac{i}{c}N & w \end{pmatrix} \qquad (1.8.7)$$

Symmetric tensors of rank two in three dimensions reduce from six to three components by transforming to the principal axes and aligning one axis with the source field intensity. For example, if there is no magnetic field and if the electric field intensity is directed along the x-axis the tensor reduces to:

$$\left(T_{ij}\right) = \frac{\varepsilon}{2}\begin{pmatrix} E^2 & 0 & 0 \\ 0 & -E^2 & 0 \\ 0 & 0 & -E^2 \end{pmatrix} \qquad (1.8.8)$$

To determine the significance of the stress tensor, let an observer be at rest in the unprimed frame and let charge q be fixed in position in the primed frame, which moves with speed v with respect to the observer. We multiply Eq. 1.8.2 by the primed Lorentz transformation matrix as observed in the moving frame and then integrate over the primed coordinates:

$$\iiiint c'_{\sigma\alpha} F^{\prime v}_{\alpha} \, dX'_1 \, dX'_2 \, dX'_3 \, dX'_4 = \iiiint c'_{\sigma\alpha} \frac{\partial T'_{\alpha\beta}}{\partial X'_{\beta}} \, dX'_1 \, dX'_2 \, dX'_3 \, dX'_4$$

$$(1.8.9)$$

Working with the left side of Eq. 1.8.9 and using results of Eq. 1.3.3 the integral in moving coordinates is equated to that in stationary coordinates:

$$\iiiint c'_{\sigma\alpha} F^{\prime v}_{\sigma} \, dX'_1 \, dX'_2 \, dX'_3 \, dX'_4 = \iiiint F^{v}_{\sigma} \, dX'_1 \, dX'_2 \, dX'_3 \, dX'_4$$

$$= \iiiint F^{v}_{\sigma} \, dX_1 \, dX_2 \, dX_3 \, dX_4$$

The four-dimensional integral on the right side of Eq. 1.8.9 reduces to three dimensions in the primed coordinates with aid of the four dimensional generalization of Gauss's divergence theorem:

$$\iiiint c'_{\sigma\alpha} \frac{\partial T'_{\alpha\beta}}{\partial X'_{\beta}} dX'_1 \, dX'_2 \, dX'_3 \, dX'_4$$

$$= \iiiint \frac{\partial \left(c'_{\sigma\alpha} T'_{\alpha\beta} \right)}{\partial X'_{\beta}} dX'_1 \, dX'_2 \, dX'_3 \, dX'_4 = \iiint c'_{\sigma\alpha} T'_{\alpha 4} \, dX'_1 \, dX'_2 \, dX'_3$$

The last equality results since all integrals vanish at the limits of the spatial integrals. Working with the last integral, note the Lorentz transformations of each tensor index of $T_{\sigma\alpha}$ and then simplify:

$$c'_{\alpha\beta} T_{\sigma\alpha} = c'_{\lambda\beta} c'_{\sigma\alpha} c'_{\lambda\gamma} T'_{\alpha\gamma} \tag{1.8.10}$$

Since $c'_{\lambda\beta} c'_{\lambda\gamma} = \delta_{\beta\lambda}$ it follows that $c'_{\alpha\beta} T_{\sigma\alpha} = c'_{\sigma\alpha} T'_{\alpha\beta}$, from which $c'_{\sigma\alpha} T_{\alpha 4} = c'_{\alpha 4} T'_{\sigma\alpha}$. This leaves the equality:

$$\iiiint F''_{\sigma} \, dX'_1 \, dX'_2 \, dX'_3 \, dX'_4 = \iiint c'_{\alpha 4} T_{\sigma\alpha} dX'_1 \, dX'_2 \, dX'_3 \tag{1.8.11}$$

Since $c'_{\alpha 4} = U_\alpha / ic$ this may be written:

$$\iiiint F''_{\sigma} \, dX'_1 \, dX'_2 \, dX'_3 \, dX'_4 = \frac{1}{ic} \iiint T_{\sigma\alpha} U_\alpha dX_1 dX_2 dX_3 \tag{1.8.12}$$

To change the 4-integral of Eq. 1.8.12 into a three-dimensional one, differentiate by (ict) to obtain:

$$\iiint F'^{v}_{1\sigma} \, dX'_1 \, dX'_2 \, dX'_3 = F_\sigma = -\frac{1}{c^2} \frac{\partial}{\partial t} \iiint T_{\sigma\alpha} U_\alpha dX_1 dX_2 dX_3 \tag{1.8.13}$$

Since all time integrals are zero at time $t = -\infty$, time integration has a value only at present time, t.

To illustrate results of these equations, consider a charge moving in the z-direction with speed small enough so $\gamma \approx 1$. For this case

$\boldsymbol{B} = -v \times \boldsymbol{E}/c^2$ and terms quadratic in velocity are negligible. Consider a charge moving with low speed in the z-direction. With the axis in the direction of motion sum $T_{\sigma\alpha}U_\alpha$ has the form:

$$T_{3\alpha}U_\alpha = \frac{\varepsilon}{2}E^2 \boldsymbol{v} \tag{1.8.14}$$

We next generalize Eq. 1.8.14 at slow velocity, $\boldsymbol{v}$, and evaluate integro-differential Eq. 1.8.13. Combining gives:

$$\boldsymbol{F} = \int \boldsymbol{F}^v dV = \frac{d}{dt}\left\{ \frac{\boldsymbol{v}}{c^2} \int \left(\frac{\varepsilon}{2}E^2 \right) dV \right\} \tag{1.8.15}$$

The sign was changed to represent reaction of the field on its source, rather than *vice versa*.

For a low speed particle undergoing differential acceleration Eq. 1.8.15 has the form:

$$\boldsymbol{F} = \frac{d}{dt}(m\boldsymbol{v}) = \frac{d\boldsymbol{p}}{dt} \tag{1.8.16}$$

Comparison shows the mass is calculated by taking the volume integral of the energy density:

$$m = \frac{1}{c^2} \int \left(\frac{\varepsilon}{2}E^2 \right) dV \tag{1.8.17}$$

With low-speed mass given by Eq. 1.8.17, Eq. 1.8.16 may be viewed either as a statement of Newton's law for electromagnetic mass or as confirmation that $\boldsymbol{F}$ is a force. Using an inverse process, the coupled field equations for particles and fields may be obtained by operating on the energy with variational calculus.

Newton's force equation, Eq. 1.8.16, has a significant philosophical ramification. In Sec. 1.1 based upon the cosmological principle we concluded the speed of light is independent of an observer's rest frame; deductions to this point are based upon that axiom. But if positions are independent of all functions of speed what mechanism supports Newton's law? For example, as an airplane accelerates its occupants feel the force. Placing the reference frame on the plane and imagining everything outside accelerates the other way affects nothing! The plane accelerates, its local environment does not, and hence the environment is a privileged frame. We conclude that although the vacuum of space does not create a preferred inertial frame, it does create a preferred acceleration frame.

1.9 Kinematic Properties of Fields

To further develop the kinematic properties of fields, begin with the four-dimensional force equation, Eq. 1.6.14:

$$F^{\nu} = \rho E + J \times B; \quad -icF_4^{\nu} = E \cdot J \tag{1.9.1}$$

To express these equalities in a way that depends only upon fields it is necessary to substitute for ρ and J from the nonhomogeneous electromagnetic equations, Eq. 1.6.8:

$$F^{\nu} = \varepsilon E \left(\nabla \cdot E \right) - B \times \left(\frac{1}{\mu} \nabla \times B - \varepsilon \frac{\partial E}{\partial t} \right)$$

$$-icF_4^{\nu} = E \cdot \left(\frac{1}{\mu} \nabla \times B - \varepsilon \frac{\partial E}{\partial t} \right) \tag{1.9.2}$$

It is helpful to add zero to each equation in the form of terms proportional to the homogeneous Maxwell equations, Eq. 1.6.11. The added terms are:

$$\frac{1}{\mu} B \left(\nabla \cdot B \right) - \varepsilon E \times \left(\nabla \times E + \frac{\partial B}{\partial t} \right)$$

$$\text{and } -B \cdot \left(\nabla \times E + \frac{\partial B}{\partial t} \right) \tag{1.9.3}$$

Combining gives:

$$\boldsymbol{F}^{v} = \varepsilon\left\{\boldsymbol{E}\left(\nabla\boldsymbol{\cdot}\boldsymbol{E}\right) - \boldsymbol{E}\times\left(\nabla\times\boldsymbol{E}\right)\right\} + \frac{1}{\mu}\left\{\boldsymbol{B}\left(\nabla\boldsymbol{\cdot}\boldsymbol{B}\right) - \boldsymbol{B}\times\left(\nabla\times\boldsymbol{B}\right)\right\} - \frac{1}{c^{2}}\frac{\partial\boldsymbol{N}}{\partial t}$$

$$icF_{4}^{v} = \frac{\partial}{\partial t}\left(\frac{\varepsilon}{2}E^{2} + \frac{1}{2\mu}B^{2}\right) + \nabla\boldsymbol{\cdot}\boldsymbol{N}$$

$$(1.9.4)$$

Writing the first of Eq. 1.9.4 in tensor form gives:

$$F_{i}^{v} = \frac{\partial}{\partial x_{j}}\left\{\varepsilon\left(E_{i}E_{j} - \frac{1}{2}\delta_{ij}E_{k}E_{k}\right) + \frac{1}{\mu}\left(B_{i}B_{j} - \frac{1}{2}\delta_{ij}B_{k}B_{k}\right)\right\}$$
$$- \frac{1}{c^{2}}\frac{\partial N_{i}}{\partial t}$$

$$(1.9.5)$$

Integrating over a closed three-dimensional volume gives:

$$\oint\left\{\varepsilon\left(E_{i}E_{j} - \frac{1}{2}\delta_{ij}E_{k}E_{k}\right) + \frac{1}{\mu}\left(B_{i}B_{j} - \frac{1}{2}\delta_{ij}B_{k}B_{k}\right)\right\}dS_{j}$$
$$= \int\left(\frac{1}{c^{2}}\frac{\partial N_{i}}{\partial t} + F_{i}^{v}\right)dV$$

$$(1.9.6)$$

By Eq. 1.8.16 the last term on the right is the rate of change of momentum of all charges contained within the volume. Therefore the first term on the right is the rate of change of field momentum, $\boldsymbol{p}_{\text{field}}$. It follows that the left side of the equation is equal to the force on the charges and fields within the volume of integration. This conclusion may be written as:

$$\boldsymbol{p}_{field} = \frac{1}{c^{2}}\int\boldsymbol{N}dV \quad \text{and} \quad \boldsymbol{F}^{v} = \rho\boldsymbol{E} + \boldsymbol{J}\times\boldsymbol{B} = \frac{d}{dt}\boldsymbol{p}_{charge}$$

$$(1.9.7)$$

Since F^{v} is a force density, it follows from Eq. 1.9.7 that the electric field intensity is a force per unit charge. Since the wave travels at speed c, by the first of Eq. 1.9.7 the momentum passing through a planar surface is:

$$\boldsymbol{p}_{field} = \frac{1}{c}\int\boldsymbol{N}\boldsymbol{\cdot}d\boldsymbol{S}$$

$$(1.9.8)$$

By definition dS is a differential vector area directed normally outward from the surface.

Integrating the second of Eq. 1.9.1 and 1.9.4 over a three dimensional volume gives:

$$\int (\boldsymbol{E}\boldsymbol{\cdot}\boldsymbol{J})\,\mathrm{d}V = \frac{d}{dt}\int\left(\frac{\varepsilon}{2}E^2 + \frac{1}{2\mu}B^2\right)dV + \oint \boldsymbol{N}\boldsymbol{\cdot}\mathrm{d}\boldsymbol{S} \tag{1.9.9}$$

Since the field intensity is a force per unit charge it follows that the left side of Eq. 1.9.9 is the rate at which energy enters the volume of integration. Therefore the volume integral on the right side is the rate at which energy increases in the interior, and the surface integral is the rate at which energy exits through the surface. It follows that the energy in the electromagnetic fields is equal to:

$$W = \int\left(\frac{\varepsilon}{2}E^2 + \frac{1}{2\mu}B^2\right)dV \tag{1.9.10}$$

It also follows that the rate at which energy exits the volume through the surface is:

$$P = \oint \boldsymbol{N}\boldsymbol{\cdot}\mathrm{d}\boldsymbol{S} \tag{1.9.11}$$

This Poynting vector-to-power relationship is Poynting's theorem.

A sometimes-useful formulation of Eq. 1.9.10 is obtained by rewriting it in terms of the potentials. Combining Eq. 1.9.10 with Eq. 1.6.8 and 1.6.17 gives:

$$W = \int\left[\rho\Phi + \boldsymbol{J}\boldsymbol{\cdot}\boldsymbol{A} - \frac{\varepsilon}{2}\boldsymbol{E}\boldsymbol{\cdot}\frac{\partial A}{\partial t}\right]dV + \oint\left[-\varepsilon(\phi E) + \frac{1}{\mu}(\boldsymbol{A}\times\boldsymbol{B})\right]\boldsymbol{\cdot}d\boldsymbol{S} \tag{1.9.12}$$

For slowly varying fields the third term in the integrand of the volume integral is negligibly small. The surface integral may be evaluated at limitlessly large range where, for small accelerations, it is also negligibly small. Therefore for this restricted case the total field energy is:

$$W = \int \left[\rho \Phi + \boldsymbol{J} \boldsymbol{\cdot} \boldsymbol{A} \right] dV \tag{1.9.13}$$

The integrals of Eq. 1.9.10 and 1.9.13 give the same result for slowly varying fields even though they are evaluated over quite different portions of space. The former includes all space but the latter includes only regions containing charges and currents. The two equations are convenient starting points to calculate properties, respectively, of fields and particles.

As an example, consider properties of a particle of radius r_0 that supports charge q and moves at nonrelativistic velocity $\boldsymbol{v}$. Using Eq. 1.5.6, and through first order in v/c, the potentials, the electromagnetic portion of the rest mass, and the linear momentum are:

$$\Phi_0 = \frac{q}{4\pi\varepsilon r_0}; \quad A_0 = \frac{\mu q}{4\pi r_0}\boldsymbol{v}; \quad m = \frac{q\Phi_0}{c^2}; \quad \boldsymbol{p} = q A_0 \tag{1.9.14}$$

The zero subscripts on the potentials indicate values on the surface of the particle; the particle momentum is a result of the vector potential produced by its own motion.

The energy of a moving particle with rest mass m and charge q is expressed by Eq. 1.4.7 where both its mass and its momentum include contributions from all aspects of the particle, electromagnetic and otherwise. The mass and momentum expressed by Eq. 1.9.14 are included. To apply these equations to a moving, charged particle let it lie within external electromagnetic potential fields Φ and $\boldsymbol{A}$. The magnitudes of the electromagnetic energy and momentum due to field-particle interactions are respectively $q\Phi$ and $q\boldsymbol{A}$. This momentum differs from that of Eq. 1.9.14 as a source-sink difference: since the particle creates the field of Eq. 1.9.14 and is acted on by this field the effect has the opposite sign. Substituting these values into Eq. 1.4.7 gives the energy-momentum relationship:

$$W = q\Phi + \sqrt{\left(\boldsymbol{p} - q\boldsymbol{A} \right)^2 c^2 + m^2 c^4} \tag{1.9.15}$$

1.10 Wave Equations, Potential Gauges, and Uniqueness

The wave properties of electromagnetic fields E and B follow from Maxwell's equations. To examine them note that away from sources Eq. 1.6.8 and 1.6.11 are:

$$\nabla \times B - \frac{1}{c^2}\frac{\partial E}{\partial t} = 0; \quad \varepsilon \nabla \cdot E = 0$$
$$\nabla \times E + \frac{\partial B}{\partial t} = 0; \quad \nabla \cdot B = 0 \tag{1.10.1}$$

Taking the curl of both vector functions in Eq. 1.10.1 and then substituting back gives:

$$\nabla \times (\nabla \times B) = \nabla(\nabla \cdot B) - \nabla^2 B = \frac{1}{c^2}\frac{\partial}{\partial t}(\nabla \times E) = -\frac{1}{c^2}\frac{\partial^2 B}{\partial t^2}$$
$$\nabla \times (\nabla \times E) = \nabla(\nabla \cdot E) - \nabla^2 E = \frac{\partial}{\partial t}(\nabla \times B) = -\frac{1}{c^2}\frac{\partial^2 E}{\partial t^2} \tag{1.10.2}$$

The first equality in each row of Eq. 1.10.2 is a vector identity and the other two come directly from the Maxwell equations. Entering the null values of divergence from Eq. 1.10.1 yields:

$$\nabla^2 E - \frac{1}{c^2}\frac{\partial^2 E}{\partial t^2} = 0; \quad \nabla^2 B - \frac{1}{c^2}\frac{\partial^2 B}{\partial t^2} = 0 \tag{1.10.3}$$

These are the wave equations in the absence of sources. It was shown in Sec. 1.6 that $c^2 = 1/(\mu\varepsilon)$.

To examine the wave behavior of the potentials, note from Eq. 1.6.17 the relationship between the force field vectors E and B is:

$$B = \nabla \times A; \quad E = -\left(\nabla\Phi + \partial A/\partial t\right) \tag{1.10.4}$$

Combining the first parts of Eq. 1.10.1 and 1.10.4 gives:

$$\nabla \times (\nabla \times A) = \nabla \times B = \frac{1}{c^2}\frac{\partial E}{\partial t} + \mu J \tag{1.10.5}$$

Using the identity for the curl of the curl of a vector in Eq. 1.10.5 and substituting in the second part of Eq. 1.10.1 gives:

$$\nabla^2 A - \nabla\left[\nabla\cdot A + \frac{1}{c^2}\frac{\partial\Phi}{\partial t}\right] - \frac{1}{c^2}\frac{\partial^2 A}{\partial t^2} = -\mu J \qquad (1.10.6)$$

Taking the divergence of the second part of Eq. 1.10.4 and then substituting $\nabla\cdot E = \rho/\varepsilon$ from Eq. 1.10.1 gives:

$$\nabla^2\Phi + \frac{\partial}{\partial t}\nabla\cdot A = -\frac{\rho}{\varepsilon} \qquad (1.10.7)$$

Rewriting Eq. 1.5.5 with three-dimensional notation gives:

$$\nabla\cdot A + \frac{1}{c^2}\frac{\partial\Phi}{\partial t} = 0 \qquad (1.10.8)$$

Substituting Eq. 1.10.8 into Eq. 1.10.6 and 1.10.7 gives:

$$\nabla^2 A - \frac{1}{c^2}\frac{\partial^2 A}{\partial t^2} = -\mu J; \quad \nabla^2\Phi - \frac{1}{c^2}\frac{\partial^2\Phi}{\partial t^2} = -\frac{\rho}{\varepsilon} \qquad (1.10.9)$$

Comparing Eq. 1.10.3 and 1.10.9 shows the fields and potentials obey the same free space wave equation.

A philosophical point is that we began the discussion of relativity with the postulate that the speed of light is independent of the speed of an observer's platform. We note in Eq. 1.10.3 and 1.10.9 that speed is determined by the properties of space and there is no dependence upon the speed of the observer. This result is expected, of course, and implies only that the constraint remains ongoing throughout the analysis.

A sometimes-significant point is the fields that follow from Eq. 1.10.4 do not uniquely require the electromagnetic potentials. Since the curl of the gradient of any continuous function is zero, the gradient of any continuous function ψ may be added to potential A without altering the magnetic force field derived from it. Adding the negative partial with respect to time of the same function to the scalar potential leaves the electric force field unaltered. Therefore altering the potential functions in the manner of Eq. 1.10.10 leaves force fields unaltered:

$$A' = A + \nabla \psi; \quad \Phi' = \Phi - \partial A / \partial t \tag{1.10.10}$$

Although functions so altered are no longer the electromagnetic potentials, they are commonly referred to as such and the specific functions used is the gauge. That the gauge of Eq. 1.10.8 has special significance is shown by the Aharonov-Bohm effect and is necessary when dealing with sources, including antennas, which require potentials that satisfy Eq. 1.5.6. In the absence of sources gauge changes are often convenient.

Uniqueness

For any spatial region of interest one may construct virtual surfaces snugly about all the contained sources. The surfaces support electric fields, possibly with both normal and tangential directional components. We wish to show that if a specific electrostatic potential meets the boundary conditions on all virtual surfaces and if it satisfies Laplace's equation $\nabla^2 \phi = 0$ in charge-free regions then it is the uniquely correct potential.

Taking $\phi \nabla \phi$ as a function of interest and applying to it the divergence theorem of Table A.4.2, No. 10, gives:

$$\oint (\phi \nabla \phi) \cdot dS = \int \nabla \cdot (\phi \nabla \phi) dV = \int \left[(\nabla \phi)^2 + \phi \nabla^2 \phi \right] dV \tag{1.10.11}$$

Since $\nabla^2 \phi = 0$ the last term on the right vanishes. Next let ϕ_1 and ϕ_2 represent two different potential functions that meet the surface boundary condition that either the potential or the normal derivative is zero. Since both ϕ_1 and ϕ_2 satisfy the Laplace equation so does their difference. Therefore the difference between the two potentials satisfy the equation:

$$\oint \Delta \phi \nabla (\Delta \phi) \cdot dS = 0 = \int (\nabla \phi)^2 dV \tag{1.10.12}$$

Since either the potentials or the normal fields of the two functions are equal on all surfaces the surface integral vanishes and the positive real integrand of the volume integral also vanishes. The functions are, therefore, equal everywhere in space and therefore the same function.

To extend this conclusion to magnetostatics within the spatial region of interest again construct virtual surfaces snugly about all sources. Then introduce vector functions U and W and define the new vector field $V = U \times (\nabla \times W)$. Taking the volume integral using Table A.4.2, No. 12, yields:

$$\oint U \times (\nabla \times W) \cdot dS = \int \left[(\nabla \times U) \cdot (\nabla \times W) - U \cdot \nabla \times (\nabla \times W) \right] dV \quad (1.10.13)$$

Similarly to electrostatic fields, construct virtual surfaces snugly about all sources. The virtual surfaces support magnetic fields, possibly with both normal and tangential directional components, and then put $U = W = A$. In free space and using Eq. 1.6.17:

$$\nabla \times B = \nabla \times (\nabla \times A) = 0 \quad (1.10.14)$$

Next let A_1 and A_2 represent different potentials that are equal on all surfaces and produce respectively magnetic fields B_1 and B_2. Combining gives the equality:

$$\oint [\Delta A \times \Delta B] \cdot \hat{n} dS = 0 = \int (\Delta B)^2 dV \quad (1.10.15)$$

The integrand of the surface integral may be written in the following forms either of which makes the integral vanish:

$$\hat{n} \times \Delta A = 0 \text{ or } \hat{n} \times \Delta B = 0 \quad (1.10.16)$$

Therefore the positive real integrand of the volume integral of Eq. 1.10.15 must also vanish, and therefore the two fields are the same.

To extend the argument to time-varying fields again draw virtual surfaces snugly about all sources and confine attention to intervening empty space. For that case Poynting's theorem, Eq. 1.9.11, reduces to:

$$-\frac{1}{\mu} \oint (E \times B) \cdot \hat{n} dS = \frac{d}{dt} \int \left(\frac{\varepsilon}{2} E^2 + \frac{1}{2\mu} B^2 \right) dV \quad (1.10.17)$$

The surfaces support electric and magnetic fields with both normal and tangential field components. Suppose two spatial fields match the boundary conditions but have different values E_1 and E_2 in space. The surface integral vanishes leaving:

$$-\frac{1}{\mu}\oint\left[\Delta E \times \Delta B\right]\cdot\hat{n}dS = \frac{d}{dt}\int\left[\frac{\varepsilon}{2}\left(\Delta E\right)^2 + \frac{1}{2\mu}\left(\Delta B\right)^2\right]dV \qquad (1.10.18)$$

Since the difference between the two sets of fields vanishes on all surfaces the surface integral vanishes. Therefore the volume integral vanishes and the energy difference between the two field sets is constant. If the fields are initially zero and then increase proportionally in magnitude there is no change in the surface integral and hence none in the volume integral and the difference between the two field energies remains equal to zero. Since the integral is zero and the integrand is positive real it follows that the two field sets are equal everywhere in space.

We conclude that if a set of fields satisfies Eq. 1.10.3 or 1.10.9 within a source-free region and if all boundary conditions are met on surfaces surrounding sources then that field set is uniquely correct.

1.11 A Lemma for Field Calculation

A lemma is needed to assist the systematic calculation of electromagnetic fields about known sources. To obtain it begin with the general form for fields in a source-free region that contains time-dependent fields:

$$\nabla \times B - \varepsilon\mu\frac{\partial E}{\partial t} = 0 = \nabla \times E + \frac{\partial B}{\partial t} \qquad (1.11.1)$$

Taking the curl of Eq. 1.11.1 then substituting back and forth as needed gives:

$$\nabla \times \left(\nabla \times B\right) + \varepsilon\mu\frac{\partial^2 B}{\partial t^2} = 0 = \nabla \times \left(\nabla \times E\right) + \varepsilon\mu\frac{\partial^2 E}{\partial t^2} \qquad (1.11.2)$$

This shows that, away from sources, $\boldsymbol{E}$ and $\boldsymbol{B}$ satisfy the same vector partial differential equation.

$$\nabla^2\Psi - \varepsilon\mu\,\partial^2\Psi/\partial t^2 = 0 \tag{1.11.3}$$

It would be most helpful to be able to derive the vectors from Eq. 1.11.3. To do so, we turn to a lemma that begins with the vector field $\boldsymbol{F}(\boldsymbol{r},t)$, defined by

$$\boldsymbol{F} = \nabla\times\left(\boldsymbol{r}\Psi\right) \tag{1.11.4}$$

The lemma is that if Ψ satisfies Eq. 1.11.3 then $\boldsymbol{F}$ satisfies the differential equation:

$$\nabla\times\left(\nabla\times\boldsymbol{F}\right)+\varepsilon\mu\,\partial^2\boldsymbol{F}/\partial t^2 = 0 \tag{1.11.5}$$

To verify that Eq. 1.11.5 is correct, multiply Eq. 1.11.3 by $(-\boldsymbol{r})$ then take the curl:

$$-\nabla\times\left(\boldsymbol{r}\nabla^2\Psi\right)+\varepsilon\mu\frac{\partial^2}{\partial t^2}\left[\nabla\times\left(\boldsymbol{r}\Psi\right)\right]=0 \tag{1.11.6}$$

Comparing Eq. 1.11.4 through 1.11.6 shows that Eq. 1.11.5 is satisfied if:

$$\nabla\times\left\{\nabla\times\left[\nabla\times\left(\boldsymbol{r}\Psi\right)\right]\right\}=-\nabla\times\left(\boldsymbol{r}\nabla^2\Psi\right) \tag{1.11.7}$$

To confirm Eq. 1.11.7, begin with the identity for the curl of a scalar-vector product:

$$\nabla\times\left(\boldsymbol{r}\Psi\right)\equiv\Psi\left(\nabla\times\boldsymbol{r}\right)-\boldsymbol{r}\times\nabla\Psi \tag{1.11.8}$$

Since $\nabla\times\boldsymbol{r}\equiv 0$, it follows that:

$$\nabla\times\left[\nabla\times\left(\boldsymbol{r}\Psi\right)\right]=-\nabla\times\left(\boldsymbol{r}\times\nabla\Psi\right) \tag{1.11.9}$$

Combining Eq. 1.11.7 and 1.11.9 gives:

$$\nabla \times \left[\nabla \times \left(\mathbf{r} \times \nabla \Psi \right) \right] - \nabla \times \left(\mathbf{r} \nabla^2 \Psi \right) = 0 \tag{1.11.10}$$

Two identities from vector analysis are:

$$\nabla \left(\mathbf{A} \cdot \mathbf{B} \right) \equiv \mathbf{A} \times \left(\nabla \times \mathbf{B} \right) + \mathbf{B} \times \left(\nabla \times \mathbf{A} \right) + \left(\mathbf{B} \cdot \nabla \right) \mathbf{A} + \left(\mathbf{A} \cdot \nabla \right) \mathbf{B}$$
$$\nabla \times \left(\mathbf{A} \times \mathbf{B} \right) \equiv \mathbf{A} \left(\nabla \cdot \mathbf{B} \right) - \mathbf{B} \left(\nabla \cdot \mathbf{A} \right) + \left(\mathbf{B} \cdot \nabla \right) \mathbf{A} - \left(\mathbf{A} \cdot \nabla \right) \mathbf{B} \tag{1.11.11}$$

Putting $\mathbf{A} = \mathbf{r}$ and $\mathbf{B} = \nabla \Psi$:

$$\nabla \left(\mathbf{r} \cdot \nabla \Psi \right) \equiv \left(\mathbf{r} \cdot \nabla \right) \nabla \Psi + \left(\nabla \Psi \cdot \nabla \right) \mathbf{r} = \left(\mathbf{r} \cdot \nabla \right) \nabla \Psi + \nabla \Psi$$
$$\nabla \times \left(\mathbf{v} \times \nabla \Psi \right) \equiv \mathbf{r} \nabla^2 \Psi - 2 \nabla \Psi + \left(\mathbf{r} \cdot \nabla \right) \nabla \Psi \tag{1.11.12}$$

Combining Eq. 1.11.10 and 1.11.12:

$$\nabla \times \left(\mathbf{r} \times \nabla \Psi \right) - \mathbf{r} \nabla^2 \Psi + \nabla \Psi + \nabla \left(\mathbf{r} \cdot \nabla \Psi \right) = 0 \tag{1.11.13}$$

Since the curl of the gradient vanishes, taking the curl of Eq. 1.11.13 yields Eq. 1.11.10 and completes the proof.

Eq. 1.11.3 is the wave equation, and from which comes its speed and wavelength. Substitution back into the equation shows that the trigonometric functions are solutions. Dropping to two dimensions, the equation simplifies to:

$$\frac{\partial^2 \Psi}{\partial z^2} - \mu \varepsilon \frac{\partial^2 \Psi}{\partial t^2} = 0 \tag{1.11.14}$$

Solutions are the trigonometric functions; consider the special case:

$$\Psi \left(z, t \right) = \Psi_0 e^{i \left(\omega t - kz \right)} \tag{1.11.15}$$

Substitution back into the differential equation shows that:

$$k = \omega \sqrt{\mu \varepsilon}; \quad c = 1 / \sqrt{\mu \varepsilon} \tag{1.11.16}$$

The speed of the wave is c, wave number is k, and the radian frequency is ω.

1.12 The Scalar Differential Equation

To solve Eq. 1.11.3 it is helpful to begin by removing the time-dependent portion. For that purpose use the Fourier integral expansion:

$$\Psi(r,t) = \int_{-\infty}^{\infty} \psi(r,\omega)e^{i\omega t}\,d\omega \tag{1.12.1}$$

Substituting Eq. 1.12.1 into Eq. 1.11.3 leads to:

$$\int_{-\infty}^{\infty} (\nabla^2\psi + k^2\psi)e^{i\omega t}\,d\omega = 0 \tag{1.12.2}$$

By definition $k^2 = \omega^2\varepsilon\mu$; this is a linear frequency-wave number dispersion relationship. For the integral of Eq. 1.12.2 to equal zero for all values of ω the integrand itself must equal zero:

$$\nabla^2\psi + k^2\psi = 0 \tag{1.12.3}$$

This is the Helmholtz equation; it leads to the full range of solutions of time-varying electromagnetic fields.

Certain differential vector operations in spherical coordinates are listed in Table 1.12.1. Using spherical coordinates with angle θ the polar angle measured from the positive z-axis, ϕ the azimuth angle measured from the positive x-axis, and r the radial distance measured from the origin, by Table 1.12.1 the Helmholtz equation is:

$$\frac{1}{r^2\sin\theta}\frac{\partial}{\partial\theta}\left[\sin\theta\frac{\partial\psi}{\partial\theta}\right] + \frac{1}{r^2\sin^2\theta}\frac{\partial^2\psi}{\partial\phi^2} + \frac{1}{r^2}\frac{\partial}{\partial r}\left[r^2\frac{\partial\psi}{\partial r}\right] + k^2\psi = 0 \tag{1.12.4}$$

Dividing the equation by k^2 shows that the radial dependence of the solution is a function only of the product $\sigma = kr$, and therefore ψ may be written as $\psi(\sigma,\theta,\phi)$. A theorem applicable to problems using spherical coordinates is that the complete solution of Eq. 1.12.4 is obtained by summing over all possible product functions $\psi(\sigma,\theta,\phi)$:

Table 1.12.1. Differential vector operations, spherical coordinates.

Orthogonal Line Elements:	$dr, r d\theta,\ r\sin\theta\ d\phi$
Gradient	$\left\{ (\nabla\psi)_r = \dfrac{\partial\psi}{\partial r};\quad (\nabla\psi)_\theta = \dfrac{1}{r}\dfrac{\partial\psi}{\partial\theta};\quad (\nabla\psi)_\phi = \dfrac{1}{r\sin\theta}\dfrac{\partial\psi}{\partial\phi} \right\}$
Divergence of Vector **A**:	$\dfrac{1}{r^2}\dfrac{\partial}{\partial r}\left(r^2 A_r\right) + \dfrac{1}{r\sin\theta}\dfrac{\partial}{\partial\theta}\left(\sin\theta A_\theta\right) + \dfrac{1}{r\sin\theta}\dfrac{\partial}{\partial\phi}A_\phi$
Components of Curl **A**:	$\left\{ \begin{aligned} (\nabla\times A)_r &= \dfrac{1}{r\sin\theta}\left[\dfrac{\partial\left(\sin\theta A_\phi\right)}{\partial\theta} - \dfrac{\partial A_\theta}{\partial\phi}\right] \\[2mm] (\nabla\times A)_\theta &= \dfrac{1}{r\sin\theta}\dfrac{\partial A_r}{\partial\phi} - \dfrac{1}{r}\dfrac{\partial\left(rA_\phi\right)}{\partial r} \\[2mm] (\nabla\times A)_\phi &= \dfrac{1}{r}\left[\dfrac{\partial\left(rA_\theta\right)}{\partial r} - \dfrac{\partial A_r}{\partial\theta}\right] \end{aligned} \right\}$
Laplacian of $\Psi = \nabla^2\Psi$:	$\left\{ \dfrac{1}{r^2}\dfrac{\partial}{\partial r}\left(r^2\dfrac{\partial\Psi}{\partial r}\right) + \dfrac{1}{r^2\sin\theta}\dfrac{\partial}{\partial\theta}\left(\sin\theta\dfrac{\partial\Psi}{\partial\theta}\right) + \dfrac{1}{r^2\sin^2\theta}\dfrac{\partial^2\Psi}{\partial\phi^2} \right\}$

$$\psi(r,\theta,\phi) = R(\sigma)\Theta(\theta)\Phi(\phi) \tag{1.12.5}$$

To obtain $\psi(\sigma,\theta,\phi)$, it is necessary to obtain solutions of Eq. 1.12.5 that involve only one independent variable. After obtaining the functional forms, all possible products are formed and weighted by a constant multiplying coefficient. The coefficient for any specific case is determined by matching boundary conditions. Finally, all individual product functions with appropriate coefficients are summed. Substituting Eq. 1.12.5 into Eq. 1.12.4 and multiplying by $r^2/\psi(r,\theta,\phi)$ gives:

$$\frac{1}{\Theta\sin\theta}\frac{d}{d\theta}\left(\sin\theta\frac{d\Theta}{d\theta}\right) + \frac{1}{\Phi\sin^2\theta}\frac{d^2\Phi}{d\phi^2} + \frac{1}{R}\frac{d}{d\sigma}\left(\sigma^2\frac{dR}{d\sigma}\right) + \sigma^2 = 0 \tag{1.12.6}$$

The first two terms are independent of the radius and the last two terms are independent of the angles, yet the two sets equal each other's negative. This requires both pairs of terms to be constant. The constant is known as the separation constant. A convenient choice of separation constant is for the radial terms to equal $\nu(\nu+1)$ and the angular terms to equal $-\nu(\nu+1)$. This definition results in the separated, complete differential equations:

$$\frac{1}{\sigma^2}\frac{d}{d\sigma}\left(\sigma^2\frac{dR}{d\sigma}\right)+\left(1-\frac{\nu(\nu+1)}{\sigma^2}\right)R=0 \tag{1.12.7}$$

$$\frac{\Phi}{\sin\theta}\frac{d}{d\theta}\left[\sin\theta\frac{d\Theta}{d\theta}\right]+\frac{\Theta}{\sin^2\theta}\frac{d^2\Phi}{d\phi^2}+\nu(\nu+1)\Theta\Phi=0 \tag{1.12.8}$$

The radial equation is in the desired form of a differential equation with one independent variable. Spherical Bessel, Neumann, and Hankel functions, respectively, $j_\nu(\sigma)$, $y_\nu(\sigma)$, and $h_\nu(\sigma)$, are solutions; the functions together with detailed properties and interrelationships are given in Sec. A.18 through A.20.

The angular equation may be written:

$$\frac{\sin\theta}{\Theta}\frac{d}{d\theta}\left[\sin\theta\frac{d\Theta}{d\theta}\right]+\nu(\nu+1)\sin^2\theta+\frac{1}{\Phi}\frac{d^2\Phi}{d\phi^2}=0 \tag{1.12.9}$$

The first two terms of Eq. 1.12.9 are functions of θ only and the third is a function of ϕ only, yet the terms equal each other's negative. Again, both sets are constant. Putting the first two terms equal to m^2, where m is the second separation constant, results in two separated equations, each involving only one independent variable:

$$\frac{1}{\sin\theta}\frac{d}{d\theta}\left(\sin\theta\frac{d\Theta}{d\theta}\right)+\left(\nu(\nu+1)-\frac{m^2}{\sin^2\theta}\right)\Theta=0 \tag{1.12.10}$$

$$\frac{d^2\Phi}{d\phi^2}+m^2\Phi=0 \tag{1.12.11}$$

Trigonometric functions are solutions of the azimuth angle differential equation; values are shown and discussed in Sec. A.10. Legendre functions are solutions of the zenith angle differential equation; the functions together with integrals and tabulated values of these differential equations are given in Sec. A.10 through A.17.

A particularly important set of solutions of the radial field is Hankel functions of the second kind and integer order: $h_\ell(\sigma)$ where "ℓ" represents an integer value of "v". Solutions of the zenith angle equation are associated Legendre functions; solutions are, in some instances, of integer order and in others of noninteger order. In all cases, the orders of the radial and zenith angle solutions are the same. Trigonometric functions form the solutions of the azimuth angle equation: $\sin\phi$, $\cos\phi$, or $\exp(\pm im\phi)$. Since all solutions to be considered extend over the full range of azimuth angle, zero through 2π, only integer values of degree m, are present. With exponential notation, the exponent may have either sign.

With symbol $z_v(\sigma)$ representing a linear combination of radial solution forms, rather than writing the solution as separate sums we write it as:

$$\psi_v^m(r,\theta,\phi) = z_v(\sigma)\,\Theta_v^m(\theta)\,e^{-im\phi} \tag{1.12.12}$$

With this notation, completeness requires m to include the full set of positive and negative integers but the degree of the Legendre function is always positive.

1.13 Radiation Fields in Spherical Coordinates

Replacing $\boldsymbol{B}$ by $\mu\boldsymbol{H}$ more closely matches common usage. Since, for what lies ahead, we are concerned only with free space properties and there μ is but a unit-determining parameter that measures the magnetic field in amperes per meter the move is not particularly significant.

The calculation procedure for periodically time varying fields is due to Hansen, and begins with the vector theorem that a field with zero divergence is completely specified by its curl. It is, therefore, helpful to introduce the two independent field sets:

$$\eta \boldsymbol{H}_1 = \boldsymbol{r} \times \nabla \Psi_1 \quad \text{and} \quad \boldsymbol{E}_2 = \boldsymbol{r} \times \nabla \Psi_2 \tag{1.13.1}$$

Symbol $\eta = \sqrt{\mu/\varepsilon}$ indicates the wave impedance.

Since the free space divergences of both vectors are zero, solutions of Eq. 1.13.1 provide the complete set of possible values for vectors $\boldsymbol{H}_1$ and $\boldsymbol{E}_2$. The remaining field solutions, $\boldsymbol{H}_2$ and $\boldsymbol{E}_1$, may be obtained from Eq. 1.13.1 using the Maxwell curl equations. The total fields, $(\boldsymbol{E}_1 + \boldsymbol{E}_2)$ and $(\boldsymbol{H}_1 + \boldsymbol{H}_2)$, are then complete. If the boundary conditions are matched, the fields are also unique.

In what follows we use $\exp(i\omega t)$ time dependence and $\exp(-jm\phi)$ azimuth angle dependence, where $i^2 = j^2 = -1$. The reasons for separate notation are that it permits separation of polarization and time dependencies and it permits restriction of separation constant m to the field of positive integers, without loss of generality. Using suppressed time dependent notation and with Hansen's method the defining terms for phasor fields are, see Eq. 1.11.4:

$$\eta \tilde{\boldsymbol{H}}_1 = \boldsymbol{r} \times \nabla \psi_1 \quad \text{and} \quad \tilde{\boldsymbol{E}}_2 = \boldsymbol{r} \times \nabla \psi_2 \tag{1.13.2}$$

A tilde over a vector indicates that it is a phasor. It is required that the scalar function ψ satisfies the Helmholtz equation, Eq. 1.12.3. For integer modes, the results are solutions in the form of Eq. 1.12.12:

$$\psi_1 = F(\ell,m) z_\ell(\sigma) \Theta_\ell^m e^{-jm\phi}$$
$$\psi_2 = jG(\ell,m) z_\ell(\sigma) \Theta_\ell^m e^{-jm\phi} \tag{1.13.3}$$

Generally speaking, the order is not restricted to integers and the radial function $z_\ell(\sigma)$ may be any linear combination of spherical Bessel and Neumann functions. The zenith angle function may be any linear combination of associated Legendre functions. Both the applicable functions and the constant multiplying coefficients $F(\ell,m)$ and $G(\ell,m)$ are ultimately determined by boundary conditions.

Applying the operation of Eq. 1.13.2 to Eq. 1.13.3 gives:

$$\boldsymbol{r} \times \nabla \psi = -\frac{\hat{\theta}}{\sin\theta} \frac{\partial \psi}{\partial \phi} + \hat{\phi} \frac{\partial \psi}{\partial \theta} \tag{1.13.4}$$

Combining gives:

$$\eta\tilde{\boldsymbol{H}}_1 = F\left(\ell,m\right)z_\ell\left(\sigma\right)\left[j\hat{\theta}\frac{m\Theta_\ell^m}{\sin\theta}+\hat{\phi}\frac{d\Theta_\ell^m}{d\theta}\right]e^{-jm\phi}$$

$$\tilde{\boldsymbol{E}}_2 = jG\left(\ell,m\right)z_\ell\left(\sigma\right)\left[j\hat{\theta}\frac{m\Theta_\ell^m}{\sin\theta}+\hat{\phi}\frac{d\Theta_\ell^m}{d\theta}\right]e^{-jm\phi}$$

$$(1.13.5)$$

Taking the curl of the second of Eq. 1.13.5 then using the Maxwell curl equation leads to:

$$\eta\tilde{\boldsymbol{H}}_2 = -iG\left(\ell,m\right)e^{-jm\phi}\left\{j\ell\left(\ell+1\right)\frac{z_\ell}{\sigma}\Theta_\ell^m\hat{r}+z_\ell^{\bullet}\left(j\frac{d\Theta_\ell^m}{d\theta}\hat{\theta}+\frac{m\Theta_\ell^m}{\sin\theta}\hat{\phi}\right)\right\}$$

$$(1.13.6)$$

The carat indicates a unit vector and the dot superscript symbolizes:

$$z_\ell^{\bullet}\left(\sigma\right)=\frac{1}{\sigma}\frac{d}{d\sigma}\left[\sigma z_\ell\left(\sigma\right)\right]$$

$$(1.13.7)$$

Taking the curl of the first of Eq. 1.13.5 then applying the Maxwell curl equation leads to:

$$\tilde{\boldsymbol{E}}_1 = iF\left(\ell,m\right)e^{-jm\phi}\left\{\ell\left(\ell+1\right)\frac{z_\ell}{\sigma}\Theta_\ell^m\hat{r}+z_\ell^{\bullet}\left(\frac{d\Theta_\ell^m}{d\theta}\hat{\theta}-j\frac{m\Theta_\ell^m}{\sin\theta}\hat{\phi}\right)\right\} \qquad (1.13.8)$$

The total fields are the sum of Eq. 1.13.5, 1.13.6, and 1.13.8. They may be written as:

with $j=\pm i$

$$\tilde{E}_r = i\sum_{\ell=1}^{\infty}\sum_{m=0}^{\ell}i^{-\ell}F\left(\ell,m\right)\ell\left(\ell+1\right)\frac{z_\ell\left(\sigma\right)}{\sigma}\Theta_\ell^m\left(\cos\theta\right)e^{-jm\phi}$$

$$\eta\tilde{H}_r = -ij\sum_{\ell=1}^{\infty}\sum_{m=0}^{\ell}i^{-\ell}G\left(\ell,m\right)\ell\left(\ell+1\right)\frac{z_\ell\left(\sigma\right)}{\sigma}\Theta_\ell^m\left(\cos\theta\right)e^{-jm\phi}$$

$$\tilde{E}_\theta = \sum_{\ell=1}^{\infty} \sum_{m=0}^{\ell} i^{-\ell} \left[iF(\ell,m) z_\ell^\bullet \frac{d\Theta_\ell^m}{d\theta} - G(\ell,m) z_\ell \frac{m\Theta_\ell^m}{\sin\theta} \right] e^{-jm\phi}$$

$$\eta\tilde{H}_\phi = \sum_{\ell=1}^{\infty} \sum_{m=0}^{\ell} i^{-\ell} \left[F(\ell,m) z_\ell \frac{d\Theta_\ell^m}{d\theta} - iG(\ell,m) z_\ell^\bullet \frac{m\Theta_\ell^m}{\sin\theta} \right] e^{-jm\phi}$$

$$\tilde{E}_\phi = -j \sum_{\ell=1}^{\infty} \sum_{m=0}^{\ell} i^{-\ell} \left[iF(\ell,m) z_\ell^\bullet \frac{m\Theta_\ell^m}{\sin\theta} - G(\ell,m) z_\ell \frac{d\Theta_\ell^m}{d\theta} \right] e^{-jm\phi}$$

$$\eta\tilde{H}_\theta = j \sum_{\ell=1}^{\infty} \sum_{m=0}^{\ell} i^{-\ell} \left[F(\ell,m) z_\ell \frac{m\Theta_\ell^m}{\sin\theta} - iG(\ell,m) z_\ell^\bullet \frac{d\Theta_\ell^m}{d\theta} \right] e^{-jm\phi}$$

$$(1.13.9)$$

The radial equations are the basis for defining multipoles. Electric multipoles of order ℓ and degree m are proportional to the infinite set of constants $F(\ell,m)$; since the fields include a radial electric field component but not a magnetic field component they are known TM (transverse magnetic) fields. The dual set of magnetic multipoles of order ℓ and degree m are proportional to the infinite set of constants $G(\ell,m)$; since the fields include a radial magnetic field component but not an electric field component they are known as TE (transverse electric) fields. Section A.22 contains a discussion of electric multipolar moments as commonly defined and their relationship to the field constants.

Without loss of generality, the phases of constants $F(\ell,m)$ and $G(\ell,m)$ and multiplying factor $i^{-\ell}$ have been picked for later convenience.

Keeping only the real or only the imaginary part with respect to "j" provides, respectively, x- or y-polarization of the electric field intensity. The fields are right or left circularly polarized, respectively, with $j = i$ or $j = -i$. Since this result applies to all time-dependent outgoing waves, it follows that it also applies when the rate of change is arbitrarily small. Hence, it describes fields in the limit as the frequency goes to zero: a static charge distribution. Because of this general result, it is helpful to obtain a physical view of what constitutes field sources. The sources of coefficients $F(\ell,m)$ and $G(\ell,m)$ for static fields are discussed in Sec. A.22. Static sources of $G(\ell,m)$ may be viewed as either separation of magnetic poles, in a way parallel with the separation of electric charges, or to circulating currents.

Consider a few special cases of Eq. 1.13.9. If the described fields are contained within a source-free region of space, and if that space is loss free, solutions have positive, integer values of orders and integer values of degrees. Spherical Bessel functions, which have no singularities, form the radial portion of the solution; spherical Neumann functions, which have singularities, are not present. Associated Legendre functions of the first kind, and of integer order, which have no singularities, form the angular portion of the solution; fractional order associated Legendre functions and those of the second kind, which have singularities, are not present.

If the fields originate at a point and support an outward flow of energy from that point, the radial portion of the solution consists of spherical Hankel functions of the second kind. A solution within an enclosed space that excludes the z-axis, but has rotational symmetry, is described by associated Legendre functions of both the first and second kind, with noninteger, or positive-real order and integer degree.

In all cases, if the medium in which the fields exist is lossy the separation constants are complex numbers with a positive real part. Since all cases of interest in this book concern lossless media and a full 2π spatial rotation about the z-axis, both the order and degree are real and degrees have only integer values.

References

R. Becker, *Electromagnetic Fields and Interactions*, Blaisdell Publishing Co. and Blackie and Son (1964), reprinted by Dover Publications (1982)

P. Graneau, N. Graneau, *In the Grip of the Distant Universe*, World Scientific (2006)

W.W. Hansen, "A New Type of Expansion in Radiation Problems," *Phys. Rev.*, vol. 47, pp. 139–143 (1935)

J.D. Jackson, *Classical Electrodynamics*, 2nd ed., John Wiley (1975)

L.D. Landau, E.M. Lifshitz, *The Classical Theory of Fields*, trans. by H. Hamermesh, Addison-Wesley (1951)

R.B. Leighton, *Principles of Modern Physics*, McGraw-Hill, New York (1959)

W.K.H. Panofsky, M. Phillips, *Classical Electricity and Magnetism*, 2nd ed., Addison-Wesley (1961)

A. Sommerfeld, *Electrodynamics*, Academic Press (1952)

J.A. Stratton, *Electromagnetic Theory*, McGraw-Hill (1941)

J.B. Westgard, *Electrodynamics: A Concise Introduction*, Springer-Verlag (1997)

Chapter 2

Properties of Radiation Fields

$\mathcal{I}$n this chapter we examine fundamental techniques and tools used for critical analyses of radiation properties. We begin with the radiation Q of dipolar sources and then extend the argument to higher order multipolar sources. To determine Q it is necessary to examine and determine techniques for calculating power in radiation fields. We also examine techniques for determining radiation reaction forces, momentum, and associated stresses. Of special importance are forces associated with reactive power, that is power that does not travel permanently outbound from the source but remains in the local region as a standing energy field. Textbooks calculate radiation reaction force on the basis of power radiated into the far field; the force brakes the charge and is significant in many cases even though it is much less than that of a binding Coulomb force. In stark contrast the reactive radiation reaction force is about the magnitude of the Coulomb binding force but acts to distort the source, not brake it.

The rest of the chapter is devoted to characterizing radiation properties of isolated multipoles and scattering by spherical objects made of different materials. The chapter concludes with a discussion of stresses on the surface of a source producing a dipolar radiation field.

2.1 Dipoles in Continuous Media

Both electric and current charge densities are useful aids for the analysis of electromagnetic fields within continuous media. The densities are defined to equal respectively the charge-to-volume and current-to-volume ratio in small, but not limitlessly small, regions. To illustrate the size restriction, consider a point within a material of interest and then

43

construct a volume that contains the point. To be an acceptable volume: (a) the smallest dimension of interest to the problem, wavelength for example, is many times larger than each of the three volume dimensions, (b) the volume contains a significantly large number of individual charges, and (c) the charge-to-volume or current-to-volume ratio varies continuously with position. If N represents the number of charges within the volume and v_j the velocity of charge 'j' the average velocity is:

$$v = \sum_{j=1}^{N} v_j / N \tag{2.1.1}$$

Under these conditions with a, b, and c the linear dimensions of the volume and with δ much smaller than anything else of interest in the system, the densities at each point are:

$$\rho = \lim_{a,b,c \to \delta} \frac{Nq}{abc}; \quad J = \sum_{j=1}^{N} \lim_{a,b,c \to \delta} \frac{Nq}{abc} v \tag{2.1.2}$$

Electric Dipole Sources

Although the continuous media of interest contain an equal balance of positive and negative charges, consider separately all negative charges within a small volume as a negative charge density; all positive charges within the same volume form a positive charge density of equal magnitude. If an external electric field is applied the two arrays of charge densities move slightly in opposite directions and the moving, bound charges produce a current density at each point. The slight shift in position also affects the edge of the material by creating a thin layer of unbalanced charge on either side of the media. Local differences in magnitudes of the positive and negative charge densities result from either nonuniform material or field gradients.

For each atom, a slight displacement of the center of charge creates an atomic dipole. The dipole moment of separated charges is discussed in Sec. A.22 and the potential it creates is shown Table A.22.1. Let p_1 represent an atomic dipole moment and symbol ∇' indicate the del operator acting at the point of origin of the vector, as opposed to the termination. From Sec. A.22 the potential about a z-directed electric dipole is:

$$\Phi(r,\theta,\phi) = -\frac{p_1}{4\pi\varepsilon}\nabla\left(\frac{1}{r}\right) = \frac{p_1}{4\pi\varepsilon}\nabla'\left(\frac{1}{r}\right) \tag{2.1.3}$$

Within a small volume the electric polarization P is by definition the ratio of the total dipole moment to the volume:

$$\boldsymbol{P} = \lim_{a,b,c\to\delta}\sum_{j=1}^{N}\boldsymbol{p}_j\bigg/abc = N\boldsymbol{p} \tag{2.1.4}$$

With ξ representing a position vector within a dielectric the total dipole moment is:

$$\boldsymbol{p} = \int \boldsymbol{P}dV' = \int \rho_P\xi dV' \tag{2.1.5}$$

The prime on dV' indicates the integration is over all source positions and subscript P indicates the charge density is composed of bound charges. The time dependence of the charge density at a fixed position is:

$$\frac{\partial\boldsymbol{p}}{\partial t} = \int \xi\frac{\partial\rho_P}{\partial t}dV' = -\int(\nabla\cdot\rho_P\boldsymbol{v})\xi dV' \tag{2.1.6}$$

The second equality uses the continuity of charge. Using tensor notation the integrand in the last term may be written:

$$\xi_x\nabla\cdot(\rho_P\boldsymbol{v}) = \nabla\cdot(\xi_x\rho_P\boldsymbol{v}) - \rho_P\boldsymbol{v}\cdot\nabla\xi_x = \nabla\cdot(\xi_x\rho_P\boldsymbol{v}) - \rho_P v_x \tag{2.1.7}$$

Integrating Eq. 2.1.7 over the material and transforming the integral over the first term of the last equality to a surface integral that is evaluated just outside the material yields a null result. Combining that null with Eq. 2.1.5 to 2.1.7 leads to the equality chain:

$$\frac{\partial\boldsymbol{p}}{\partial t} = \int \frac{\partial\boldsymbol{P}}{\partial t}dV = \int \rho_P\boldsymbol{v}dV' = \int \boldsymbol{J}_P dV' \tag{2.1.8}$$

Equating integrands and combining with the continuity of bound charge gives:

$$\frac{\partial \boldsymbol{P}}{\partial t} = \rho_P \boldsymbol{v} = \boldsymbol{J}_P = -v\nabla\bullet\boldsymbol{P}; \quad \nabla\bullet\boldsymbol{P} = -\rho_P \tag{2.1.9}$$

In words, polarization current exists where the divergence of the polarization is different from zero.

Integrating Eq. 2.1.9 over the material creates the equality chain:

$$\int \nabla\bullet\boldsymbol{P}dV' = -\int \rho_P dV' = \oint \boldsymbol{P}\bullet d\boldsymbol{S} = -\oint \rho_{PS}dS \tag{2.1.10}$$

In summary, with subscript 'n' indicating the normal component at the surface the polarization-created surface charge density ρ_{PS} and volume charge density ρ_P are given by:

$$\rho_{PS} = -\hat{n}\bullet\boldsymbol{P}; \quad \nabla\bullet\boldsymbol{P} = -\rho_P \tag{2.1.11}$$

Magnetic Dipole Sources

Continuous media may also contain solenoidal currents; examples are the angular momenta of both the spin and angular motion of atomic electrons. Such solenoidal currents create magnetic moments. With ξ defined as in Eq. 2.1.5 by definition the magnetic dipole moment of a circulating current is:

$$\boldsymbol{m} = \frac{1}{2}\int (\xi \times \boldsymbol{J})dV \tag{2.1.12}$$

For a planar current Eq. 2.1.12 is equal to the product of the circulatory current and the subscribed area. Consider the magnetic properties of a planar array of such dipoles. With uniform media each dipole is surrounded by identical neighbors and in the space between them the dipolar currents are oppositely directed and hence cancel. Therefore interior effective solenoidal currents cancel in the interior, leaving only surface currents. Each surface dipole contributes its surface current over the length of its diameter and the continuous chain of surface dipoles creates an effective surface sheet current density.

To calculate the resulting vector potential we begin with the static version of Eq. 1.5.7, where r is the field position, r' is the source position, and R is the distance between the source and field points:

$$A(r) = \frac{\mu}{4\pi} \int \frac{J(r')}{R} dV'$$

(2.1.13)

To determine large-scale effects of small-scale solenoidal currents we do a Taylor series expansions of Eq. 2.1.13. The zero order term vanishes and the first order term is:

$$A_k(r) = -\frac{\mu}{4\pi}\left[\frac{\partial}{\partial x_j}\left(\frac{1}{r}\right)\right]_R \int \xi_j J_k dV = \frac{\mu}{4\pi}\frac{R_j}{R^3} \int \xi_j J_k dV$$

(2.1.14)

Adding and subtracting $\xi_j J_k$ gives:

$$A_k = \frac{\mu}{4\pi}\frac{R_j}{2R^3} \int \left[\left(\xi_j J_k + \xi_k J_j\right) + \left(\xi_j J_k - \xi_k J_j\right)\right] dV$$

(2.1.15)

To analyze the first term of Eq. 2.1.15 we turn to Eq. 2.1.16 where, with a static field the divergence of the current density is zero and hence the line of equalities is zero. Taking the volume integral of the two terms on the right, changing the volume integral of the divergence term to a surface integral, and evaluating it just outside the magnetic material shows the current and hence the integral is zero; therefore the volume integral of the second term, also the first integrand of Eq. 2.1.15, is also zero.

$$\xi_j \xi_k \nabla \cdot J = 0 = \nabla \cdot \left(\xi_j \xi_k J\right) - J \cdot \nabla \xi_j \xi_k$$

$$0 = \int J \cdot \nabla \xi_j \xi_k dV = \int \left(\xi_j J_k + \xi_k J_j\right) dV$$

(2.1.16)

Returning to three dimensional vector notation, the integrand of the final term of Eq. 2.1.15 is $(\xi \times J) \times R$. Substituting the definition of magnetic dipole moment m from Eq. 2.1.12 into the equation gives the desired expression for the vector potential of a magnetic dipole:

$$A(r,t) = \frac{1}{R^3} m \times R = -m \times \nabla\left(\frac{1}{r}\right) \tag{2.1.17}$$

The magnetization of a material is defined from microscopic magnetic dipole moments in a way that parallels the definition of polarization using microscopic electric dipole moments:

$$M = \lim_{a,b,c \to \delta} \sum_{j=1}^{N} m_j \bigg/ abc = Nm; \quad m = MdV \tag{2.1.18}$$

We may, therefore, write:

$$\begin{aligned} A &= \frac{\mu}{4\pi} \int M \times \nabla'\left(\frac{1}{r}\right) dV' \\ &= \frac{\mu}{4\pi} \int \frac{1}{r} \nabla' \times M dV' - \frac{\mu}{4\pi} \int \nabla' \times \left(\frac{M}{r}\right) dV' \end{aligned} \tag{2.1.19}$$

With the aid of the vector form of Gauss' law, the expression for A as a function of M is shown in the first line of Eq. 2.1.20; the second line follows by comparing that result with Eq. 2.1.14:

$$A = \frac{\mu}{4\pi} \int \frac{1}{r} \nabla' \times M dV' + \frac{\mu}{4\pi} \oint \frac{1}{r} M \times dS' \tag{2.1.20}$$
$$J_m = \nabla \times M; \quad I_m = M \times \hat{n}$$

This equation shows that the magnetic current density is equal to the curl of the magnetization, $J_m = \text{curl } M$, and the surface current density is $I_m = M \times n$.

Maxwell Source Equations

The two source vector equations applicable both within and without material are listed in Eq. 1.6.8, and repeated in the top line of Eq. 2.1.21, where subscript 'tot' indicates the sum of free and bound charges.

$$\nabla \times \boldsymbol{B} - \frac{1}{c^2}\frac{\partial \boldsymbol{E}}{\partial t} = \mu \boldsymbol{J}_{tot}; \quad \varepsilon \nabla \cdot \boldsymbol{E} = \rho_{tot}$$

$$\boldsymbol{J}_{tot} = \boldsymbol{J}_{fr} + \frac{\partial \boldsymbol{P}}{\partial t} + \nabla \times \boldsymbol{M}; \quad \rho_{tot} = \rho_{fr} - \nabla \cdot \boldsymbol{P}$$

$$(2.1.21)$$

Collecting terms gives:

$$\nabla \cdot (\varepsilon \boldsymbol{E} + \boldsymbol{P}) = \rho_{fr}$$

$$\nabla \times \boldsymbol{B}/\mu = \left(\boldsymbol{J}_{fr} + \nabla \times \boldsymbol{M} + \frac{\partial}{\partial t}(\varepsilon \boldsymbol{E} + \boldsymbol{P}) \right)$$

$$(2.1.22)$$

It aids in the analysis of many problems to define vectors that arise from free charges and currents only. For that reason vectors $\boldsymbol{H}$ and $\boldsymbol{D}$ are defined as:

$$\boldsymbol{D} = \varepsilon \boldsymbol{E} + \boldsymbol{P}; \quad \boldsymbol{H} = \boldsymbol{B}/\mu - \boldsymbol{M}$$

$$\nabla \times \boldsymbol{H} - \partial \boldsymbol{D}/\partial t = J_{fr}; \quad \nabla \cdot \boldsymbol{D} = \rho_{fr}$$

$$(2.1.23)$$

Of course, as a matter of convenience, we anticipated this result and defined $\boldsymbol{H}$ in Sec. 1.13.

The volume integral of the divergence equation is:

$$\int \nabla \cdot \boldsymbol{D} dV \equiv \oint \boldsymbol{D} \cdot d\boldsymbol{S} = \int \rho_{fr} dV = Q$$

$$(2.1.24)$$

The equality of the second and last terms shows the surface integral of $\boldsymbol{D}$ is equal to the contained charge. The vector is, therefore, defined as the electric flux density.

The volume integral of the curl equation is:

$$\int \nabla \times \boldsymbol{H} dV \equiv \oint d\boldsymbol{S} \times \boldsymbol{H} = \int \boldsymbol{J} dV$$

$$(2.1.25)$$

It is convenient, but not restrictive, to break the volume of integration into differential right circular cylinders with the axis being the direction of the local current. Within a differential volume let ρ equal the radial distance from z-axis and ϕ be the angle about that axis. Differential area dS is ρ-directed and has magnitude $\rho\,d\phi\,dz$. Since H is ϕ-directed and the last two integrands of Eq. 2.1.25 are z-directed, integrating over the z-coordinate gives:

$$\oint H \cdot d\ell = I \tag{2.1.26}$$

This shows that currents create magnetic field intensity H circumferentially about themselves. The magnitude varies inversely with distance from the current axis.

Due to the historical accident of how magnetic vectors H and B were first understood the nomenclature was, and remains, that H is called the magnetic field intensity and B the magnetic flux density.

Boundary Conditions

How do field vectors change between adjacent continuous media with different electromagnetic properties? To address the issue select a point of contact between two media and construct a virtual differential right circular cylinder with its axis normal to and centered on the interface. It follows from Eq. 1.6.11 and 2.1.23 that the divergences of the flux densities within the volume are:

$$\nabla \cdot D = \rho_{fr}; \quad \nabla \cdot B = 0 \tag{2.1.27}$$

Let the volume be electrically small and shaped such that the height is much smaller than the radius. Upon taking the volume integral of Eq. 2.1.27 over such a volume in the limit as the height becomes vanishingly small, the integral over the electric flux is equal to the free charge enclosed; an equivalent integral over the magnetic field is zero. Therefore, with subscripts '1' and '2' indicating cylindrical surfaces in the different media and ρ_s the free surface charge density at the interface, the integrated results are:

$$\hat{n} = normal\ unit\ vector$$
$$\hat{n} \cdot (D_2 - D_1) = \rho_s; \quad \hat{n} \cdot (B_2 - B_1) = 0 \tag{2.1.28}$$

These equations show the normal component of the **B** field is continuous through a media interface and the normal component of the **D** field differs by the surface charge density.

Construct an area that extends across the interface between two media with its normal in the interface and consider the curls of the field vectors from Eq. 1.6.3, 2.1.23:

$$\nabla \times \boldsymbol{E} = -\partial \boldsymbol{B}/\partial t; \quad \nabla \times \boldsymbol{H} = \boldsymbol{J} + \partial \boldsymbol{D}/\partial t \tag{2.1.29}$$

Evaluate the line integral of each vector around the surface area in the limit as the interfacial lengths become arbitrarily small. With I_s representing the sheet surface current density, in that limit the line integrals are:

$$\hat{n} = normal\, unit\, vector$$
$$\hat{n} \times \left(\boldsymbol{E}_2 - \boldsymbol{E}_1 \right) = 0; \quad \hat{n} \times \left(\boldsymbol{H}_2 - \boldsymbol{H}_1 \right) = \boldsymbol{I}_s \tag{2.1.30}$$

These equations show the tangential components of the **E** field are continuous through a media interface and the tangential components of the **H** field differ by the sheet current density.

2.2 Electromagnetic Fields in Continuous Media

Constitutive Relationships

Although different media support quite different responses to applied fields many useful materials respond in proportion to the force. For such linearly responding crystalline material and with ε_{jk} and μ_{jk} representing second rank tensors the response has the form:

$$D_j = \varepsilon_{jk} E_k; \quad B_j = \mu_{jk} H_k \tag{2.2.1}$$

In many cases the crystals are much smaller than the object of interest. For such cases the tensors become constitutive scalars:

$$\boldsymbol{D} = \varepsilon\varepsilon_r \boldsymbol{E}; \quad \boldsymbol{B} = \mu\mu_r \boldsymbol{H} \tag{2.2.2}$$

With this notation ε and μ are free space values and ε_r and μ_r are relative constitutive values.

In many significant cases involving a nonlinear response to an electric field, the response may be described by the expansion:

$$D = \gamma_1 E + \gamma_3 E^2 E + \gamma_5 E^4 E + ...$$

(2.2.3)

The gammas are constant, experimentally determined coefficients. Materials of this class are useful for frequency mixing.

The electromagnetic power that emanates from any closed volume is given by Poynting's theorem, Eq. 1.9.11. Use of the theorem is sometimes aided by recasting it as a function of free currents only. For that purpose and referring to Eq. 2.1.29 we take the scalar product between H and the first term, the scalar product between E and the second term, and then subtract the second from the first. After using the vector theorem for the divergence of the cross product the sum is:

$$\nabla \cdot (E \times H) = -H \cdot \partial B / \partial t - E \cdot \partial D / \partial t - E \cdot J_{fr}$$

(2.2.4)

Integrating Eq. 2.2.4 over the surface of a closed volume and then replacing the integral of the divergence by the surface integral equivalent gives:

$$-\int \left(E \cdot \frac{\partial}{\partial t} D + H \cdot \frac{\partial}{\partial t} B \right) dV = \oint N \cdot dS + \int E \cdot J_{fr} dV$$

(2.2.5)

For linear, scalar media this may be written:

$$-\frac{1}{2} \frac{\partial}{\partial t} \int \left(\varepsilon \varepsilon_r E^2 + \mu \mu_r H^2 \right) dV = \oint N \cdot dS + \int E \cdot J_{fr} dV$$

(2.2.6)

The surface integral is power that exits the system through the surface and the left side is the rate at which energy goes into the fields. It follows from energy conservation that the final term on the right is rate at which energy is converted into Joule heat.

Of special interest is the behavior of electromagnetic fields in insulating, continuous media. Returning to Eq. 2.1.29 and taking the curl of the first term, substituting it into the second, and then using the vector identity Table A.4.1 No. 9 gives:

$$\nabla^2 E - \mu\varepsilon\varepsilon_r \frac{\partial^2 E}{\partial t^2} = \frac{1}{\varepsilon}\left(\nabla\rho_{tot} + \frac{1}{c^2}\frac{\partial J_{tot}}{\partial t}\right) \tag{2.2.7}$$

Within ideal insulating material the positive and negative free charges densities are equal; there are neither free charges nor free currents and, except for ferrimagnetic material, magnetic moment M is negligibly small. Incorporating these results into Eq. 2.2.7 gives:

$$\nabla^2 E - \mu\varepsilon\varepsilon_r \frac{\partial^2}{\partial t^2} E = \mu\frac{\partial^2}{\partial t^2} P \tag{2.2.8}$$

This is the wave equation within the media. Since the positive charge density does not significantly respond to applied electric forces, we reassign symbol ρ as the negative bound charge density and:

$$\mu\frac{\partial P}{\partial t} = -N\mu e\frac{\partial^2 r}{\partial t^2} \tag{2.2.9}$$

These are point equations that describe events at a specific point. If a single charge exists in otherwise free space with the definition $q = Ne$, the wave equation becomes:

$$\nabla^2 E - \mu\varepsilon \frac{\partial^2}{\partial t^2} E = \mu q\frac{\partial^2}{\partial t^2} r \tag{2.2.10}$$

This equation describes the electric force on individual charges in an applied electromagnetic field.

2.3 Boxed, Discrete Electromagnetic Fields

It is helpful in analyses of confined radiation to know possible modes within a cavity, the energy associated with each mode, and the number of independent modes that can exist. For that purpose consider all possible electric field modes inside an otherwise empty, rectangular cavity that is confined by walls of infinite conductivity. From Eq. 1.12.2 and 1.12.3 wave number k is, by definition:

$$k = \omega/c \tag{2.3.1}$$

This is most easily analyzed by using monochromatic radiation and then a Fourier integral to extend that result to the desired overall time dependence. For each frequency, see Eq. 1.6.8 and 1.6.11, the Maxwell equations in an empty hollow chamber are:

$$\eta\tilde{H} = \frac{i}{k}\nabla\times\tilde{E} \qquad \tilde{E} = -\frac{i}{k}\nabla\times\left(\eta\tilde{H}\right)$$
$$\nabla\cdot\tilde{H} = 0 \qquad\qquad \nabla\cdot\tilde{E} = 0 \tag{2.3.2}$$

Let the cavity be a rectangular box that extends from 0 to a along the x-axis, 0 to b along the y-axis, and 0 to d along the z-axis. Boundary conditions applied to perfectly conducting walls require all parallel electric field components to be zero at the surface. Since solutions of Eq. A.7.3 are both time and spatial sinusoids the general form of electric field components is:

$$E_x = E_1 \cos\left(k_x x\right)\sin\left(k_y y\right)\sin\left(k_z z\right)e^{i\omega t}$$
$$E_y = E_2 \sin\left(k_x x\right)\cos\left(k_y y\right)\sin\left(k_z z\right)e^{i\omega t} \tag{2.3.3}$$
$$E_z = E_3 \sin\left(k_x x\right)\sin\left(k_y y\right)\cos\left(k_z z\right)e^{i\omega t}$$

Constants E_1, E_2, and E_3 are specific to a specific problem. Since k satisfies Eq. A.5.17 it is also a vector, and since by Eq. 2.3.2 the divergence is equal to zero it follows that:

$$\boldsymbol{k}\cdot\tilde{E} = 0 = k_x E_1 + k_y E_2 + k_z E_3 \tag{2.3.4}$$

This condition permits expressing two of the field constants as functions of the other, from which an acceptable form is:

$$E_x = -\frac{k_x k_z}{k_x^2 + k_y^2} E_3 \cos(k_x x)\sin(k_y y)\sin(k_z z)\cos\omega t$$

$$E_y = -\frac{k_y k_z}{k_x^2 + k_y^2} E_3 \sin(k_x x)\cos(k_y y)\sin(k_z z)\cos\omega t \qquad (2.3.5)$$

$$E_z = E_3 \sin(k_x x)\sin(k_y y)\cos(k_z z)\cos\omega t$$

Use of Maxwell's equations shows the accompanying set of magnetic field components are:

$$\eta H_x = \frac{i}{k}\left[k_y E_3 - k_z E_2\right]\sin(k_x x)\cos(k_y y)\cos(k_z z)e^{i\omega t}$$

$$\eta H_y = \frac{i}{k}\left[k_z E_1 - k_x E_3\right]\cos(k_x x)\sin(k_y y)\cos(k_z z)e^{i\omega t} \qquad (2.3.6)$$

$$\eta H_z = \frac{i}{k}\left[k_x E_2 - k_y E_1\right]\cos(k_x x)\cos(k_y y)\sin(k_z z)e^{i\omega t}$$

Substituting the field coefficients of Eq. 2.3.5 into 2.3.6 shows that:

$$\eta H_x = iE_3\left(\frac{k_y k}{k_x^2 + k_y^2}\right)\sin(k_x x)\cos(k_y y)\cos(k_z z)e^{i\omega t}$$

$$\eta H_y = -iE_3\left(\frac{k_x k}{k_x^2 + k_y^2}\right)\cos(k_x x)\sin(k_y y)\cos(k_z z)e^{i\omega t} \qquad (2.3.7)$$

$$\eta H_z = 0$$

Inspection of the above equations shows that the electric and magnetic fields are out of time phase. Since the ideal cavity is lossless the energy is constant, and since the electric and magnetic energies are equal the total energy is twice either of them. Solving for the total energy by integrating the energy density over the volume gives:

$$W = \frac{\varepsilon}{16}E_3^2\left(1 + \frac{k_z^2}{k_x^2 + k_y^2}\right)V \qquad (2.3.8)$$

With integer values of ℓ,m,n the conducting boundary conditions give:

$$k_x = \frac{\ell\pi}{a} \quad k_y = \frac{m\pi}{b} \quad k_z = \frac{n\pi}{d} \tag{2.3.9}$$

Combining and introducing w as the energy density gives:

$$k_x = \frac{\ell\pi}{a} \quad k_y = \frac{m\pi}{b} \quad k_z = \frac{n\pi}{d} \tag{2.3.10}$$

$$w = \frac{\varepsilon}{16}E_3^{\,2}\left(1 + \frac{(n/d)^2}{(\ell/a)^2 + (m/b)^2}\right) \tag{2.3.11}$$

To find the number of independent solutions within volume V it is convenient to analyze a cubic cavity. A moment's consideration shows the number of available states equals the number of spatial points in the positive quadrant of k-space. A special case of interest is with integers ℓ,m,n all much greater than one. For that case the number of points nearly equals the volume of that quadrant and, in k-space, the unit length is π/a. The total number of points is therefore 1/8 the volume in phase space:

$$N = \frac{1}{8}\left(\frac{a}{\pi}\right)^3 \int_0^{2\pi} d\phi \int_0^{\pi} \sin\theta\, d\theta \int_0^{k} k^2\, dk = \frac{k^3 V}{6\pi^2} \tag{2.3.12}$$

The above argument yields the number of possible field solutions that exist with $H_z = 0$. A parallel argument gives the number of possible field solutions that exist with $E_z = 0$. The sum is the total number of possible solutions:

$$N = \frac{k^3 V}{3\pi^2} \tag{2.3.13}$$

The number of states between frequencies ω and $\omega + d\omega$ is:

$$\frac{1}{V}dN = \frac{\omega^2}{\pi^2 c^3}d\omega \qquad (2.3.14)$$

This becomes the expression for the number of energy states available in free space by imagining all space to be in an enclosed system and letting the dimensions of the system become limitlessly large.

2.4 Q of Time Varying Systems

By definition the Q of a radiation field is the radian frequency times the standing energy-to-emitted power ratio. Although relatively small for antennas on the order of half wavelength or larger, for electrically small and decreasingly sizes Q rises precipitously.

Defining $W(t)$ to be the energy stored in the system, the magnitude of the Q of any system is defined to be the dimensionless ratio:

$$Q = \frac{\omega W(t)}{dW(t)/dt} \qquad (2.4.1)$$

This shows Q is a measure of how rapidly a system decays. Rewriting Eq. 2.4.1 gives:

$$\frac{dW(t)}{dt} = -\frac{\omega}{Q}W(t) \qquad (2.4.2)$$

The solution of Eq. 2.4.2 is:

$$W(t) = W_0 e^{-\omega t/Q} \qquad (2.4.3)$$

W_0 is the initial energy. As a non-electrical example, consider a ball bouncing on a smooth, horizontal, surface. In a uniform gravitational field, the energy is proportional to the height and at maximum height the energy is entirely due to position. The maximum height reached by the ball follows the exponential decay of Eq. 2.4.3 and the ratio of heights on successive bounces is $e^{-2\pi/Q}$.

As a circuit example consider a series electric circuit containing both resistive and passive circuit elements. It is driven by time-dependent voltage $v(t)$ that produces current flow $i(t)$. The voltage and current satisfy the integro-differential equation:

$$L\frac{di(t)}{dt} + Ri(t) + \frac{1}{C}\int i(t)\,dt = v(t) \tag{2.4.4}$$

The homogeneous part is the damped harmonic oscillator equation:

$$\frac{d^2 i(t)}{dt^2} + \frac{R}{L}\frac{di(t)}{dt} + \frac{1}{LC}i(t) = 0 \tag{2.4.5}$$

Current as a function of time satisfies the equation:

$$i(t) = I_0 e^{st} \tag{2.4.6}$$

Substituting Eq. 2.4.6 into 2.4.5 shows that

$$s = -\frac{R}{2L} \pm \sqrt{\frac{R^2}{4L^2} - \frac{1}{LC}} \tag{2.4.7}$$

Introduce the notation:

$$\alpha = R/2L \quad \text{and} \quad \omega_0 = 1/\sqrt{LC} \tag{2.4.8}$$

Combining shows the solution of the homogeneous equation is:

$$i(t) = I_0 e^{-\alpha t} e^{\pm t\sqrt{\alpha^2 - \omega_0^2}} \tag{2.4.9}$$

The character of the solution depends upon the relative sizes of α and ω_0. Consider the special case:

$$\omega_0 > \alpha \tag{2.4.10}$$

Combining Eq. 2.4.10 with Eq. 2.4.9 gives:

$$i(t) = I_0 e^{-\alpha t} e^{\pm it \sqrt{\omega_0^2 - \alpha^2}} \tag{2.4.11}$$

The energy of the system is proportional to:

$$W(t) \sim i(t) i^*(t) = I_0 I_0^* e^{-2\alpha t} \tag{2.4.12}$$

The rate of energy decay, *i.e.*, the power out, is:

$$P(t) \sim \frac{d}{dt}\left[i(t) i^*(t)\right] = -2\alpha I_0 I_0^* e^{-2\alpha t} \tag{2.4.13}$$

Combining the definition of Q with the above gives:

$$Q = \left|\frac{\omega W_{pk}}{P_{av}}\right| = \frac{\omega}{2\alpha} \tag{2.4.14}$$

With this series circuit the Q-bandwidth relationship follows using the input impedance of the circuit:

$$Z(\omega) = R + i\omega L\left(1 - \omega_0^2 / \omega^2\right) \tag{2.4.15}$$

Bandwidth is defined as the frequency difference between half-power points. The impedance is real and equal to R at frequency ω_0; half power points have equal real and reactive parts and hence the impedance is equal to $\sqrt{2} R$. With ω_1 representing the frequency of a half power point:

$$R = \omega_1 L\left(1 - \omega_0^2 / \omega_1^2\right) \tag{2.4.16}$$

Combining gives:

$$(\omega_1 - \omega_0)(\omega_1 + \omega_0) - \frac{\omega_1 R}{L} \tag{2.4.17}$$

For the important special case of low frequency modulation of high frequency carriers the bandwidth is small and, for that case, Eq. 2.4.17 is approximately $\Delta\omega = R/2L$ where $\Delta\omega = \pm(\omega_1 - \omega_0)$, the difference between half-power and resonant frequencies. Combining the definitions of Eq. 2.4.8 and 2.4.14 shows the bandwidth, B, normalized to the actual frequency is:

$$B = \Delta\omega/\omega_0 = 1/Q \tag{2.4.18}$$

Consider the case of a lossy inductor. The input impedance is:

$$Z = R + i\omega L \tag{2.4.19}$$

If the current is $I_0\cos(\omega t)$ the energy stored in the inductance and the power loss in the resistance are:

$$W(t) = \frac{1}{4}LI_0^2\left[1+\cos(2\omega t)\right]; \quad P(t) = \frac{1}{2}RI_0^2\left[1+\cos(2\omega t)\right] \tag{2.4.20}$$

Combining the definition of Eq. 2.4.14 with Eq. 2.4.20 gives:

$$Q = \frac{\omega L}{R} = \tan\zeta \tag{2.4.21}$$

Angle ζ is the phase angle of the impedance. A similar expression holds for lossy capacitors. This expression is particularly useful at frequencies far removed from the resonant frequency.

Returning to antennas, the relationship between Q and the peak standing energy about an antenna is:

$$W_{pk} = \frac{P_{av}}{\omega}Q = \frac{Q}{\omega}\frac{dW_{pk}}{dt} \tag{2.4.22}$$

An isolated oscillator decays as:

$$W_{pk} = W_{pk0}e^{-\omega t/Q} \tag{2.4.23}$$

A fundamental difficulty with calculating antenna Q is: if an antenna has been producing radiation since time $t = -\infty$, a requirement for steady state analysis, an infinite amount of field energy exists and as the radius of integration increases without limit so does the energy. Satisfactory results ensue only if the field energy in the expression for Q excludes energy that has permanently left the local system; separating traveling from standing energies is a critical step in the calculation of antenna Q.

Problems associated with high-Q antennas include large surface currents and associated ohmic losses, the reactive energy returns to the source twice each field cycle and must be somehow accommodated, and if a limited energy supply is available the required standing field energy for a desired power output may exceed the supply.

2.5 Instantaneous and Complex Power in Fields

It is generally convenient to use complex notation to describe electromagnetic fields. Phasors are useful primarily because they transform the differential and integral operations of calculus into arithmetic ones. Although that is helpful enough to justify using phasors, there are difficulties. Since those difficulties arise in subsequent chapters we detail the use of phasors in this one. The real and reactive powers in a radiation field are of particular significance. By definition, real power exits the emitting region but reactive power merely oscillates in position.

Although all local energy contributes to Q, only energy within a half wavelength of an antenna returns to it during steady state operation and hence only it affects an antenna's input impedance. At the driving terminals of an antenna, energy radiated permanently away from the system is indistinguishable from energy captured by an absorber and absorption creates a radiation resistance. Energy oscillating to and from the source produces a radiation reactance. To a prime mover driving the source the effective impedance is indistinguishable from the input impedance of a properly synthesized closed circuit. Hence, from the point of view of the driving source, an antenna may be replaced by and analyzed as if it were an electric circuit.

Written in phasor form, but keeping the retarded time phase dependence, in terms of the letter functions of Sec. A.19, and with $t_R = t - \sigma/\omega$, the retarded time, the general form of field expansion Eq. 1.13.9 for fields with rotational symmetry is:

$$\sigma^2 \tilde{E}_r = \sum_{\ell=1}^{\infty} F_\ell \ell(\ell+1)\left[B_\ell(\sigma)+iA_\ell(\sigma)\right]P_\ell(\cos\theta)e^{i\omega t_R}$$

$$\sigma^2 \eta \tilde{H}_r = -\sum_{\ell=1}^{\infty} G_\ell \ell(\ell+1)\left[B_\ell(\sigma)+iA_\ell(\sigma)\right]P_\ell(\cos\theta)e^{i\omega t_R}$$

$$\sigma \tilde{E}_\theta = \sum_{\ell=1}^{\infty} F_\ell\left[D_\ell(\sigma)+iC_\ell(\sigma)\right]\frac{dP_\ell(\cos\theta)}{d\theta}e^{i\omega t_R}$$

$$\sigma \eta \tilde{H}_\phi = \sum_{\ell=1}^{\infty} F_\ell\left[A_\ell(\sigma)-iB_\ell(\sigma)\right]\frac{dP_\ell(\cos\theta)}{d\theta}e^{i\omega t_R} \qquad (2.5.1)$$

$$\sigma \tilde{E}_\phi = \sum_{\ell=1}^{\infty} G_\ell\left[A_\ell(\sigma)-iB_\ell(\sigma)\right]\frac{dP_\ell(\cos\theta)}{d\theta}e^{i\omega t_R}$$

$$\sigma \eta \tilde{H}_\theta = -\sum_{\ell=1}^{\infty} G_\ell\left[D_\ell(\sigma)+iC_\ell(\sigma)\right]\frac{dP_\ell(\cos\theta)}{d\theta}e^{i\omega t_R}$$

Each and every emission field with rotational symmetry about the z-axis may be fully described by making appropriate choices of multiplying coefficients F_ℓ and G_ℓ. As noted in Sec. 1.13 since there are no radial magnetic field terms proportional to F_ℓ and no electric field terms proportional to G_ℓ. Hence these fields are denoted, respectively, as transverse magnetic, TM, and transverse, TE, electric fields.

Using Eq. 2.5.1 and Table A.16.1.6 the surface integral of the complex Poynting vector evaluated on a circumscribing, spherical surface of radius σ/k is:

$$P_c(\sigma) = \oint \mathbf{N}_c \cdot d\mathbf{S}$$

$$= \frac{\pi}{\eta k^2}\sum_{\ell=1}^{\infty}\frac{\ell(\ell+1)}{(2\ell+1)}\left\{\begin{array}{l}\left(F_\ell F_\ell^* + G_\ell G_\ell^*\right)\left[A_\ell(\sigma)D_\ell(\sigma)-B_\ell(\sigma)C_\ell(\sigma)\right] \\ +i\left(F_\ell F_\ell^* - G_\ell G_\ell^*\right)\left[A_\ell(\sigma)C_\ell(\sigma)+B_\ell(\sigma)D_\ell(\sigma)\right]\end{array}\right\} \qquad (2.5.2)$$

The absence of cross product terms between TM and TE modes shows that the two modal types act independently. The sign of the imaginary term depends upon whether the field is TM or TE; the net is zero if the coefficients have equal magnitudes. Since each modal coefficient is multiplied by its own complex conjugate a phase difference between sources has no effect and all modal phase factors are suppressed.

Inspection of Eq. 2.5.2 shows the two numbers needed to evaluate modal power are weighted sums over $(A_\ell D_\ell - B_\ell C_\ell)$, and $(A_\ell C_\ell + B_\ell D_\ell)$. By Table A.20.2.1, $(A_\ell D_\ell - B_\ell C_\ell)$, is equal to one for all orders. We define the second term to be:

$$\gamma_\ell(\sigma) = A_\ell(\sigma) C_\ell(\sigma) + B_\ell(\sigma) D_\ell(\sigma) \tag{2.5.3}$$

Values of $\gamma_\ell(\sigma)$ are listed in Table 2.5.1. That table shows the magnitude of $\gamma_\ell(\sigma)$ increases precipitously with small and decreasing values of σ and with increasing modal number ℓ. All signs in Table 2.5.1 are the same and hence $\gamma_\ell(\sigma)$ is a monotone decreasing function of σ. Since $\gamma_\ell(\sigma)$ is a measure of reactive power we note the reactance of the surface of a circumscribing virtual surface has the same sign for all radii: capacitive for TM modes and inductive for TE modes. This is in marked contrast with the numerical analysis of center-driven biconical antennas where the sign of the reactance of a TM antenna at the input terminals is primarily a function of normalized cone length. Changes in the sign of the input reactance versus antenna arm length for TM sources are due to the transmission line character of the antenna arms and not to intrinsic properties of an external, virtual radiating surface.

Table 2.5.1. Radial dependence of $\gamma_1(s)$ for modes 1 through 5.

$$\gamma_1(\sigma) = -\frac{1}{\sigma^3}$$

$$\gamma_2(\sigma) = -\frac{18}{\sigma^5} - \frac{3}{\sigma^3}$$

$$\gamma_3(\sigma) = -\frac{675}{\sigma^7} - \frac{90}{\sigma^5} - \frac{6}{\sigma^3}$$

$$\gamma_4(\sigma) = -\frac{44\,100}{\sigma^9} - \frac{4\,725}{\sigma^7} - \frac{270}{\sigma^5} - \frac{10}{\sigma^3}$$

$$\gamma_5(\sigma) = -\frac{4\,465\,125}{\sigma^{11}} - \frac{396\,900}{\sigma^9} - \frac{18\,900}{\sigma^7} - \frac{630}{\sigma^5} - \frac{15}{\sigma^3}$$

2.6 Time Varying Power in Actual Radiation Fields

The actual fields, from which the phasor fields of Eq. 2.5.1 follow, are shown in Eq. 2.6.1 and vary trigonometrically with time. Since adjustment of the time origin adjusts the absolute phase of the full field set results of analyzing any particular one are general. As will be noted, however, for this case the phase difference between TE and TM modes is significant.

$$\sigma^2 E_r = \sum_{\ell=1}^{\infty} F_\ell \ell(\ell+1)\left[B_\ell \cos(\omega t_R - \phi) - A_\ell \sin(\omega t_R - \phi)\right] P_\ell(\cos\theta)$$

$$\sigma^2 \eta H_r = -\sum_{\ell=1}^{\infty} G_\ell \ell(\ell+1)\left[B_\ell \cos(\omega t_R - \phi) - A_\ell \sin(\omega t_R - \phi)\right] P_\ell(\cos\theta)$$

$$\sigma E_\theta = \sum_{\ell=1}^{\infty} F_\ell \left[D_\ell \cos(\omega t_R - \phi) - C_\ell \sin(\omega t_R - \phi)\right]\frac{d}{d\theta} P_\ell(\cos\theta)$$

$$\sigma \eta H_\phi = \sum_{\ell=1}^{\infty} F_\ell \left[A_\ell \cos(\omega t_R - \phi) + B_\ell \sin(\omega t_R - \phi)\right]\frac{d}{d\theta} P_\ell(\cos\theta) \tag{2.6.1}$$

$$\sigma E_\phi = \sum_{\ell=1}^{\infty} G_\ell \left[A_\ell \cos(\omega t_R - \phi) + B_\ell \sin(\omega t_R - \phi)\right]\frac{d}{d\theta} P_\ell(\cos\theta)$$

$$\sigma \eta H_\theta = -\sum_{\ell=1}^{\infty} G_\ell \left[D_\ell \cos(\omega t_R - \phi) - C_\ell \sin(\omega t_R - \phi)\right]\frac{d}{d\theta} P_\ell(\cos\theta)$$

Using Eq. 2.6.1 to evaluate the radial component of the time-dependent Poynting vector then integrating over a circumscribing, constant radius surface centered at the origin gives the surface power:

$$p(\sigma, t_R) = \oint \mathbf{N} \cdot d\mathbf{S} =$$

$$\frac{\pi}{\eta k^2}\sum_{\ell=1}^{\infty}\frac{\ell(\ell+1)}{(2\ell+1)}\left[F_\ell^2 + G_\ell^2\right]\left\{ \begin{array}{l}(A_\ell D_\ell - B_\ell C_\ell) + (A_\ell D_\ell + B_\ell C_\ell)\cos 2(\omega t_R - \phi) \\ -(A_\ell C_\ell - B_\ell D_\ell)\sin 2(\omega t_R - \phi)\end{array}\right\} \tag{2.6.2}$$

This equation, in contrast with Eq. 2.5.2, depends upon the relative phases of the driving modes.

Equation 2.6.2 contains three separate parameters: $(A_\ell D_\ell - B_\ell C_\ell)$, $(A_\ell D_\ell + B_\ell C_\ell)$, and $(A_\ell C_\ell - B_\ell D_\ell)$. Since in what follows we seek to work with functions that vanish at limitlessly large ranges, we define $\alpha_\ell(\sigma)$ and $\beta_\ell(\sigma)$ as:

$$\alpha_\ell(\sigma) = \left(A_\ell D_\ell + B_\ell C_\ell\right) - (-1)^\ell; \quad \beta_\ell(\sigma) = \left(A_\ell C_\ell - B_\ell D_\ell\right) \tag{2.6.3}$$

Combining gives:

$$
\begin{aligned}
p(\sigma, t_R) = \frac{\pi}{\eta k^2} \sum_{\ell=1}^{\infty} & \frac{\ell(\ell+1)}{(2\ell+1)} \left[F_\ell^2 + G_\ell^2\right] \\
\times & \left\{ \begin{array}{l} \left[1+(-1)^\ell \cos 2\left(\omega t_R - \phi\right)\right] \\ + \left[\alpha_\ell(\sigma)\cos 2\left(\omega t_R - \phi\right) - \beta_\ell(\sigma)\sin 2\left(\omega t_R - \phi\right)\right] \end{array} \right\}
\end{aligned} \tag{2.6.4}
$$

Within the curly brackets of Eq. 2.6.4, the envelope of the first term is independent of range. Functions $\alpha_\ell(\sigma)$ and $\beta_\ell(\sigma)$ are alternating series that constitute the reactive power density. Each is an oscillating function of range; values are listed in Tables 2.6.1 and 2.6.2 for $\ell = 1$ through 5.

The first term in Eq. 2.6.4 is the real power $p_r(\sigma, t_R)$ where:

$$p_r(\sigma, t_R) = \frac{\pi}{\eta k^2} \sum_{\ell=1}^{\infty} \frac{\ell(\ell+1)}{(2\ell+1)} \left[F_\ell^2 + G_\ell^2\right]\left[1+(-1)^\ell \cos 2\left(\omega t_R - \phi\right)\right] \tag{2.6.5}$$

This power travels ever outward at speed c in the form of periodic, trigonometric pulses; there is no range dependence and it does not go to a limit at infinity.

The distance-dependent power terms in Eq. 2.6.4 are given by $p_i(\sigma, t_R)$ where:

$$p_i(\sigma, t_R) = \frac{\pi}{\eta k^2} \sum_{\ell=1}^{\infty} \frac{\ell(\ell+1)}{(2\ell+1)} \left[F_\ell^2 + G_\ell^2\right]\left[\begin{array}{l} \alpha_\ell(\sigma)\cos 2\left(\omega t_R - \phi\right) \\ -\beta_\ell(\sigma)\sin 2\left(\omega t_R - \phi\right) \end{array} \right] \tag{2.6.6}$$

As may be seen from Tables 2.6.1 and 2.6.2, the envelope of each term is largest at the emitter surface and goes asymptotically to zero at infinite radius.

Table 2.6.1. Radial dependence of $\alpha_\ell(\sigma)$.

$$\alpha_1(\sigma) = \frac{2}{\sigma^2}$$

$$\alpha_2(\sigma) = \frac{36}{\sigma^4} - \frac{18}{\sigma^2}$$

$$\alpha_3(\sigma) = \frac{1\,350}{\sigma^6} - \frac{720}{\sigma^4} + \frac{72}{\sigma^2}$$

$$\alpha_4(\sigma) = \frac{88\,200}{\sigma^8} - \frac{49\,350}{\sigma^6} + \frac{6\,000}{\sigma^4} - \frac{200}{\sigma^2}$$

$$\alpha_5(\sigma) = \frac{8\,930\,250}{\sigma^{10}} - \frac{5\,159\,700}{\sigma^8} + \frac{699\,300}{\sigma^6} - \frac{31\,500}{\sigma^4} + \frac{450}{\sigma^2}$$

Table 2.6.2. Radial dependence of $\beta_\ell(\sigma)$.

$$\beta_1(\sigma) = -\frac{1}{\sigma^3} + \frac{2}{\sigma}$$

$$\beta_2(\sigma) = -\frac{18}{\sigma^5} + \frac{33}{\sigma^3} - \frac{6}{\sigma}$$

$$\beta_3(\sigma) = -\frac{675}{\sigma^7} + \frac{1\,250}{\sigma^5} - \frac{276}{\sigma^3} + \frac{12}{\sigma}$$

$$\beta_4(\sigma) = -\frac{44\,100}{\sigma^9} + \frac{83\,475}{\sigma^7} - \frac{20\,220}{\sigma^5} + \frac{1\,300}{\sigma^3} - \frac{20}{\sigma}$$

$$\beta_5(\sigma) = -\frac{4\,465{,}125}{\sigma^{11}} + \frac{8\,533\,350}{\sigma^9} - \frac{2\,201\,850}{\sigma^7} + \frac{169\,470}{\sigma^5} - \frac{4\,425}{\sigma^3} + \frac{30}{\sigma}$$

2.7 Comparison of Complex and Instantaneous Powers

Although only TM modes are analyzed in this section, TE modes obey the same laws with the sole exception that the sign is different for the imaginary.

The time-dependent field power of Eq. 2.5.1 leads to the complex power of Eq. 2.5.2. With ξ as a constant phase factor it may be put in the form:

$$P_c(\sigma, t_R)$$

$$= \frac{\pi}{\eta k^2} \sum_{\ell=1}^{\infty} \frac{\ell(\ell+1)}{(2\ell+1)} F_\ell^2 \left\{ \left[1 \pm \cos(2\omega t_R - 2\xi_\ell) \right] + \gamma_\ell(\sigma)\sin(2\omega t_R - 2\xi_\ell) \right\} \quad (2.7.1)$$

The power expression from the time-dependent fields is:

$$p(\sigma,t_R) = \frac{\pi}{\eta k^2} \sum_{\ell=1}^{\infty} \frac{\ell(\ell+1)}{(2\ell+1)} F_\ell^2 \left\{ \begin{array}{l} \left[1 + (-1)^\ell \cos 2(\omega t_R - \phi) \right] \\ + \left[\alpha_\ell(\sigma)\cos 2(\omega t_R - \phi) - \beta_\ell(\sigma)\sin 2(\omega t_R - \phi) \right] \end{array} \right\}$$

$$(2.7.2)$$

Equations 2.7.1 and 2.7.2 describe the same energy flow; each curly bracket is multiplied by the same factor but there are, respectively, two and three time-dependent terms.

The first term of Eq. 2.7.1 describes a unidirectional energy flow away from the source. This, the real part of the complex power, does not go to a limit at infinite radius, it is equal to zero twice each field cycle, and it is never negative. The second term describes radially directed, alternately directed power in time quadrature with the first and hence is the reactive power; it vanishes in the limit of infinite radius. The first term of Eq. 2.7.2 is the real power. Like its counterpart in Eq. 2.7.1, it does not go to a limit at infinite radius, it is equal to zero twice each field cycle, and it is never negative. It, too, describes a unidirectional energy flow away from the source. The real power and $\alpha_\ell(\sigma)$ powers are in time phase and both are in time quadrature with $\beta_\ell(\sigma)$ power. Both $\alpha_\ell(\sigma)$ and $\beta_\ell(\sigma)$ powers vanish at limitlessly large distance; at each point both oscillate between equal negative and positive parts and hence both describe radially directed, alternating power. The phases of the real part of the complex power and the real power differ by a radius-dependent phase factor.

A quantitative expression for phase angle $\xi_\ell(\sigma)$ follows by equating Eq. 2.7.1 and 2.7.2:

$$\tan(2\xi_\ell) = \frac{A_\ell B_\ell}{A_\ell^2 - B_\ell^2} \tag{2.7.3}$$

It follows from Eq. 2.7.1 that the group velocity of the real part of the complex power is:

$$v_{gp} = \frac{c}{1 + d\xi_\ell/d\sigma} \tag{2.7.4}$$

It may be verified using Table A.20.2 No. 4 that:

$$\frac{d}{d\sigma}\left(\frac{A_\ell B_\ell}{A_\ell^2 - B_\ell^2}\right) \leq 0 \qquad (2.7.5)$$

Combining Eq. 2.7.3 and 2.7.5 with functional properties of the tangent gives:

$$d\xi_\ell / d\sigma \leq 0 \qquad (2.7.6)$$

Combining Eq. 2.7.4 and 2.7.6 shows that the real part of the complex power, Eq. 2.7.1, propagates faster than the speed of light. A basic tenet of physics is that the speed of electromagnetic energy is never greater than c. This confirms the complex power is not a physical entity and it does not correctly describe an actual energy flow. In contrast, the first term of Eq. 2.7.2 does travel at the speed of light and does describe an actual energy flow.

It follows from Eq. 2.7.1 that if the calculus operations of differentiating or integrating complex power with respect to the radius is done, the calculation must include operations on the function $\xi_\ell(\sigma)$. Yet with complex power, knowledge of $\xi_\ell(\sigma)$ is suppressed and unavailable. Therefore, it is not possible to carry out such operations from knowledge of only complex power.

If suppressed phase angle $\xi_\ell(\sigma)$ of mode ℓ is assigned a value of zero at a vanishingly small radius, the value decreases with increasing radius

Table 2.7.1. Radius for which selected values of phase angle occur, three lowest modes.

$\xi_\ell(\sigma)$	$\ell = 1$	$\ell = 2$	$\ell = 3$
0	0	0	0
$-\pi/2$	0.618	0.777	0.785
$-\pi$	1	1.414	1.566
$-3\pi/2$	1.618	1.882	2.421
-2π	∞	2.149	2.439
$-5\pi/2$		4.104	3.289
-3π		∞	4.310
$-7\pi/2$			2.852
-4π			∞

to equal $-(\ell+1)\pi$ at infinite radius. Table 2.7.1 lists values of σ for which the phase angle reaches selected values as a function of radius and modal number.

Since the use of complex power is uncompromised in electric circuits, the complex power expression of Eq. 2.7.1, re-expressed as Eq. 2.7.7, applies to the driving circuitry, including the input side of the radiating surface, $\sigma = ka$:

$$P_c(\sigma, t_R) = \frac{\pi}{\eta k^2} \sum_{\ell=1}^{\infty} \frac{\ell(\ell+1)}{(2\ell+1)} F_\ell^2 \left\{ \begin{array}{l} \left[A_\ell D_\ell - B_\ell C_\ell\right]\left[1 + \cos 2\left(\omega t_R - \xi_\ell\right)\right] \\ -\left[A_\ell C_\ell + B_\ell D_\ell\right]\sin 2\left(\omega t_R - \xi_\ell\right) \end{array} \right\}$$

$$(2.7.7)$$

The time-dependent power expression of Eq. 2.7.2, re-expressed as Eq. 2.7.8, applies to the external region, including the output side of the radiating surface $\sigma = ka$:

$$p(\sigma, t_R) = \frac{\pi}{\eta k^2} \sum_{\ell=1}^{\infty} \frac{\ell(\ell+1)}{(2\ell+1)} F_\ell^2 \left\{ \begin{array}{l} \left[A_\ell D_\ell - B_\ell C_\ell\right] \pm \left[A_\ell D_\ell + B_\ell C_\ell\right]\cos\left(2\omega t_R\right) \\ \mp \left[A_\ell C_\ell - B_\ell D_\ell\right]\sin\left(2\omega t_R\right) \end{array} \right\}$$

$$(2.7.8)$$

The mean square value of the time varying portions are respectively given by:

$$\left[A_\ell D_\ell - B_\ell C_\ell\right]^2 + \left[A_\ell C_\ell + B_\ell D_\ell\right]^2 = \left[A_\ell D_\ell + B_\ell C_\ell\right]^2 + \left[A_\ell C_\ell - B_\ell D_\ell\right]^2$$

$$(2.7.9)$$

It follows by inspection that Eq. 2.7.9 is an identity. Therefore, the total power is continuous through the interface. The left side terms are the magnitudes of the real plus imaginary parts of the input complex power on the source side. On the right side, the first term applies to the time variation of the real power and the in-phase oscillatory power. The second term represents the out-of-phase oscillatory power.

The first two terms inside the curly brackets of Eq. 2.7.8 may be written as:

$$\left[A_\ell D_\ell - B_\ell C_\ell\right] + \left[A_\ell D_\ell + B_\ell C_\ell\right]\cos\left(2\omega t_R\right)$$
$$= \left[A_\ell D_\ell - B_\ell C_\ell\right]\left[1 + (-1)^\ell \cos\left(2\omega t_R\right)\right] + \left[A_\ell D_\ell + B_\ell C_\ell - (-1)^\ell\right]\cos\left(2\omega t_R\right)$$

$$(2.7.10)$$

Comparison of Eq. 2.7.7 and 2.7.8 as modified by Eq. 2.7.10 at $\sigma = ka$ shows that the real power undergoes a phase discontinuity of $2\xi_\ell$ as it passes through the antenna. The absolute phase is determined by the phase of the source and the antenna circuit impedances.

In summary, although the total time-dependent power is continuous through the interface between the source and field regions, the separation of that power into constituent parts is different. On the source side, the power separates into real and reactive parts the time varying portions of which are in time quadrature. Power that is in phase with the input power represents power loss from the system. On the field side, power that is in phase with the real power does not represent power loss; some oscillatory power is in phase with the real power and some is in phase quadrature.

At the surface of a radiating sphere, it is correct to write the complex power in the form of Eq. 2.5.2 as:

$$P_c(ka) = \oint N_c \cdot dS$$
$$= \frac{\pi}{\eta k^2} \sum_{\ell=1}^{\infty} \frac{\ell(\ell+1)}{(2\ell+1)} \left\{ \begin{array}{l} \left(F_\ell F_\ell^* + G_\ell G_\ell^*\right)\left[A_\ell(ka)D_\ell(ka) - B_\ell(ka)C_\ell(ka)\right] \\ + i\left(F_\ell F_\ell^* - G_\ell G_\ell^*\right)\left[A_\ell(ka)C_\ell(ka) + B_\ell(ka)D_\ell(ka)\right] \end{array} \right\}$$

$$(2.7.11)$$

The equation is correct only at radius a and the equality of the imaginary part does not extend to larger radii. The equality also shows another important mathematical feature of electromagnetic radiation: The upper and lower terms, respectively $(A_\ell D_\ell - B_\ell C_\ell)$ and $(A_\ell C_\ell + B_\ell D_\ell)$, relate to the Bessel and Neumann functions as:

$$\left(A_\ell D_\ell - B_\ell C_\ell\right) = \left(j_\ell \dot{y}_\ell - y_\ell \dot{j}_\ell\right)\sigma^2$$
$$\left(A_\ell C_\ell + B_\ell D_\ell\right) = \left(j_\ell \dot{j}_\ell + y_\ell \dot{y}_\ell\right)\sigma^2$$

$$(2.7.12)$$

The upper part of Eq. 2.7.12, a quantity proportional to the real output power, consists of the products of terms proportional to products of one spherical Bessel function versus one spherical Neumann function. We conclude that electromagnetic energy can only be radiated away from a source if fields proportional to both functions are present. The lower part of Eq. 2.7.12, a quantity proportional to the imaginary part of the surface power, consists of products of terms proportional to Bessel functions plus those proportional to Neumann functions. This term, therefore, is present with all fields.

Since outward modal speeds differ, consider complex power notation for radiation emitted from radius a. At radius $b > a$ the intermodal phase differences differ from the emitted ones and wave reconstruction does not accurately reproduce the emitted one.

2.8 Traveling Waves

With suppressed $\exp(i\omega t)$ time dependence, the electric and magnetic phasor fields of a unit magnitude, x-polarized, z-directed plane wave expressed in rectangular coordinates are:

$$\tilde{E} = \hat{x}e^{-ikz} \quad \text{and} \quad \eta\tilde{H} = \hat{y}e^{-ikz} \tag{2.8.1}$$

The same fields expressed in spherical coordinates are:

$$\tilde{E} = \left\{\sin\theta\cos\phi\hat{r} + \cos\theta\cos\phi\hat{\theta} - \sin\phi\hat{\phi}\right\}e^{-i\sigma\cos\theta}$$

$$\eta\tilde{H} = \left\{\sin\theta\sin\phi\hat{r} + \cos\theta\sin\phi\hat{\theta} + \cos\phi\hat{\phi}\right\}e^{-i\sigma\cos\theta} \tag{2.8.2}$$

Since the fields of Eq. 1.13.9 may be used to describe all steady state fields, including plane waves. For a plane wave it is necessary to obtain appropriate values of the field coefficients $F(\ell,m)$ and $G(\ell,m)$. For that purpose it is convenient to consider the radial field components. Again referring to Eq. 1.13.9, the radial component of the electric field intensity is:

$$\tilde{E}_r = i \sum_{\ell=0}^{\infty} \sum_{m=0}^{\ell} i^{-\ell} F(\ell,m) \ell(\ell+1) \frac{z_\ell(\sigma)}{\sigma} \Theta_\ell^m (\cos\theta) e^{-jm\phi} \qquad (2.8.3)$$

Since a plane wave contains no singularities, neither does expression Eq. 2.8.3; it follows that the coefficients of spherical Neumann functions vanish leaving only spherical Bessel functions. Since the wave occupies all azimuth angles ϕ in the range $0 \le \theta \le 2\pi$, degree m is an integer. Since the z-axes are included in the solution only integer order, associated Legendre functions of the first kind, $P_\ell^m(\cos\theta)$, are present. Applying these conditions and equating Eq. 2.8.3 with the radial component of Eq. 2.8.2 gives:

$$\sin\theta \cos\phi \, e^{-i\sigma\cos\theta} = \frac{i}{\sigma} \sum_{\ell=0}^{\infty} \sum_{m=0}^{\ell} i^{-\ell} F(\ell,m) \ell(\ell+1) j_\ell(\sigma) P_\ell^m (\cos\theta) e^{-jm\phi}$$

$$(2.8.4)$$

The azimuth angular dependence of Eq. 2.8.4 shows that only the coefficients of degree one are different from zero. Also, since the left side is real with respect to 'j' so is the right side. This leaves the residual terms:

$$e^{-i\sigma\cos\theta} = \frac{i}{\sigma} \sum_{\ell=1}^{\infty} i^{-\ell} F(\ell,1) \ell(\ell+1) j_\ell(\sigma) \frac{P_\ell^1(\cos\theta)}{\sin\theta} \qquad (2.8.5)$$

Another expansion for the exponential is listed in Table A.16.1 No. 2:

$$e^{-i\sigma\cos\theta} = \frac{i}{\sigma} \sum_{\ell=1}^{\infty} i^{-\ell} (2\ell+1) j_\ell(\sigma) \frac{P_\ell^1(\cos\theta)}{\sin\theta} \qquad (2.8.6)$$

Equating the two expressions gives:

$$F(\ell,1) = \frac{(2\ell+1)}{\ell(\ell+1)} \qquad (2.8.7)$$

Combining gives the radial component of a unit magnitude electric field in the plane wave:

$$\tilde{E}_r = \frac{i}{\sigma} \sum_{\ell=1}^{\infty} i^{-\ell} (2\ell+1) j_\ell(\sigma) P_\ell^1(\cos\theta) \cos\phi \tag{2.8.8}$$

Working with TE modes in a similar way gives:

$$G(\ell,1) = -\frac{(2\ell+1)}{\ell(\ell+1)} \tag{2.8.9}$$

$$\eta \tilde{H}_r = \frac{i}{\sigma} \sum_{\ell=1}^{\infty} i^{-\ell} (2\ell+1) j_\ell(\sigma) P_\ell^1(\cos\theta) \sin\phi \tag{2.8.10}$$

Angular field components follow from the radial components and the form of Eq. 1.13.9:

$$\begin{aligned}
\tilde{E}_\theta &= \sum_{\ell=1}^{\infty} i^{-\ell} \frac{(2\ell+1)}{\ell(\ell+1)} \left[ij_\ell^\bullet(\sigma) \frac{dP_\ell^1}{d\theta} + j_\ell(\sigma) \frac{P_\ell^1}{\sin\theta} \right] \cos\phi \\
\eta \tilde{H}_\phi &= \sum_{\ell=1}^{\infty} i^{-\ell} \frac{(2\ell+1)}{\ell(\ell+1)} \left[j_\ell(\sigma) \frac{dP_\ell^1}{d\theta} + ij_\ell^\bullet(\sigma) \frac{P_\ell^1}{\sin\theta} \right] \cos\phi \\
\tilde{E}_\phi &= -\sum_{\ell=1}^{\infty} i^{-\ell} \frac{(2\ell+1)}{\ell(\ell+1)} \left[j_\ell(\sigma) \frac{dP_\ell^1}{d\theta} + ij_\ell^\bullet(\sigma) \frac{P_\ell^1}{\sin\theta} \right] \sin\phi \\
\eta \tilde{H}_\theta &= \sum_{\ell=1}^{\infty} i^{-\ell} \frac{(2\ell+1)}{\ell(\ell+1)} \left[ij_\ell^\bullet(\sigma) \frac{dP_\ell^1}{d\theta} + j_\ell(\sigma) \frac{P_\ell^1}{\sin\theta} \right] \sin\phi
\end{aligned} \tag{2.8.11}$$

Equations 2.8.8, 2.8.10, and 2.8.11 are the time suppressed electric and magnetic fields of a unit magnitude, x-polarized, z-directed plane wave expressed in spherical coordinates.

2.9 Scattering by a Sphere, General Aspects

In the remaining sections of this chapter we discuss responses of passive, linear systems to electromagnetic fields. Interactions between objects and electromagnetic fields are most easily analyzed by imagining the scatterer immersed in a plane wave. Even though plane waves don't exist, spherical waves do exist and, if the radius of the sphere is much larger than other dimensions of interest, approach being a plane wave as the distance from the source increases without limit. It is only necessary for the radius of the sphere to be much larger than any other spatial dimension of interest to the problem.

We analyze scattering by ideally conducting spheres that respond linearly to the imposed field. A conclusion is that a significant portion of the incoming energy is scattered permanently away from the scatterer. Detailed summations over known functions are obtained as problem solutions, although the field expressions are infinite sums over products of associated Legendre functions with spherical Bessel and Neumann functions. The functions can be numerically evaluated at all orders in the far field and at small orders in the near field but near field evaluation is difficult at high orders.

A spherical scatterer of radius a is immersed in the plane wave described by Eq. 2.8.8, 2.8.10, and 2.8.11. To obtain all fields about the scatterer, a step-wise procedure is used to analyze the interaction between a sphere and a plane wave. In the first step energy and momentum is extracted from the wave and applied to the sphere; these are extinction values of energy and momentum. In the second step the extinction values separate into absorbed and scattered parts. The scatterer permanently retains the absorbed part and the scattered part goes back into space. To determine the scattered fields, note again that all possible fields are expressible in the form of Eq. 1.13.9. Problem solution is simplified if three characteristics of the scattered field are noted: First, since scattered fields exist on the z-axis and since only associated Legendre polynomials of integer order converge exist on that axis the zenith angle dependence varies as associated Legendre polynomials of integer order. Second, the total scattered power is constant in the limit of infinite radius and constant power requires outgoing fields to vary with distance as $e^{-i\sigma}/\sigma$. Only spherical Hankel functions of the second kind have this limiting form and satisfy the spherical Bessel differential equation. Third, the scattered field possesses only the symmetries of the scatterer and the input field. Therefore, only fields of degree one are present.

To solve for the magnitudes and phases of the scattered modes it is convenient to multiply each set of modes by sets of complex scattering coefficients: α_ℓ for TE modes and β_ℓ for TM modes. With this definition the complete expression for the radial components of the scattered fields are:

$$\tilde{E}_r = i\sum_{\ell=1}^{\infty} i^{-\ell}\beta_\ell(2\ell+1)\frac{h_\ell(\sigma)}{\sigma}P_\ell^1(\cos\theta)\cos\phi$$

$$\eta\tilde{H}_r = i\sum_{\ell=1}^{\infty} i^{-\ell}\alpha_\ell(2\ell+1)\frac{h_\ell(\sigma)}{\sigma}P_\ell^1(\cos\theta)\sin\phi$$

(2.9.1)

Problem solution requires evaluation of α_ℓ and β_ℓ for each order ℓ.

The sum of the incident and scattered TM and TE fields are:

$$\tilde{E}_{TM}^{ex} = \sum_{\ell=1}^{\infty} i^{-\ell}\left\{\begin{array}{l}\dfrac{i}{\sigma}(2\ell+1)\left(j_\ell(\sigma)+\beta_\ell h_\ell(\sigma)\right)P_\ell^1(\cos\theta)\cos\phi\hat{r}\\[2ex] +\dfrac{(2\ell+1)}{\ell(\ell+1)}i\left(j_\ell^{\bullet}(\sigma)+\beta h_\ell^{\bullet}(\sigma)\right)\left[\dfrac{dP_\ell^1}{d\theta}\cos\phi\hat{\theta}-\dfrac{P_\ell^1}{\sin\theta}\sin\phi\hat{\phi}\right]\end{array}\right\}$$

$$\tilde{H}_{TM}^{ex} = \frac{1}{\eta}\sum_{\ell=1}^{\infty} i^{-\ell}\frac{(2\ell+1)}{\ell(\ell+1)}\left(j_\ell(\sigma)+\beta_\ell h_\ell(\sigma)\right)\left[\frac{P_\ell^1}{\sin\theta}\sin\phi\hat{\theta}+\frac{dP_\ell^1}{d\theta}\cos\phi\hat{\phi}\right]$$

(2.9.2)

$$\tilde{H}_{TE}^{ex} = \frac{1}{\eta}\sum_{\ell=1}^{\infty} i^{-\ell}{}_\ell\left\{\begin{array}{l}\dfrac{i}{\sigma'}(2\ell+1)\left(j_\ell(\sigma)+\alpha_\ell h_\ell(\sigma)\right)P_\ell^1(\cos\theta)\sin\phi\hat{r}\\[2ex] +\dfrac{(2\ell+1)}{\ell(\ell+1)}i\left(j_\ell^{\bullet}(\sigma)+\alpha_\ell h_\ell^{\bullet}(\sigma)\right)\left[\dfrac{dP_\ell^1}{d\theta}\sin\phi\hat{\theta}+\dfrac{P_\ell^1}{\sin\theta}\cos\phi\hat{\phi}\right]\end{array}\right\}$$

$$\tilde{E}_{TE}^{ex} = \sum_{\ell=1}^{\infty} i^{-\ell}\frac{(2\ell+1)}{\ell(\ell+1)}\left(j_\ell(\sigma)+\alpha_\ell h_\ell(\sigma)\right)\left[\frac{P_\ell^1}{\sin\theta}\cos\phi\hat{\theta}-\frac{dP_\ell^1}{d\theta}\sin\phi\hat{\phi}\right]$$

By Poynting's theorem the time-average power, P_{av}, on a spherical, virtual surface of radius σ/k that circumscribes the scatterer is equal to the real part of the surface integral of the radial component of the complex Poynting vector, see Eq. 1.9.11:

$$P_{av} = \frac{\sigma^2}{\eta k^2}\int_0^{2\pi} d\phi\int_0^{\pi}\sin\theta d\theta\,\text{Re}\left(N_{cr}\right)$$

(2.9.3)

Using Eq. 2.9.2 to evaluate the radial component, and with an asterisk indicating complex conjugate, the complex Poynting vector is:

$$
N_{cr} = \mathrm{Re}\,\frac{\sigma^2}{2\eta k^2} \sum_{\ell=1}^{\infty} \sum_{n=1}^{\infty} i^{n-\ell} \left(\frac{(2\ell+1)}{\ell(\ell+1)}\right)\left(\frac{(2n+1)}{n(n+1)}\right) \times
$$

$$
\left\{
\begin{array}{l}
i\left[
\begin{array}{l}
\left(j_n + \beta_n^* h_n^*\right)\left(\dot{j}_\ell + \beta_\ell \dot{h}_\ell\right)\left(\dfrac{dP_\ell^1}{d\theta}\dfrac{dP_n^1}{d\theta}\cos^2\phi + \dfrac{P_\ell^1 P_n^1}{\sin^2\theta}\sin^2\phi\right) \\[3mm]
-\left(j_\ell + \alpha_\ell h_\ell\right)\left(\dot{j}_n + \alpha_n^* \dot{h}_n^*\right)\left(\dfrac{dP_\ell^1}{d\theta}\dfrac{dP_n^1}{d\theta}\sin^2\phi + \dfrac{P_\ell^1 P_n^1}{\sin^2\theta}\cos^2\phi\right)
\end{array}
\right] \\[10mm]
+\left[
\begin{array}{l}
\left(j_\ell + \alpha_\ell h_\ell\right)\left(j_n + \beta_n^* h_n^*\right)\left(\dfrac{dP_\ell^1}{d\theta}\dfrac{P_n^1}{\sin\theta}\sin^2\phi + \dfrac{P_\ell^1}{\sin\theta}\dfrac{dP_n^1}{d\theta}\cos^2\phi\right) \\[3mm]
+\left(\dot{j}_n + \alpha_n^* \dot{h}_n^*\right)\left(\dot{j}_\ell + \beta_\ell \dot{h}_\ell\right)\left(\dfrac{dP_\ell^1}{d\theta}\dfrac{P_n^1}{\sin\theta}\cos^2\phi + \dfrac{P_\ell^1}{\sin\theta}\dfrac{dP_n^1}{d\theta}\sin^2\phi\right)
\end{array}
\right]
\end{array}
\right\}
\quad (2.9.4)
$$

Inserting Eq. 2.9.4 into the integral of Eq. 2.9.3 and integrating over the azimuth angle gives:

$$
P_{av} = \int_0^\pi \sin\theta\,d\theta\,\mathrm{Re}\,\frac{\pi\sigma^2}{2\eta k^2} \sum_{\ell=1}^{\infty} \sum_{n=1}^{\infty} i^{n-\ell} \left(\frac{(2\ell+1)}{\ell(\ell+1)}\right)\left(\frac{(2n+1)}{n(n+1)}\right)
$$

$$
\times\left\{
\begin{array}{l}
i\left[\left(j_n + \beta_n^* h_n^*\right)\left(\dot{j}_\ell + \beta_\ell \dot{h}_\ell\right) - \left(j_\ell + \alpha_\ell h_\ell\right)\left(\dot{j}_n + \alpha_n^* \dot{h}_n^*\right)\right] \\[3mm]
\times\left[\dfrac{dP_\ell^1}{d\theta}\dfrac{dP_n^1}{d\theta} + \dfrac{P_\ell^1 P_n^1}{\sin^2\theta}\right] \\[5mm]
+\left[\left(\dot{j}_\ell + \beta_\ell \dot{h}_\ell\right)\left(\dot{j}_n + \alpha_n^* \dot{h}_n^*\right) + \left(j_\ell + \alpha_\ell h_\ell\right)\left(j_n + \beta_n^* h_n^*\right)\right] \\[3mm]
\times\left[\dfrac{1}{\sin\theta}\dfrac{d\left(P_n^1 P_\ell^1\right)}{d\theta}\right]
\end{array}
\right\}
\quad (2.9.5)
$$

Using integrals in Table A.16.1 No. 3 and A.16.1 No. 6 to evaluate the integrals of Eq. 2.9.5 gives:

$$P_{av} = \frac{\pi\sigma^2}{\eta k^2}\,\mathrm{Re}\sum_{\ell=1}^{\infty} i(2\ell+1)\left[\begin{array}{c}\left(j_\ell + \beta_\ell^* h_\ell^*\right)\left(j_\ell^{\cdot} + \beta_\ell h_\ell^{\cdot}\right) \\ -\left(j_\ell + \alpha_\ell h_\ell\right)\left(j_\ell^{\cdot} + \alpha_\ell^* h_\ell^{\cdot*}\right)\end{array}\right] \qquad (2.9.6)$$

Use of the identities of Sec. A.20 shows the equation simplifies to:

$$P_{av} = \frac{\pi}{\eta k^2}\sum_{\ell=1}^{\infty}(2\ell+1)\left[\mathrm{Re}\left(\alpha_\ell + \beta_\ell\right) + \left(\alpha_\ell\alpha_\ell^* + \beta_\ell\beta_\ell^*\right)\right] \qquad (2.9.7)$$

Energy and momentum are carried into the system by the plane wave and both are transferred to the scatterer. The input power is equal to the term proportional to $\mathrm{Re}\left(\alpha_\ell + \beta_\ell\right)$. The power extracted from the beam is defined as extinction power, and is always positive. Changing the sign to conform to this usage, the extinction power is:

$$P_{EX} = -\frac{\pi}{\eta k^2}\,\mathrm{Re}\sum_{\ell=1}^{\infty}(2\ell+1)\left(\alpha_\ell + \beta_\ell\right) \qquad (2.9.8)$$

The power scattered back into the field is equal to:

$$P_{SC} = \frac{\pi}{\eta k^2}\sum_{\ell=1}^{\infty}(2\ell+1)\left(\alpha_\ell\alpha_\ell^* + \beta_\ell\beta_\ell^*\right) \qquad (2.9.9)$$

Absorbed power equals the difference between the scattered and extinction powers. The absorbed power is zero for lossless scatterers and hence Eq. 2.9.7 is equal to zero.

Critical scattering parameters are commonly normalized to a value that is independent of the magnitude of the incoming plane wave. Define scattering cross section, C_{SC}, to equal the scattered power-to-incoming power density ratio. For unit magnitude electric field intensity the incoming power density is $1/(2\eta)$. Values are sometimes also normalized by the geometric cross section. Cross section has the dimensions of an area, and normalization with respect to the geometric cross sectional area measures the size the scatterer appears to be versus the size it would appear with zero-wavelength optics. The geometric cross section C_{GE} is the geometric area the scatterer presents to the plane wave. The geometric cross sectional area of a spherical scatterer of radius a, for example, is $C_{GE} = \pi a^2$.

Combining definitions with Eq. 2.8.11 and Eq. 2.9.9 shows the scattering-to-geometric cross section ratio is:

$$\frac{C_{SC}}{C_{GE}} = \frac{2}{k^2 a^2} \sum_{\ell=1}^{\infty} (2\ell+1)\left[\alpha_\ell \alpha_\ell^* + \beta_\ell \beta_\ell^*\right] \qquad (2.9.10)$$

Similarly, the extinction cross section, C_{EX}, is defined to equal the extinction power-to-incoming power density ratio. Combining the definition with Eq. 2.8.11 and Eq. 2.9.8 shows the extinction-to-geometric cross section ratio is:

$$\frac{C_{EX}}{C_{GE}} = -\frac{2}{k^2 a^2} \sum_{\ell=1}^{\infty} (2\ell+1)\operatorname{Re}\left(\alpha_\ell + \beta_\ell\right) \qquad (2.9.11)$$

Another significant number is the radar cross section. Define the radar cross section, C_{RCS}, to equal the quotient of the power that would be scattered if the power density were everywhere equal to its value at $\theta = \pi$ divided by the incoming power density. It is a measure of the power returned towards a single interrogating source. By definition, the power scattered in direction $\theta = \pi$ is the back-scattered power. To determine the radar cross section evaluate Eq. 2.9.4 at $\theta = \pi$. The angular functions at that angle are equal to:

$$\frac{dP_\ell^1}{d\theta} = -\frac{P_\ell^1}{\sin\theta} = \frac{1}{2}\ell(\ell+1)(-1)^\ell \qquad (2.9.12)$$

Carrying out the calculation then normalizing by both the incoming power density and the geometric cross section results in the normalized radar cross section:

$$\frac{C_{RCS}}{C_{GE}} = \frac{1}{k^2 a^2} \sum_{\ell=1}^{\infty} \sum_{n=1}^{\ell} (2\ell+1)(2n+1)(-1)^{\ell+n} U(\ell-n)\left[(\alpha_\ell - \beta_\ell)(\alpha_n^* - \beta_n^*)\right]$$

$$(2.9.13)$$

Function $U(\ell-n)$ is the step function:

$$U(\ell-n) = \begin{vmatrix} 1 & \ell > n \\ 1/2 & \ell = n \\ 0 & \ell < n \end{vmatrix} \qquad (2.9.14)$$

As shown by Eq. 1.9.8 the fields carry momentum as well as energy, and momentum transfer from the field to the scatterer constitutes an applied force. The momentum transferred to the scatterer by the extinction energy is in the direction of the incoming wave and, by Eq. 1.9.8, is equal to the energy divided by c. The scattered power transfers momentum in proportion to the cosine of the angle between the incident and scattering directions. The back-scattered and forward-scattered portions of the power produce momentum respectively into or away from the direction of the beam. The sign of the total transferred momentum depends upon which type dominates, and that depends upon details of specific scatterers. Although the resulting force is too small to be significant in most macro-scale applications, nonetheless it exists, it affects all scatterers and receiving antennas, and, like energy, it is conserved.

It is also possible to calculate the force on a scatterer because of the scattered field, F_{SC}. With incident x-directed electric and y-directed magnetic fields, surface currents and magnetic field intensity are, respectively, at least partially x- and y-directed. By ampere's law their interaction creates a z-directed force. To calculate the force note that the momentum density is directly proportional to the power density: they differ only by a factor of c. The force arises from the z-component of the scattered power. It, in turn, is equal to the integral of the product of the Poynting vector and the cosine of the scattering angle. With the help of Eq. 1.9.7 the expression for the force in the direction of the plane wave is:

$$F_{SC} = -\frac{\sigma^2}{2ck^2} \int_0^{2\pi}\int_0^{\pi} \sin\theta\cos\theta\, d\theta \operatorname{Re}(N_r) \qquad (2.9.15)$$

Substituting the scattered fields of Eq. 2.9.5 into Eq. 2.9.15 and integrating

$$F_{SC} = -\int_0^\pi \sin\theta d\theta \frac{\pi\varepsilon\sigma^2}{2k^2} \mathrm{Re} \sum_{\ell=1}^\infty \sum_{n=1}^\infty i^{n-\ell} \left(\frac{(2\ell+1)}{\ell(\ell+1)}\right)\left(\frac{(2n+1)}{n(n+1)}\right)$$

$$\times \left\{ \begin{array}{l} i\left[\beta_\ell\beta_n^* h_n^* \dot{h}_\ell - \alpha_\ell\alpha_n^* h_\ell \dot{h}_n^*\right]\left[\dfrac{dP_\ell^1}{d\theta}\dfrac{dP_n^1}{d\theta} + \dfrac{P_\ell^1 P_n^1}{\sin^2\theta}\right]\cos\theta \\[4mm] + \left[\alpha_\ell\beta_n^* h_\ell \dot{h}_n^* + \alpha_n^*\beta_\ell \dot{h}_\ell h_n^*\right]\dfrac{1}{\sin\theta}\dfrac{d\left(P_n^1 P_\ell^1\right)}{d\theta}\cos\theta \end{array} \right\} \qquad (2.9.16)$$

Inserting the integrals of Table A.16.1 No. 4 and A.16.1 No. 7 into Eq. 2.9.16 gives:

$$F_{SC} = \left\{ \begin{array}{l} \dfrac{\pi\varepsilon\sigma^2}{k^2}\mathrm{Re}\sum_{\ell=1}^\infty \left(\dfrac{\ell(\ell+2)}{(\ell+1)}\right)\left[\alpha_\ell\alpha_{\ell+1}^* h_{\ell+1}^* \dot{h}_\ell - \beta_\ell\beta_{\ell+1}^* h_\ell \dot{h}_{\ell+1}^*\right] \\[5mm] -\dfrac{\pi\varepsilon\sigma^2}{k^2}\mathrm{Re}\sum_{\ell=1}^\infty \left(\dfrac{(\ell-1)(\ell+1)}{\ell}\right)\left[\beta_\ell\beta_{\ell-1}^* h_{\ell-1}^* \dot{h}_\ell - \alpha_\ell\alpha_{\ell-1}^* h_\ell \dot{h}_{\ell-1}^*\right] \\[5mm] -\dfrac{\pi\varepsilon\sigma^2}{k^2}\mathrm{Re}\sum_{\ell=1}^\infty \left(\dfrac{(2\ell+1)}{\ell(\ell+1)}\right)\left[\alpha_\ell\beta_\ell^* \dot{h}_\ell h_\ell^* + \alpha_\ell^*\beta_\ell h_\ell \dot{h}_\ell^*\right] \end{array} \right\}$$

$$(2.9.17)$$

In the far field Eq. 2.9.17 goes to:

$$F_{SC} = -\frac{\varepsilon\pi}{k^2}\sum_{\ell=1}^\infty \left\{ \begin{array}{l} \dfrac{\ell(\ell+2)}{(\ell+1)}\left(\alpha_\ell\alpha_{\ell+1}^* + \alpha_\ell^*\alpha_{\ell+1} + \beta_\ell\beta_{\ell+1}^* + \beta_\ell^*\beta_{\ell+1}\right) \\[4mm] + \dfrac{(2\ell+1)}{\ell(\ell+1)}\left(\alpha_\ell\beta_\ell^* + \alpha_\ell^*\beta_\ell\right) \end{array} \right\} \qquad (2.9.18)$$

Using Eq. 2.9.11, the force caused by the reception of the extinction power, the extinction force, F_{EX} is in the direction of the incoming field. Normalizing F_{EX} by the incoming power density determines the normalized force, f_{EX}. Normalizing it by the geometric cross section gives:

$$\frac{f_{EX}}{C_{GE}} = -\frac{2}{ck^2 a^2} \sum_{\ell=1}^{\infty} (2\ell+1)\,\mathrm{Re}\!\left(\alpha_\ell + \beta_\ell\right) \tag{2.9.19}$$

Summing gives the normalized total force on the scatterer:

$$\frac{\left(f_{SC} + f_{EX}\right)}{C_{GE}} = -\frac{2}{ck^2 a^2} \sum_{\ell=1}^{\infty} \left\{ \begin{array}{l} (2\ell+1)\,\mathrm{Re}\!\left(\alpha_\ell + \beta_\ell\right) \\[6pt] +\dfrac{\ell(\ell+2)}{(\ell+1)}\left(\alpha_\ell \alpha_{\ell+1}^* + \alpha_\ell^* \alpha_{\ell+1} + \beta_\ell \beta_{\ell+1}^* + \beta_\ell^* \beta_{\ell+1}\right) \\[8pt] +\dfrac{(2\ell+1)}{\ell(\ell+1)}\left(\alpha_\ell \beta_\ell^* + \alpha_\ell^* \beta_\ell\right) \end{array} \right\} \tag{2.9.20}$$

Lossless scatterers do absorb momentum and are thereby pushed in the direction of an incoming plane wave. In many cases the effect is insignificant, for example the effect on atmospheric molecules is much less than disturbances due to thermal unbalance. In other cases, however, such as the impulse electromagnetic wave produced by a nuclear blast, it is very significant.

2.10 Scattering Spheres, Specific Examples

The scattered fields generated by a sphere immersed in a plane wave, and hence the descriptive scattering coefficients α_ℓ and β_ℓ, depend upon both the size-to-wavelength ratio and the constitutive properties of the material. In this section both permittivity and permeability are scalars and insulators.

Solving the scattering problem requires evaluating scattering coefficients α_ℓ and β_ℓ. We begin with the fields inside the sphere with the origin included in the region of validity. The general expressions for interior fields, separated into TM and TE parts and indicated by superscript 'in', are valid at all points $r \le a$. Since passive field points are nonsingular the radial portion of the expressions are expressed by spherical Bessel functions only. With γ_ℓ and δ_ℓ representing coefficients to be determined:

$$\tilde{E}_{TM}^{in} = \sum_{\ell=1}^{\infty} i^{-\ell} \gamma_\ell \left\{ \begin{array}{l} \dfrac{i}{\sigma'}(2\ell+1) j_\ell(\sigma') P_\ell^1(\cos\theta)\cos\phi\,\hat{r} \\[2mm] + \dfrac{(2\ell+1)}{\ell(\ell+1)} i j_\ell^\bullet(\sigma') \left[\dfrac{dP_\ell^1}{d\theta}\cos\phi\,\hat{\theta} - \dfrac{P_\ell^1}{\sin\theta}\sin\phi\,\hat{\phi} \right] \end{array} \right\}$$

$$\tilde{H}_{TM}^{in} = \frac{1}{\eta'} \sum_{\ell=1}^{\infty} i^{-\ell} \gamma_\ell \frac{(2\ell+1)}{\ell(\ell+1)} j_\ell(\sigma') \left[\frac{P_\ell^1}{\sin\theta}\sin\phi\,\hat{\theta} + \frac{dP_\ell^1}{d\theta}\cos\phi\,\hat{\phi} \right]$$

(2.10.1)

$$\tilde{H}_{TE}^{in} = \frac{1}{\eta'} \sum_{\ell=1}^{\infty} i^{-\ell} \delta_\ell \left\{ \begin{array}{l} \dfrac{i}{\sigma'}(2\ell+1) j_\ell(\sigma') P_\ell^1(\cos\theta)\sin\phi\,\hat{r} \\[2mm] + \dfrac{(2\ell+1)}{\ell(\ell+1)} i j_\ell^\bullet(\sigma') \left[\dfrac{dP_\ell^1}{d\theta}\sin\phi\,\hat{\theta} + \dfrac{P_\ell^1}{\sin\theta}\cos\phi\,\hat{\phi} \right] \end{array} \right\}$$

$$\tilde{E}_{TE}^{in} = \sum_{\ell=1}^{\infty} i^{-\ell} \delta_\ell \frac{(2\ell+1)}{\ell(\ell+1)} j_\ell(\sigma') \left[\frac{P_\ell^1}{\sin\theta}\cos\phi\,\hat{\theta} - \frac{dP_\ell^1}{d\theta}\sin\phi\,\hat{\phi} \right]$$

Both wave number k' and wave impedance η' depend upon the material of which the sphere is composed. The external fields are expressed by Eq. 2.9.2. To evaluate the four field coefficients we obtain one equation by equating the interior and exterior radial components of the **B** field across the interface, a second equation by equating the interior and exterior radial components of the **D** field across the interface, a third equation by equating the interior and exterior theta components of the TM **E** field across the interface, and a fourth equation by equating the interior and exterior phi components of the TE **H** field across the interface.

With η_r equal the relative wave impedance in the scatterer we obtain the scattering coefficients:

$$\alpha_n = -\frac{j_n(\sigma) j_n^\bullet(\sigma') - \eta_r j_n^\bullet(\sigma) j_n(\sigma')}{j_n^\bullet(\sigma') h_n(\sigma) - \eta_r j_n(\sigma') h_n^\bullet(\sigma)}$$

(2.10.2)

$$\beta_\ell = -\frac{\eta_r j_\ell(\sigma) j_\ell^\bullet(\sigma') - j_\ell^\bullet(\sigma) j_\ell(\sigma')}{\eta_r j_\ell^\bullet(\sigma') h_\ell(\sigma) - j_\ell(\sigma') h_\ell^\bullet(\sigma)}$$

The internal field coefficients are:

$$\delta_n = \frac{j_n(\sigma)h_n^\bullet(\sigma) - j_n^\bullet(\sigma)h_n(\sigma)}{j_n(\sigma')h_n^\bullet(\sigma) - j_n^\bullet(\sigma')h_n(\sigma)/\eta_r}$$

$$\gamma_\ell = \frac{j_\ell(\sigma)h_\ell^\bullet(\sigma) - j_\ell^\bullet(\sigma)h_\ell(\sigma)}{j_\ell(\sigma')h_\ell^\bullet(\sigma)/\eta_r - j_\ell^\bullet(\sigma')h_\ell(\sigma)}$$

(2.10.3)

Case 1: An important special case is a sphere of material with equal relative permittivity and permeability, values of which are real if the material is lossless and complex if it is lossy. Although the wave impedance is equal that of free space the wave numbers are not. Making that substitution into Eq. 2.10.2 shows the scattering coefficients are:

$$\alpha_n = \beta_\ell = -\frac{j_n(\sigma)j_n^\bullet(\sigma') - j_n^\bullet(\sigma)j_n(\sigma')}{j_n^\bullet(\sigma')h_n(\sigma) - j_n(\sigma')h_n^\bullet(\sigma)}$$

(2.10.4)

Scattering occurs at all radii, even for matched permittivity and permeability. The internal field coefficients are

$$\delta_n = \gamma_\ell = \frac{j_n(\sigma)h_n^\bullet(\sigma) - j_n^\bullet(\sigma)h_n(\sigma)}{j_n(\sigma')h_n^\bullet(\sigma) - j_n^\bullet(\sigma')h_n(\sigma)}$$

(2.10.5)

For the specials case of a virtual sphere, $k = k'$, the scattering coefficients vanish and the internal field coefficients are $\delta_n = \gamma_n = 1$.

Case 2: A commercially significant case is scattering by a conducting sphere. The wave impedance of an ideal conductor is $\eta' = 0$. Substituting that value into Eq. 2.10.2 shows the scattering coefficients are:

$$\alpha_n = -\frac{j_n(ka)}{h_n(ka)} \text{ and } \beta_n = -\frac{j_n^\bullet(ka)}{h_n^\bullet(ka)}$$

(2.10.6)

For the dipole case, which is typical of electrically small scatterers:

$$\alpha_1(ka) = \frac{i(ka)^3}{3} \text{ and } \beta_1 = -\frac{2i(ka)^3}{3}$$

(2.10.7)

The cross sections and normalized forces defined in Sec. 2.9 are:

$$\frac{C_{EX}}{C_{GE}} = \frac{C_{SC}}{C_{GE}} = \frac{cf_{EX}}{C_{GE}} = \frac{10(ka)^4}{3}; \quad \frac{C_{RCS}}{C_{GE}} = 9(ka)^4; \quad \frac{cf_{SC}}{C_{GE}} = \frac{4(ka)^4}{3} \qquad (2.10.8)$$

Figure 2.10.1 shows the normalized extinction cross section for a conducting scatterer as a function of ka for scatterers of any physical size. Since power in a plane wave is fully directed, Fig. 2.10.1 also shows the momentum transferred to the scatterer; Eq. 2.10.8 shows that for small scatterers it varies as $(ka)^4$. The largest normalized extinction cross section occurs at $ka \cong 1.2$, and is equal to 2.68. For large and increasing values of ka the total cross section oscillates as it approaches an asymptotic limit of twice the geometric cross section.

In the lossless case the extinction and scattering cross sections are equal and the total force on the illuminated sphere is the extinction plus scattered forces, $(F_{EX} + F_{SC})$. The force on the scatterer because of the scattered field is shown in Fig. 2.10.2. Electrically small objects scatter predominantly back into the direction from which the wave came, increasing the thrust in the direction of the wave. Electrically large objects scatter predominantly in the direction of the incoming wave, decreasing the thrust on the scatterer. The sign of the scattering force changes at about $ka \cong 1.38$. The largest forward magnitude is about 0.257 and occurs at $ka \cong 1.12$.

The extinction momentum is in the direction of the incoming wave. All interacting energy forms part of the extinction momentum but, upon re-radiation, it may either add or subtract momentum from the scatterer. Since the subtracted momentum cannot exceed the extinction momentum it follows that:

$$\frac{Absorbed\ Energy}{Absorbed\ Momentum} \leq c \qquad (2.10.9)$$

Scatterers are commonly divided into groupings by radius-to-wavelength ratio. In the Rayleigh region $ka \ll 1$, in the Mie region ka is on the order of one, and in the optical region $ka \gg 1$.

The normalized radar cross section is shown in Fig. 2.10.3. Since conducting spheres sized to fit a human hand have radar cross-sections that are convenient to measure, the curve of Fig. 2.10.3 is often used as a laboratory calibration standard.

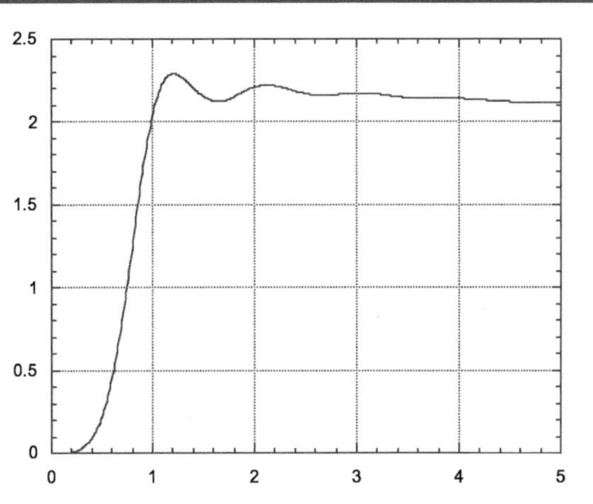

Figure 2.10.1. Ratio of extinction-to-geometric cross sections, C_{EX}/C_{GE}, versus *ka* for a conducting sphere of radius *a*.

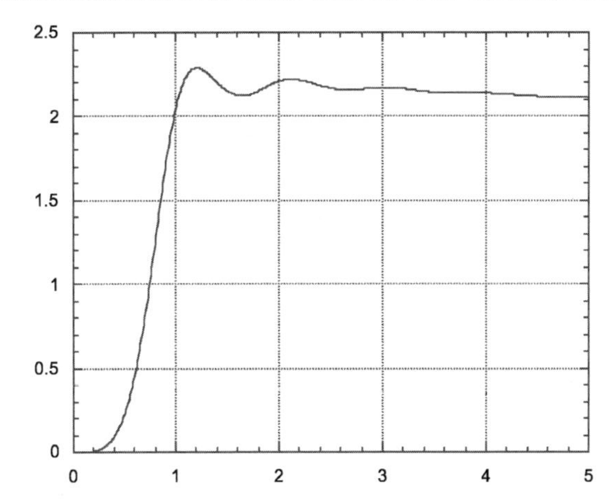

Figure 2.10.2. Ratio of scattering force-to-geometric cross section, f_{SC}/C_{GE}, versus *ka* for a conducting sphere of radius *a*.

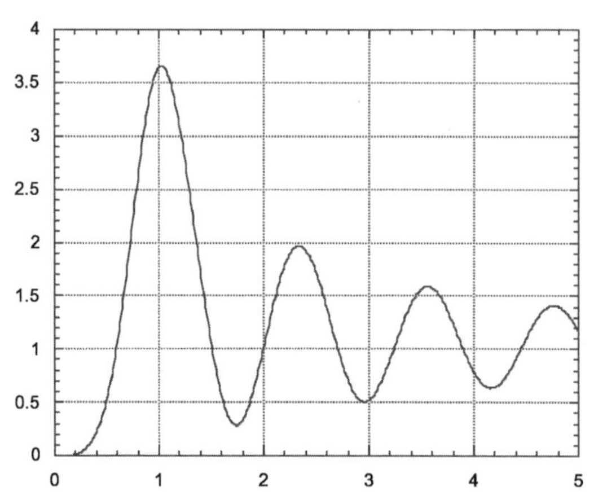

Figure 2.10.3. Ratio of radar-to-geometric cross section, C_{RCS}/C_{GE},
versus *ka* for a conducting sphere of radius *a*.

Effects due to optical scattering determine the optical properties of the sky. A clear atmosphere of gaseous nitrogen and oxygen, without suspended particulate matter, scatters a portion of the light that passes through it. Since the molecules are much smaller than a wavelength of visible light, a larger fraction of the incoming power is scattered at the shorter wavelengths than longer ones and, therefore, more blue light than red. When the sun is directly overhead the sky away from the directly incoming beam is illuminated by scattered light, which is dominantly blue. Some of the blue light is scattered to the earth and gives the sky its characteristically blue color.

Light travels farther through the atmosphere at sunrise and sunset than it does at noon and hence more light is scattered. The direct sunlight that remains, therefore, is dominantly red. If particulate matter, such as dust, about the size of an optical wavelength is present, scattering is insensitive to the wavelength and the sky appears to be dark.

References

D.M. Grimes, C.A. Grimes, "Transmission and Reception of Power by Antennas," in T. W. Barrett, D. M. Grimes, *Advanced Electromagnetism:*

Foundations, Theory and Applications, World Scientific Publishing (1995) pp. 763–791

R.F. Harrington, "Effect of Antenna Size on Gain, Bandwidth, and Efficiency," J. Research, Nat. Bureau of Standards, Vol. 64D, pp. 1–12 (1960)

H. Hertz, *Electric Waves: Researches on the Propagation of Electric Action with Finite Velocity Through Space* (1893) Translated by D. E. Jones, Dover Publications (1962)

J.D. Jackson, *Classical Electrodynamics*, 2nd ed., John Wiley (1975)

L.D. Landau, E.M. Lifshitz, *The Classical Theory of Fields*, trans. by H. Hamermesh, Addison-Wesley (1951)

R.B. Leighton, *Principles of Modern Physics*, McGraw-Hill, New York (1959)

G. Mie, "A Contribution to Optical Extinction by Metallic Colloidal Suspensions," *Ann. Physik.* vol. 25, p. 377 (1908)

W.K.H. Panofsky, M. Phillips, *Classical Electricity and Magnetism*, 2nd ed., Addison-Wesley (1961)

S. Ramo, J.R. Whinnery, T. Van Duzer, *Fields and Waves in communication Electronics, 2nd ed* John Wiley (1984)

H.C. Van de Hulst, *Light Scattering by Small Particles*, John Wiley (1957)

Chapter 3

Transmitting Biconical Antennas

In this chapter we analyze a transmitting biconical antenna located in otherwise free space. Biconical antennas are important since they are the only antenna embodiment that approximates linear dipole antennas for which exact field expansions are available. Although biconical antennas have little to do, at least directly, with photon emission by an electron they serve as a quintessential example of the power of electromagnetic analysis techniques. Expressions are derived for the input impedance and for the full set of electric and magnetic fields at all points about a biconical antenna with perfectly conducting arms and caps. Surface currents, pattern directivity, field power density, *etc.* follow from the fields. In Chapter 4 we provide a full analysis of receiving biconical antennas. Together these analyses provide a base upon which to construct an analysis of general radiators and, from there, a source that generates radiation possessing all known photon properties.

Although detailed summations over known functions are obtained as solutions, field expressions are infinite sums over products of associated Legendre functions with spherical Bessel and Neumann functions. These functions can be numerically evaluated at all orders in the far field and at small orders in the near field but near field evaluation is not feasible at high orders.

3.1 Transmitting Biconical Antennas

A biconical antenna is illustrated in Fig. 3.1.1. The input power is applied within a sphere of radius b, centered at the apices of the cones. The cones extend from radius b to radius a, the length of the cones, at angle ψ as measured from the z-axis. All surfaces are ideal

88

conductors. Source radius b is much smaller than either cone length a or wavelength λ.

Biconical antennas are unique in that they are amenable to a rigorous and complete electromagnetic analysis and are shaped similarly to many practical antennas. Any solution with fields that satisfy the Maxwell equations and for which the fields match the boundary conditions is both a unique solution, see Sec. 1.10, and a complete solution. Completeness assures that all solution terms are present, in contrast with numerical solutions that begin with an assumed symmetry and obtain an iterative answer. For those cases, the output solution contains only symmetries present in the initial input and hence the solution is only as complete as the initial input.

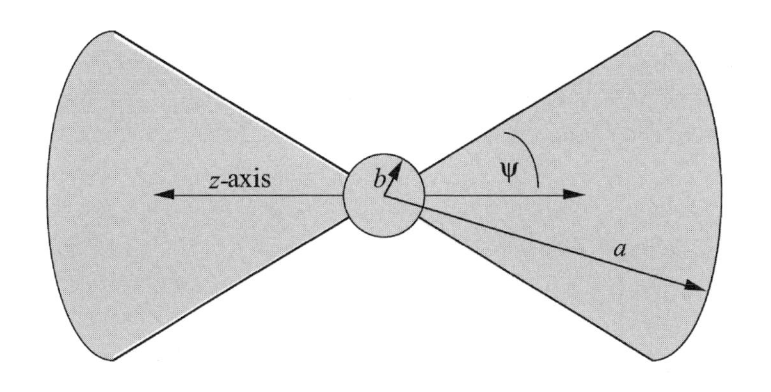

Figure 3.1.1. Schematic illustration of a biconical antenna.

The antenna arms are conical sections that extend between b and a, expansion half-angles ψ are measured from the z-axis, and the outer termination of the cone is capped by a spherical segment of radius a.

Transmitting antennas include an energy source that applies a sinusoidal steady state voltage or current to source region b. The two cones, although oppositely directed, act as a transmission line and direct the energy through the inner region, radius b to radius a, as a TEM mode. The energy then passes through the open aperture at $r = a$ and enters the outer region. All radiation has rotational symmetry about the antenna axis and many wavelengths from the antenna the electric field intensity is linearly polarized in the direction of that axis. The impedance the antenna presents to the source is determined by details of the antenna structure: cone angles, cone length, and the wavelength of the radiation. The outgoing waves undergo a discontinuity in the wave admittance (impedance) at the open aperture that results in infinite sets of TM modes in both the interior and exterior regions. Both inner and outer modes support standing energy and a steady state outward energy flow. Solution of the transmitting antenna problem requires solving for the input admittance, the coefficients of the interior TEM mode, of the infinite sets of both interior and exterior TM modes, and the radiation pattern.

The analysis is simplified by dividing space in the following way:

$$\text{SOURCE: } r < b; \quad 0 \le \theta \le \pi; \quad 0 \le \phi \le 2\pi \tag{3.1.1}$$

INTERIOR

$$Arms : b < r < a ; \quad 0 \le \theta < \psi; \quad \pi - \psi < \theta \le \pi; \ 0 \le \phi \le 2\pi \tag{3.1.2}$$
$$Free\,space : b < r < a ; \quad \psi < \theta < \pi - \psi; \quad 0 \le \phi \le 2\pi$$

INTERFACE

$$Aperture : r = a ; \quad \psi \le \theta \le \pi - \psi; \quad 0 \le \phi \le 2\pi \tag{3.1.3}$$
$$Arms : r = a; \quad 0 \le \theta < \psi, \ \pi - \psi < \theta \le \pi; \ 0 \le \phi \le 2\pi$$

EXTERIOR
$$Free\,space : \ r > a ; \ 0 \le \theta \le \pi; \ 0 \le \phi \le 2\pi \tag{3.1.4}$$

3.2 Fields

The first objective is to obtain an expression for all fields. The procedure begins with the general expansion, Eq. 1.13.9, and imposes boundary conditions specific to the biconical structure of Fig. 3.1.1. As was the case for the analysis of scatterers, field determination is greatly simplified by incorporating general field properties before matching the boundary conditions. General field properties are: (1) Since the antenna has rotational symmetry about the z-axis there is no dependence upon azimuth angle ϕ and only functions with degree m equal to zero form part of the solution. All coefficients $F(v,m)$ and $G(v,m)$ are equal to zero for m greater than zero. This changes the sums over orders and degrees of Eq. 1.13.9 to a sum over orders only. (2) The source drives straight currents that produce no current loops. Since TE coefficients are generated by current loops all coefficients $G(v,0)$ are equal to zero. (3) A source located evenly between the two cones drives the surface current density with symmetry $I(r,\psi) = I(r,\pi-\psi)$, and it drives surface charge density with symmetry $\rho(r,\psi) = -\rho(r,\pi-\psi)$. By Eq. 1.13.9, and with $v = \ell$ an integer, E_θ is proportional to $dP_\ell(\cos\theta)/d\theta$. It is shown in Sec. A.12 that Legendre functions have either even or odd symmetry as ℓ is even or odd. Consider a Legendre function of order ℓ containing terms with the symmetry of $\cos^\ell\theta$. For that term:

$$\text{If } P_\ell\left(\cos\theta\right) \approx \cos^\ell\theta \text{ then } E_\theta \approx \frac{dP_\ell}{d\theta} \approx \ell\cos^{\ell-1}\theta\sin\theta$$

$$\ell \text{ odd}: E_\theta\left(\sigma,\theta\right) = E_\theta\left(\sigma,\pi-\theta\right). \text{if } \ell \text{ even}: E_\theta\left(\sigma,\theta\right) = -E_\theta\left(\sigma,\pi-\theta\right)$$

$$(3.2.1)$$

Since the source drives only even symmetry electric fields, it follows that only odd symmetry Legendre functions appear in the field solution. Therefore, the coefficients of all even order Legendre functions are equal to zero.

The Exterior Region

(4) Since the z-axis is included in the field region all terms have null coefficients except Legendre functions of the first kind. (5) In the limit as the radius approaches infinity, energy conservation requires the radial dependence to be $exp[i(\omega t-\sigma)]/\sigma$ which, in turn, requires the coefficients of all radial functions except Hankel functions of the second kind to be zero.

After incorporating the five constraints into Eq. 1.13.9 and making the notational shift we obtain:

$$F_\ell = i^{1-\ell}F\left(\ell,0\right)$$

The most general possible set of exterior field components is:

$$\sigma E_r = \sum_{\ell=1;o}^{\infty} \ell(\ell+1) F_\ell h_\ell(\sigma) P_\ell(\cos\theta)$$

$$E_\theta = \sum_{\ell=1;o}^{\infty} F_\ell h_\ell^{\cdot}(\sigma) \frac{dP_\ell(\cos\theta)}{d\theta} \qquad (3.2.2)$$

$$\eta H_\phi = -i \sum_{\ell=1;o}^{\infty} F_\ell h_\ell(\sigma) \frac{dP_\ell(\cos\theta)}{d\theta}$$

The symbol $\ell = 1;o$ indicates the sum begins with $\ell = 1$ and is over odd integers only. The constants F_ℓ form an infinite set of unknown but constant field coefficients. Complete problem solution requires obtaining a solution for each of them.

The Interior Region

(6) Since the cones exclude fields from the z-axis modal orders need not be integers. Since symmetry requirement (3) requires null coefficients for even functions by Eq. A.12.14 the coefficients of the even parity portion of Legendre functions, $L_\nu(\cos\theta)$, are equal to zero. This restricts solutions to odd parity Legendre functions, $M_\nu(\cos\theta)$. It follows in the same way that the zero order Legendre function, $P_0(\cos\theta)$, has a null coefficient but, by Eq. A.12.14 and A.12.15, zero-order Legendre function of the second kind, $Q_0(\cos\theta)$, does not; the derivative of the zero order Legendre function of the second kind remains finite on cone surfaces. (7) Both the source voltage and the source current are finite. The voltage and current are, respectively, proportional to σ times the electric and magnetic field intensity and the radial functions approach zero as $j_\nu(\sigma) \Rightarrow \sigma^\nu$ and $y_\nu(\sigma) \Rightarrow \sigma^{-(\nu+1)}$, see Eq. A.18.7 and A.18.11. Therefore the input voltage and current values remain finite only if the coefficients of all spherical Neumann functions except $\nu = \ell = 0$ are equal to zero.

Incorporating these constraints into Eq. 1.13.9 and separately denoting the zero order TEM mode shows that the general forms of the interior field components are:

$$E_r = \sum_{v>0}^{\infty} \Gamma_v v(v+1) \frac{j_v(\sigma)}{\sigma} M_v(\cos\theta)$$

$$E_\theta = \sum_{v>0}^{\infty} \Gamma_v j_v \frac{dM_v}{d\theta} + i\left[c_0 \dot{j}_0(\sigma) + d_0 \dot{y}_0(\sigma)\right]\frac{dQ_0(\cos\theta)}{d\theta} \tag{3.2.3}$$

$$\eta H_\phi = -i\sum_{v>0}^{\infty} \Gamma_v j_v \frac{dM_v}{d\theta} + \left[c_0 j_0(\sigma) + d_0 y_0(\sigma)\right]\frac{dQ_0(\cos\theta)}{d\theta}$$

Coefficients of noninteger order modes, $F(v,0)$ of Eq. 1.13.9, are denoted by Γ_v and coefficients of zero order spherical Bessel and Neumann functions respectively by the constants c_0 and d_0.

3.3 TEM Mode

The TEM mode may be reformulated in terms of measurable antenna parameters. Consider properties of $Q_0(\cos\theta)$, see Sec. A.12:

$$Q_0(\cos\theta) = \ln\left[\cot\left(\frac{\theta}{2}\right)\right] = \frac{1}{2}\left[\ln\left(\frac{1+\cos\theta}{1-\cos\theta}\right)\right] \tag{3.3.1}$$

Differentiating:

$$\frac{dQ_0}{d\theta} = -\frac{1}{\sin\theta} \tag{3.3.2}$$

The zero order spherical Bessel, Neumann and related functions are:

$$j_0(\sigma) = \frac{\sin\sigma}{\sigma}; \quad y_0(\sigma) = -\frac{\cos\sigma}{\sigma}; \quad \dot{j}_0(\sigma) = \frac{\cos\sigma}{\sigma}; \quad \dot{y}_0(\sigma) = \frac{\sin\sigma}{\sigma}$$

$$\tag{3.3.3}$$

Substituting $v = 0$ and Eq. 3.3.2 and 3.3.3 into Eq. 1.13.9 gives:

$$E_r = 0$$

$$\sigma E_\theta = \frac{1}{i\sin\theta}\left(c_0\cos\sigma + d_0\sin\sigma\right) \qquad (3.3.4)$$

$$\sigma\eta H_\phi = \frac{1}{\sin\theta}\left(d_0\cos\sigma - c_0\sin\sigma\right)$$

The voltage difference between equal radii positions on the two antenna arms is a measurable quantity. It may be calculated from knowledge of the antenna structure and the electric field intensity using Eq. 3.3.3:

$$V(r) = \frac{\sigma}{k}\int_\psi^{\pi-\psi} E_\theta d\theta = \frac{1}{ik}\left(c_0\cos\sigma + d_0\sin\sigma\right)\int_\psi^{\pi-\psi}\frac{d\theta}{\sin\theta} \qquad (3.3.5)$$

Integrating Eq. 3.3.2 shows that:

$$\int_\psi^{\pi-\psi}\frac{d\theta}{\sin\theta} = 2\ln\left[\cot\left(\frac{\psi}{2}\right)\right]$$

It is useful in what lies ahead to define the line admittance of the transmission line formed by the two antenna arms to be $G(\psi)$ where:

$$G(\psi) = \frac{\pi}{\eta\ln\left[\cot\left(\psi/2\right)\right]} \qquad (3.3.6)$$

Combining Eq. 3.3.5 and 3.3.6 shows voltage $V(r)$ is:

$$V(r) = \frac{2\pi}{ik\eta G(\psi)}\left(c_0\cos\sigma + d_0\sin\sigma\right) \qquad (3.3.7)$$

Substituting Eq. 3.3.7 into the TEM component of the electric field intensity, Eq. 3.2.3, gives the zero order electric field intensity:

$$E_\theta = \frac{\eta k V(r) G(\psi)}{2\pi\sigma\sin\theta} \tag{3.3.8}$$

Since the magnetic field intensity is directed around the cone arms, the current on the antenna arms is radially directed. Use of Eq. 3.3.3 gives the relationship:

$$I(r) = \frac{\sigma}{k}\sin\theta \int_0^{2\pi} H_\phi d\phi = \frac{2\pi}{\eta k}\left(d_0\cos\sigma - c_0\sin\sigma\right) \tag{3.3.9}$$

Substituting Eq. 3.3.9 into the TEM component of the magnetic field intensity term of Eq. 3.2.3 gives the zero order magnetic field intensity:

$$H_\phi = \frac{k I(r)}{2\pi\sigma\sin\theta} \tag{3.3.10}$$

Defining position $r = a$ to be the terminus, the voltage and current there follow from Eq. 3.3.7 and Eq. 3.3.10:

$$V(a) = \frac{2\pi}{\eta G i k}\left\{c_0\cos(ka) + d_0\sin(ka)\right\}$$
$$I(a) = \frac{2\pi}{\eta k}\left\{d_0\cos(ka) - c_0\sin(ka)\right\} \tag{3.3.11}$$

Inverting Eq. 3.3.11 to obtain the field coefficients in terms of voltage and current at the antenna terminals gives:

$$c_0 = \frac{\eta k}{2\pi}\left\{iGV(a)\cos(ka) - I(a)\sin(ka)\right\}$$
$$d_0 = \frac{\eta k}{2\pi}\left\{I(a)\cos(ka) + iGV(a)\sin(ka)\right\} \tag{3.3.12}$$

Next, define $Y(a)$ as the terminator admittance:

$$Y(a) = I(a)/V(a)$$

Rearranging gives the voltage between, and the current on, the cone arms as a function of $Y(a)$:

$$V(r) = \frac{V(a)}{G}\left\{G\cos\left[k(a-r)\right] + iY(a)\sin\left[k(a-r)\right]\right\}$$

$$I(r) = V(a)\left\{Y(a)\cos\left[k(a-r)\right] + iG\sin\left[k(a-r)\right]\right\}$$

(3.3.13)

In terms of the terminator and line admittances, the admittance at each radius along the cones is:

$$Y(r) = G\frac{Y(a)\cos\left[k(a-r)\right] + iG\sin\left[k(a-r)\right]}{G\cos\left[k(a-r)\right] + iY(a)\sin\left[k(a-r)\right]}$$

(3.3.14)

Use Eq. 3.3.14 to define the antenna input admittance $Y(0)$ then put it equal to Y_0, the admittance at $r = b$ in the limit as b approaches zero. Also, define the input voltage, $V(0)$, and current, $I(0)$, as:

$$V(0) = \lim_{b\to 0} V(b); \quad I(0) = \lim_{b\to 0} I(b)$$

(3.3.15)

The input admittance is:

$$Y_0 = G\left\{\frac{Y(a)\cos(ka) + iG\sin(ka)}{G\cos(ka) + iY(a)\sin ka}\right\}$$

(3.3.16)

The radial dependence of the admittance as a function of the input and line admittances is:

$$Y(\sigma) = G\left\{\frac{Y_0\cos\sigma - iG\sin\sigma}{G\cos\sigma - iY_0\sin\sigma}\right\}$$

(3.3.17)

The line admittance equations have the exact form of admittance transfer along a TEM transmission line and show that the cone arms jointly act as a constant admittance line guiding the TEM mode from the source to the terminus. Quite differently from a parallel wire transmission line in which the guiding conductors remain equally spaced along the length of the line, here the guiding conductors are oppositely directed on either side of the source. Like many transmission lines the line impedance is constant, see Eq. 3.3.6. Voltage is measured between equal radius points on the cone arms and the current is measured along each arm.

3.4 Boundary Conditions

Putting the TEM results of Sec. 3.3 as part of the interior field equations shows that the general interior field form is:

$$E_r = \sum_{v>0}^{\infty} \Gamma_v v(v+1) \frac{j_v(\sigma)}{\sigma} M_v(\cos\theta)$$

$$E_\theta = \sum_{v>0}^{\infty} \Gamma_v j_v^\bullet \frac{dM_v(\cos\theta)}{d\theta} + \frac{\eta k G V(\sigma)}{2\pi\sigma\sin\theta} \qquad (3.4.1)$$

$$\eta H_\phi = -i\sum_{v>0}^{\infty} \Gamma_v j_v \frac{dM_v(\cos\theta)}{d\theta} + \frac{kI(\sigma)}{2\pi\sigma\sin\theta}$$

The infinite set of multiplying coefficients Γ_v and the input admittance $Y(0)$ are unknown and to be determined.

Since the magnetic field is entirely ϕ-directed, all currents on the cones are directed along the length of the cones. The total current consists of the sum of currents associated with the TM modes and the TEM mode. Define the TM modal current $I'(\sigma)$ to be the complementary current and the TEM modal current $I(\sigma)$ to be the principal current. The total current is the sum:

$$I_T(\sigma) = I'(\sigma) + I(\sigma) \qquad (3.4.2)$$

The first term in the expression for H_ϕ shows that the complementary current, in amperes, is:

$$I'(\sigma,\psi) = \frac{2\pi\sigma}{i\eta k} \sum_{v>0}^{\infty} \Gamma_v j_v(\sigma) \frac{dM_v(\cos\theta)}{d\theta}\bigg|_{\theta=\psi} \qquad (3.4.3)$$

Since $j_\nu(\sigma)$ varies as σ^ν for small radii, where $\nu > 0$, it follows that the complementary current vanishes in that limit:

$$\lim_{\sigma \to 0} I'(0) = 0 \tag{3.4.4}$$

The principal current at the origin follows from Eq. 3.3.9 and 3.3.12, and is:

$$\lim_{\sigma \to 0} I(0) = I(a)\cos(ka) + iGV(a)\sin(ka) \tag{3.4.5}$$

Since only the principal current exists at the source, only it can support the energy flow away from the source. Since the time-average power supported by the TEM mode does not depend upon the radius, it follows that the time-average power is guided through the region by the principal current.

Applying conducting boundary conditions to the exterior fields of Eq. 3.2.2 shows that the field intensities on the caps are related to the surface charges and currents as:

$$0 \le \theta < \psi \quad \text{and} \quad \pi - \psi < \theta \le \pi;$$

$$\varepsilon E_r(ka,\theta,\phi) = \frac{\varepsilon}{\sigma} \sum_{\ell=1;o}^{\infty} \ell(\ell+1)F_\ell h_\ell(ka) P_\ell(\cos\theta) = \rho(ka,\theta,\phi)$$

$$E_\theta(ka,\theta,\phi) = \sum_{\ell=1;o}^{\infty} F_\ell h_\ell'(ka) \frac{dP_\ell(\cos\theta)}{d\theta} = 0 \tag{3.4.6}$$

$$H_\phi(ka,\theta,\phi) = -\frac{i}{\eta} \sum_{\ell=1;o}^{\infty} F_\ell h_\ell(ka) \frac{dP_\ell(\cos\theta)}{d\theta} = -I_\theta(ka,\theta,\phi)$$

Symbol $\rho(ka,\theta,\phi)$ indicates the surface charge density on the caps in coulombs per square meter and symbol $I_\theta(ka,\theta,\phi)$ indicates surface current density on the caps in amperes per meter. Applying conducting boundary conditions to the interior field components of Eq. 3.4.1 shows that the interior field intensities on the arms are subject to the constraints:

$$kb \le \sigma \le ka$$

$$E_r(\sigma,\psi,\phi) = \sum_v v(v+1)F_v \frac{j_v(\sigma)}{\sigma} M_v(\cos\psi) = 0$$

$$\varepsilon E_\theta(\sigma,\psi,\phi) = \frac{\eta k G V(\sigma)}{2\pi\sigma\sin\psi} + \varepsilon \sum_{v>0} \Gamma_v \dot{j}_v \left.\frac{dM_v(\cos\theta)}{d\theta}\right|_{\theta=\psi} = \rho(\sigma,\psi,\phi) + \rho'(\sigma,\psi,\phi)$$

$$\eta H_\phi(\sigma,\psi,\phi) = -i\sum_{v>0}\Gamma_v j_v \left.\frac{dM_v(\cos\theta)}{d\theta}\right|_{\theta=\psi} + \frac{kI(\sigma)}{2\pi\sigma\sin\theta} = I_r(\sigma,\psi,\phi) + I'_r(\sigma,\psi,\phi)$$

$$(3.4.7)$$

Symbols with and without the primes indicate, respectively, principal and complimentary surface charge and current densities on the cone arms.

The null value of the radial field component on the conical surfaces is only satisfied by a nontrivial solution if for every value of v:

$$M_v(\cos\psi) = 0 \qquad\qquad (3.4.8)$$

Equation 3.4.8 determines an infinite and unique set of positive-real eigenvalues of v. Plots of v versus angle ψ, for which Eq. 3.4.8 is satisfied, are shown in Fig. 3.4.1 for the first through the fifth sequence of roots. Function $M_v(\cos\theta)$ is plotted versus v in Fig. 3.4.2, showing the first 23 zeros. Plots of $M_v(\cos\theta)$ versus θ at the first two roots of $M_v[\cos(5°)]$ are illustrated by Figure 3.4.3.

Virtual boundary conditions of Eq. 2.1.28 and 2.1.30 apply at the aperture and all field components are continuous through the boundary. Imposing these conditions on Eq. 3.2.2 and 3.4.1 gives the constraining equations:

$$\psi < \theta < \pi - \psi;$$

$$\sigma E_r(ka,\theta,\phi) = \sum_{\ell=1;o} \ell(\ell+1)F_\ell h_\ell^\cdot(ka)P_\ell(\cos\theta) = \sum_{v>0} v(v+1)\Gamma_v \dot{j}_v(ka)M_v(\cos\theta)$$

$$(3.4.9)$$

$$\psi < \theta < \pi - \psi;$$

$$E_\theta\left(ka,\theta,\phi\right) = \sum_{\ell=1;o}^{\infty} F_\ell h_\ell^{\cdot}\left(ka\right)\frac{dP_\ell\left(\cos\theta\right)}{d\theta} = \sum_{v} \Gamma_v j_v^{\cdot}\left(ka\right)\frac{dM_v\left(\cos\theta\right)}{d\theta} + \frac{\eta kGV\left(a\right)}{2\pi\sigma\sin\theta}$$

$$(3.4.10)$$

$$\psi < \theta < \pi - \psi;$$

$$\eta H_\phi\left(ka,\theta,\phi\right) = -i\sum_{\ell=1;o}^{\infty} F_\ell h_\ell\left(ka\right)\frac{dP_\ell\left(\cos\theta\right)}{d\theta} = -i\sum_{v} \Gamma_v j_v\left(ka\right)\frac{dM_v\left(\cos\theta\right)}{d\theta} + \frac{kV\left(a\right)}{2\pi\sigma\sin\theta}$$

$$(3.4.11)$$

These are the field values on the interface between interior and exterior regions. This completes the discussion of the field equations at a point as boundary conditions. Figure 3.4.1 is a plot of $M_v[\cos(5°)]$ versus order. Figure 3.4.2 shows values of order, v, versus cone angle ψ at which the three functions $M_v(\cos\psi)$, $M_v^1(\cos\psi)$, and $dL_\lambda^1(\cos\theta)/d\theta$ at $\theta = \psi$, are equal to zero; cardinal numbers indicate root order. Figure 3.4.3 is a plot of $M_v(\cos\theta)$ versus θ with the constraint $M_v[\cos(5°)] = 0$.

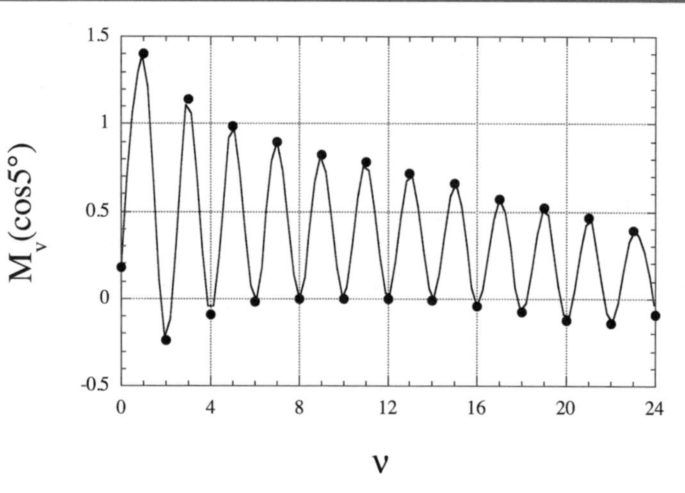

Figure 3.4.1. Function $M_v[\cos(5°)]$ plotted versus v.

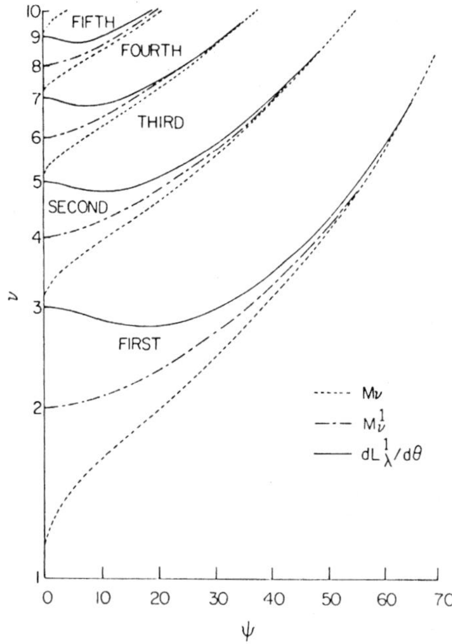

Figure 3.4.2. Root values, noninteger Legendre functions.

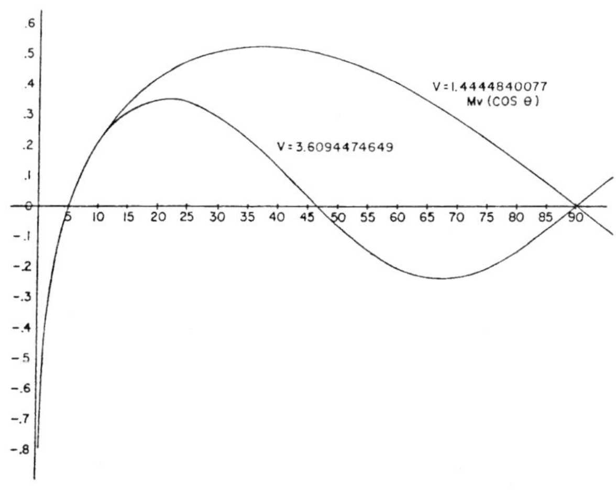

Figure 3.4.3. Two lowest order functions $M_v(\cos\theta)$ vs. θ; $M_v[\cos(5°)] = 0$.

3.5 Defining Integral Equations

In the preceding sections, the general forms of the boundary conditions are obtained as infinite sums over radial and harmonic functions. The exact form of the functions and the relationships between them is specified. In each case what remains are sums over an infinite set of modal orders and it remains to separate out the coefficients, one by one. In all but one case this is accomplished using the orthogonality of the Legendre functions. Sets of orthogonal integrals are formed and evaluated that change the equalities involving Legendre functions of the preceding sections into linear algebraic equations. The algebraic equations are used to solve for the coefficients.

The first algebraic equation is obtained without using orthogonality. Operating on Eq. 3.4.11 to evaluate the line integrals of the expression for H_ϕ around the periphery of the antenna arm on both sides of the $r = a$ boundary gives:

$$\int_\psi^{\pi-\psi} aH_\phi d\phi = -i\frac{a}{\eta}\sum_{\ell=1}^\infty F_\ell h_\ell(ka)P_\ell(\cos\theta)\Big|_\psi^{\pi-\psi}$$

$$= -i\frac{a}{\eta}\sum_v \Gamma_v j_v(ka)M_v(\cos\theta)\Big|_\psi^{\pi-\psi} + \frac{I(a)}{\eta G} \tag{3.5.1}$$

The condition $M_v(\cos\psi) = 0$ removes the sum over v. Collecting the remaining terms and making the substitution that $I(a) = Y(a)V(a)$ gives the interior line admittance at the terminus as a function of the exterior coefficients:

$$Y(a) = \frac{2ia}{V(a)}\sum_{\ell o;1}^\infty F_\ell h_\ell(ka)P_\ell(\cos\psi) \tag{3.5.2}$$

The symbol $\ell o;1$ indicates that ℓ represents the field of odd integers with lowest value of one. This equation, the first of the algebraic equations, equates the applied voltage and the admittance to a sum over odd order, exterior modes.

The next algebraic equation is obtained using the orthogonality of integer-order Legendre functions. Multiplying Eq. 3.4.10 by $\sin\theta d\theta dP_n(\cos\theta)/d\theta$ and integrating over the aperture gives:

$$\int_0^\pi \sin\theta d\theta \frac{dP_n}{d\theta} \sum_{\ell o;1}^\infty F_\ell h_\ell^\cdot(ka) \frac{dP_\ell}{d\theta} = \int_\psi^{\pi-\psi} \sin\theta d\theta \frac{dP_n}{d\theta} \left\{ \sum_{v>0}^\infty \Gamma_v j_v^\cdot(ka) \frac{dM_v}{d\theta} + \frac{\eta GV(a)}{2\pi a \sin\theta} \right\}$$

$$(3.5.3)$$

Although the limits on both integrals are from ψ to $\pi - \psi$, since the sum on the left side of Eq. 3.5.3 vanishes over the caps the range of integration is extended from 0 to π without affecting the value of the integral. Next we use the full angular range and the definitions of Tables A.16.1 and A.17.1 to obtain:

$$I_{\ell\ell} = \frac{2}{2\ell+1} \quad \text{and} \quad I_{\ell v} = \int_\psi^{\pi-\psi} P_\ell(\cos\theta) M_v(\cos\theta) \sin\theta d\theta \qquad (3.5.4)$$

Symbol 'I' with two subscripts indicates an integral and with one subscript indicates current. Evaluating Eq. 3.5.3 by incorporating Eq. 3.5.4, Table A.16.1 No. 6, and Table A.17.1 No. 1 gives:

$$\ell(\ell+1) F_\ell h_\ell^\cdot(ka) I_{\ell\ell} = \ell(\ell+1) \sum_{v>0}^\infty \Gamma_v j_v^\cdot(ka) I_{\ell v} - \frac{\eta GV(a)}{\pi a} P_\ell(\cos\psi) \qquad (3.5.5)$$

Equation 3.5.5 is the second algebraic expression that equates individual exterior modal coefficients to a sum over interior modal coefficients.

The third algebraic equation uses the orthogonality of fractional order Legendre functions. Begin by taking the product of Eq. 3.4.11 and $[dM_\mu(\cos\theta)/d\theta]\sin\theta d\theta$ and then integrating over the aperture:

$$\int_\psi^{\pi-\psi} \sin\theta d\theta \sum_{\ell o;1}^\infty F_\ell h_\ell(ka) \frac{dP_\ell}{d\theta} \frac{dM_\mu}{d\theta}$$

$$= \int_\psi^{\pi-\psi} \sin\theta d\theta \frac{dM_\mu}{d\theta} \left\{ \sum_v^\infty \Gamma_v j_v(ka) \frac{dM_v}{d\theta} + \frac{\eta GV(a)}{2\pi a \sin\theta} \right\} \qquad (3.5.6)$$

Evaluating the integrals of Eq. 3.5.6 using integrals listed in Table A.17.1, that is A.17.1 No. 5 with A.17.1 No. 1 and integral A.17.1 No. 7 with A.17.1 No. 6 gives:

$$\mu(\mu+1)I_{\mu\mu}\Gamma_\mu j_\mu(ka) = \sum_{\ell o;1}^{\infty} \ell(\ell+1)F_\ell h_\ell(ka)I_{\ell\mu} \tag{3.5.7}$$

Equation Eq. 3.5.7 is an algebraic expression that equates individual modal coefficients to a sum over exterior modes.

3.6 Solution of the Biconical Antenna Problem

The result of applying the orthogonality of Legendre functions to point equations is an expression for individual exterior or interior modal magnitudes as sums over interior or exterior modes, respectively. The equations are:

$$F_\ell h_\ell^{\bullet}(ka) = -\frac{V(a)}{a}\frac{\eta G}{\pi}\frac{P_\ell(\cos\psi)}{\ell(\ell+1)I_{\ell\ell}} + \sum_{v>0}^{\infty}\Gamma_v j_\ell^{\bullet}(ka)\frac{I_{\ell v}}{I_{\ell\ell}} \tag{3.6.1}$$

$$\Gamma_v j_v(ka) = \frac{1}{v(v+1)}\sum_{\ell o;1}^{\infty} F_\ell h_\ell(ka)\ell(\ell+1)\frac{I_{\ell v}}{I_{vv}} \tag{3.6.2}$$

Each equation contains an infinite number of linear algebraic equations. The zero order interior mode satisfies the equation:

$$Y(a) = \frac{2ia}{V(a)}\sum_{\ell o;1}^{\infty} F_\ell h_\ell(ka)P_\ell(\cos\psi) \tag{3.6.3}$$

It remains to solve the three equations for the individual coefficients. After some manipulation, including multiplying by $h_\ell(\sigma)/h_\ell^{\bullet}(\sigma)$, Eq. 3.6.1 and 3.6.2 combine to form the equality:

$$F_\ell h_\ell (ka) - \sum_{n=1}^{\infty} F_n h_n (ka) \sum_{v>0} \frac{n(n+1)}{v(v+1)} \frac{I_{\ell v} I_{nv}}{I_{\ell \ell} I_{vv}} \frac{\dot{j}_v (ka) h_\ell (ka)}{j_v (ka) h_\ell^{\cdot} (ka)}$$

$$= -\frac{\eta G}{\pi} \frac{V(a)}{a} \frac{P_\ell (\cos\psi)}{\ell(\ell+1) I_{\ell\ell}} \frac{h_\ell (ka)}{h_\ell^{\cdot} (ka)} \tag{3.6.4}$$

Equation 3.6.4 represents an infinite set of linear equations, one for each coefficient F_ℓ, and has the form:

$$x_\ell + \sum_{n=1}^{\infty} N_{\ell n} x_n = B_\ell \tag{3.6.5}$$

Because the magnitudes of coefficients F_ℓ decrease rapidly with increasing modal number, and to keep the magnitude within available computer range, it is helpful to solve the problem with the initial variable x_ℓ equal to $F_\ell h_\ell(ka)$. Next, after solving for x_ℓ and knowing $h_\ell(ka)$, solve for F_ℓ.

Although an equation of the form of Eq. 3.6.5 may be readily solved using matrix techniques, doing so requires the series to be truncated and truncation produces errors. The solution procedure is to: (1) pick an arbitrary but specific value for the ratio $V(a)/a$ and we used one, (2) use the matrix solution to solve for the product $F_\ell h_\ell(ka)$, (3) divide by $h_\ell(ka)$ to obtain F_ℓ. The procedure determines as many of the previously unknown exterior coefficients as needed, and is limited only by the capability of available computers. This completes the calculation of the exterior coefficients. Knowing $F_\ell h_\ell(ka)$, Eq. 3.6.2 may be truncated and solved for F_ℓ, and Eq. 3.6.3 may be truncated and solved for the admittance $Y(a)$:

$$Y(a) = \frac{2iG}{V(a)/a} \sum_{\ell=1}^{\infty} F_\ell h_\ell (ka) \tag{3.6.6}$$

All quantities on the right side are known. Since each mode $F_\ell h_\ell(ka)$ is proportional to $V(a)/a$, the magnitudes in the numerator and denominator of Eq. 3.6.6 cancel and it follows that the value of $Y(a)$ is correct for any applied voltage.

To change the field normalization to the more conveniently determined value $V(0)/a = 1$, enter the value into the first of Eq. 3.3.13 to obtain:

$$\frac{V(a)}{a} = \frac{G}{G\cos(ka) + iY(a)\sin(ka)} \tag{3.6.7}$$

Use of Eq. 3.6.7 to re-normalize F_ℓ completes the numerical analysis of biconical transmitting antennas.

Table 3.6.1. Constants for $\psi = 5°$, $ka = 2$.

ℓ	$F_\ell h_\ell(ka)$	v	$\Gamma_v j_v(ka)$
1	(1.50924–i2.40989)D–01	0	
3	(5.42697–i1.70419)D–02	1.444 4840	(4.33823–i17.5963)D–02
5	(21.9562–i6.98844)D–03	3.609 4475	(3.68170–i2.37137)D–02
7	(12.8283–i3.81076)D–03	5.754 8721	(2.23379–i1.17634)D–02
9	(7.18028–i2.93358)D–03	7.887 3272	(2.59971–i1.27571)D–02
11	(5.72681–i2.11622)D–03	10.016 937	(–10.8348+i5.10427)D–02
13	(4.17152–i1.57614)D–03	12.143 571	(–7.89611+i4.07774)D–03
15	(3.10771–i1.19513)D–03	14.268 228	(–3.57476+i1.60594)D–03
17	(23.4057–i9.13644)D–04	16.391 498	(–19.5095+i7.62755)D–04

Badii, Tomiyama, and Grimes did numerical analyses of several biconical antennas through 12-place accuracy. The analyses included series truncation with 17 external (maximum modal number of 33) and 16 internal modes. Table 3.6.1 lists, with six place accuracy, values of $F_\ell h_\ell(ka)$ and $\Gamma_v j_v$ (ka) for an antenna with $\psi = 5°$ and $ka = 2$, external modes one through 17 and internal modes 1.444 through 16.391.

Table 3.6.2 lists the first six figures of F_ℓ and Γ_v for the same antenna. The values illustrate that the magnitudes of F_ℓ and Γ_v respectively decrease and increase rapidly with increasing modal number. The coefficients and Eq. 3.6.7 determine the terminal admittance, $Y(a)$. $Y(a)$ and Eq. 3.3.14 determine the antenna's input admittance $Y(0)$.

Table 3.6.2. Constants for $\psi = 5°$, $ka = 2$.

ℓ	F_ℓ	v	Γ_v
1	(-6.00998–i50.5094)D–02	0	
3	(–9.96861–i36.9686)D–03	1.444484	(1.32836–i5.387966)D–01
5	(–3.75732–i11.8104)D–04	3.609448	(14.3649–i9.25205)D–01
7	(–6.96510–i20.3034)D–06	5.754872	(3.36507–i1.77209)D+01
9	(–7.74256–i21.5902)D–08	7.887327	(3.04273–i1.49310)D+03
11	(–5.7288–i15.5031)D–10	10.01694	(–16.5306+i7.78756)D+05
13	(–3.01440–i7.97810)D–12	12.14357	(–2.67294+i1.22520)D+07
15	(–1.18075–i3.07030)D–14	14.26823	(–2.98297+i1.34009)D+09
17	(–1.22586–i9.11817)D–18	16.39150	(–6.07222+i2.68528)D+11

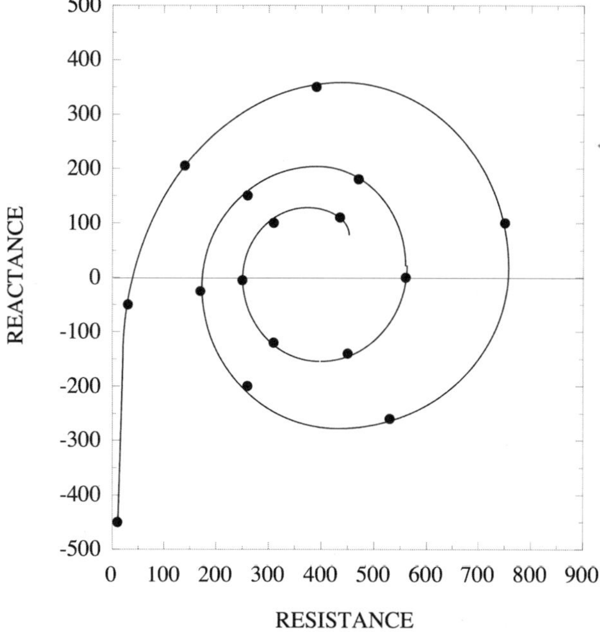

Figure 3.6.1. Input impedance of a biconical antenna, with $\psi = 5°$, as a function of arm length. Zero reactance values are at ka = 1.11, 2.59, 4.06, 5.51, 7.14. The first mark is at ka = 0.5, each succeeding mark increases ka by 0.5.

Figure 3.6.1 shows the input impedance as a function of arm length for 5° cones. The mark at $-i450$ ohms shows the input impedance of an antenna with $ka = 0.5$; succeeding marks are spaced at intervals of cones made longer by $\Delta(ka) = 0.5$. Figure 3.6.2 shows the input impedance of an antenna with $\psi = 5°$ as a function of arm length. Note that peaks are centered at about $ka \cong 1.11$, 4.06, and 7.14 and the initial resonance is much sharper than succeeding ones.

By Eq. 3.4.6, each modal contribution to the total magnetic field intensity at the aperture is equal to $F_\ell h_\ell(ka) d\,[P_\ell(\cos\theta)]/d\theta$; the magnitude of $F_\ell h_\ell(ka)$ is listed in Table 3.6.1. From the theory of Legendre polynomials, see Table A.12.1:

$$\frac{d}{d\theta} P_\ell \cos(\theta)\Big|_{\frac{\pi}{2}} = (-1)^{(\ell-1)/2} \frac{(\ell)!!}{(\ell-1)!!} \tag{3.6.8}$$

The ratios of the modal contribution to H_ϕ to that of the exterior dipole mode at $\phi = \pi/2$ are listed in Table 3.6.3. At the aperture, the modal magnitudes decrease so slowly with increasing modal number that a reasonably accurate description of interface affects requires a large number of modes. In the far field, on the other hand, only the first few modes determine the fields.

In the interior region, modal magnitudes of $H_\phi(ka,\pi/2)$ are equal to:

$$\Gamma_v j_v (ka)\left[(v)!!/(v-1)!!\right] \tag{3.6.9}$$

The ratio of factorials and the ratio-magnitude product for each mode are listed in Table 3.6.3; it is a measure of the rate of convergence of the field expressions with increasing order. Although the external modes are monotone decreasing with increasing modal number, the internal modes are not; the interior modes decrease but not monotonically with increasing modal number. The difference is because interior-to-exterior modal coupling depends upon the numerical difference between the interior modal orders and odd integers, as well as the magnitudes of the modes.

In Table 3.6.3 all ratios are normalized by the magnitude of the exterior dipole mode. The first three columns refer to exterior fields: the

Table 3.6.3. Magnetic field modal magnitudes.

ℓ	Aperture Ratio	Far Field Ratio	ν	Factorial Ratio	Interior Ratio
1	1	1			
3	2.0005D–01	7.5275D–02	1.4444840	3.2498	2.0713
5	7.1033D–02	2.4366D–03	3.6094475	3.6529	5.6258D–01
7	4.6052D–02	4.2199D–05	5.7548721	3.8350	3.4050D–01
9	3.0563D–02	4.5092D–07	7.8873272	3.9504	4.0232D–01
11	2.1471D–02	3.2493D–09	10.016937	6.8083D–02	2.8677D–02
13	1.5682D–02	1.6767D–11	12.143571	3.8859D–01	1.1845D–03
15	1.1710D–02	6.4671D–14	14.268228	9.6692D–01	1.8995D–04
17	7.8363D–03	1.8087D–17	16.391498	1.6152	3.4457D–03

first is modal number, the second the modal-to-dipole field ratio at the aperture and the third the modal-to-dipole field ratio at far field. The next three columns refer to interior fields: the first is modal number, the second the magnitude of Eq. 3.6.9, and the third modal-to-aperture field ratio.

Figure 3.6.2 plots the radiated power output for a biconical antenna with a constant input voltage. Power peaks occur at $ka = 1.11$, 4.06, and 7.14.

Field values determine the charge and current densities on the antenna surfaces. Summarized results are:

$$Cap: \quad \sigma = ka; \quad 0 \le \theta < \psi; \quad \pi - \psi < \theta \le \pi$$

$$\rho(ka,\theta) = \varepsilon \sum_{\ell o;1}^{\infty} \ell(\ell+1) F_\ell \frac{h_\ell(ka)}{ka} P_\ell(\cos\theta) \frac{coulombs}{meter^2} \tag{3.6.10}$$

$$I_\theta(ka,\theta) = \frac{i}{\eta} \sum_{\ell o;1}^{\infty} F_\ell h_\ell(ka) \frac{dP_\ell(\cos\theta)}{d\theta} \frac{amperes}{meter}$$

$$Cones: \quad b < r < a; \quad \theta = \psi$$

$$\rho(\sigma,\psi) = \left\{ \varepsilon \sum_{\nu>0}^{\infty} \Gamma_\nu \dot{j}_\nu(\sigma) \frac{dM_\nu(\cos\theta)}{d\theta}\Big|_{\theta=\psi} + \frac{kGV(r)}{2\pi c\sigma \sin\psi} \right\} \frac{coulombs}{meter^2}$$

$$I_r(\sigma,\psi) + I'_r(r,\psi) = \left\{ -\frac{kI(r)}{2\pi\sigma \sin\psi} + \frac{i}{\eta} \sum_{\nu>0}^{\infty} \Gamma_\nu j_\nu(\sigma) \frac{dM_\nu(\cos\theta)}{d\theta}\Big|_{\theta=\psi} \right\} \frac{amperes}{meter}$$

$$\tag{3.6.11}$$

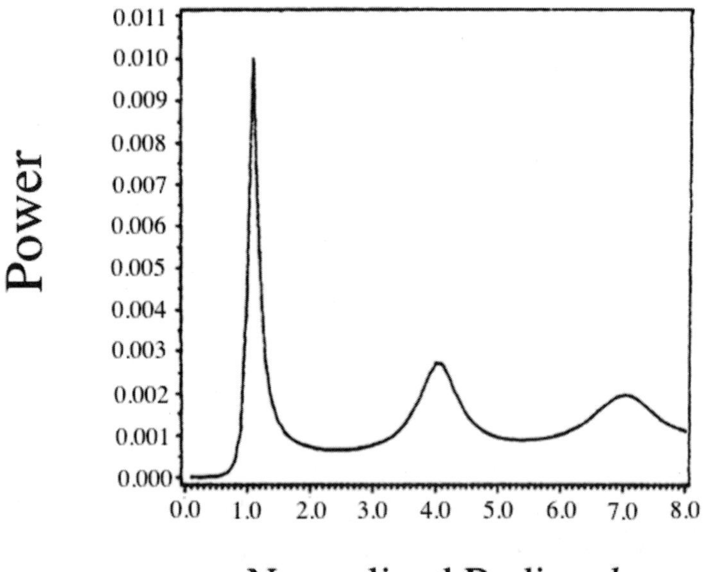

Figure 3.6.2. Output power versus ka for a 5°, biconical, transmitting antenna.

On the cones the surface current density is radially directed and on the caps it is zenith angle directed. The two currents have quite different dependencies upon radius and zenith angle and therefore the current is not continuous through the cone-cap junction. A loop of charge accumulates at the junction with a sign and magnitude that depends upon antenna structural details and the radiated wavelength. The resulting ring charge is:

$$Q(ka,\psi) = \frac{2\pi a}{i\omega}\left(\sum_{\ell o;1}^{\infty} \beta_\ell h_\ell(ka)\frac{dP_\ell(\cos\psi)}{d\psi} + \sum_{v>0}^{\infty}\Gamma_v j_v(ka)\frac{dM_v(\cos\psi)}{d\psi}\right) + \frac{iI(a)}{\omega\sin\psi}$$

$$(3.6.12)$$

Charge densities on the cap and cone have quite different dependencies upon radius and zenith angle. The electric field intensity on the cone and cap are, respectively θ and r directed, and just off an ideal 90° junction the field is directed at an angle of 45° as measured from both the cone and cap.

3.7 Power

For a transmitting antenna, the time-average power in the interior and exterior regions follow by use of the fields of Eq. 3.4.1 and 3.2.2, respectively. In the interior region, the time-average real power satisfies the transmission line rules between radii b and a. The input impedance and the radiated power are strong functions of the physical location of the standing energy wave, and it depends upon the antenna arm length.

The time-average power produced by an antenna is equal to the integral of the real part of the radial component of the complex Poynting vector at the surface of a virtual sphere, which for ease in calculation is made concentric with the antenna. The fields of Eq. 3.2.2 show that the time-average output power is:

$$P_{av} = \frac{\sigma^2}{2\eta k^2} \mathrm{Re}\left\{ \int_0^{2\pi} d\phi \int_0^{\pi} \sin\theta d\theta \sum_{\ell=1}^{\infty} \sum_{n=1}^{\infty} F_\ell F_n^* i h_\ell^\cdot(\sigma) h_n^*(\sigma) \frac{dP_\ell}{d\theta} \frac{dP_n}{d\theta} \right\} \qquad (3.7.1)$$

Replacing the Hankel functions by their far field values and evaluating the integral gives:

$$P_{av} = \frac{1}{\eta k^2} \sum_{\ell=1}^{\infty} \frac{\ell(\ell+1)}{(2\ell+1)} F_\ell F_\ell^* \qquad (3.7.2)$$

Since all terms on the right side are known, Eq. 3.7.2 is sufficient to evaluate the output power.

The time-average power input, P_{in}, to the antenna is:

$$P_{in} = \frac{1}{2}\mathrm{Re}\left[V(0)I^*(0)\right] \qquad (3.7.3)$$

By Eq. 3.3.13 the TEM voltage and current in the interior region are:

$$V(r) = \frac{V(a)}{G}\left\{G\cos\left[k(a-r)\right] + iY(a)\sin\left[k(a-r)\right]\right\}$$

$$I(r) = V(a)\left\{Y(a)\cos\left[k(a-r)\right] + iG\sin\left[k(a-r)\right]\right\} \qquad (3.7.4)$$

     Photon Creation-Annihilation

Combining Eq. 3.7.3 and 3.7.4 gives:

$$P_{in} = \frac{1}{2}\mathrm{Re}\left[V(a)V^*(a)\right]Y^*(a) \tag{3.7.5}$$

Since $V(a)$ and $Y(a)$ are known, Eq. 3.7.5 is sufficient to evaluate the input power.

In a lossless antenna:

$$P_{in} = P_{av} \tag{3.7.6}$$

The equality serves as a check on all procedures.

The complex power, P_c, on a concentric sphere of normalized radius σ is:

$$kb < \sigma < ka$$

$$P_c(\sigma) = \int_0^\pi \sin\theta\, d\theta \left\{ \frac{i\pi}{2}\sum_v \Gamma_v \Gamma_v^* j_v j_v^* \left(\frac{dM_v}{d\theta}\right)^2 + \frac{\eta k^2 G}{2\pi\sigma^2 \sin^2\theta} V(r)I^*(r) \right\} \tag{3.7.7}$$

$$\sigma > ka$$

$$P_c(\sigma) = \frac{i\pi\sigma^2}{\eta k^2}\int_0^\pi \left(\frac{dP_\ell}{d\theta}\right)^2 \sin\theta\, d\theta \sum_{\ell=1}^\infty F_\ell F_\ell^* h_\ell(\sigma)h_\ell^*(\sigma)$$

Since the electric field intensity just off the surface of the caps has only a radial component there is no normally directed Poynting vector and no energy is exchanged between the cap and the field. Since all fields are continuous through the aperture, the total complex power is a continuous function of radius between positions $a-\delta$ and $a+\delta$, where δ is a differential radial length. All fields are continuous through the aperture and, therefore, so is the energy density. Adjacent to the caps, the radial component of the electric field intensity and the azimuth component of the magnetic field intensity are not equal to zero and therefore the energy per unit length as a function of radius is discontinuous between positions $a-\delta$ and $a+\delta$. The magnitude of the discontinuity increases with increasing cone angle.

Figures 3.7.1–3.7.3 describe the complex powers within three antennas with normalized arms lengths of $ka = 0.70$, 1.28, and 2.00; all have cone angles of one degree. The antennas are, respectively, electrically short, resonant, and electrically long. In all cases the real power, P_{real}, is constant.

The normalized complex power about an electrically short antenna, $ka = 0.7$, is shown in Fig. 3.7.1; the real power is small and the terminal impedance is capacitive. The peak reactive power is capacitive and occurs at approximately $kr = 0.1$. From there, the power decreases slowly with decreasing radius until reaching the terminals, $kr = 0$. For increasing radius it decreases more rapidly until reaching $kr = ka$, where it drops abruptly, then decreases slowly to zero with increasing radius for $kr > ka$.

The normalized complex power about a resonant antenna, $ka = 1.28$ is shown in Fig. 3.7.2; the real power is large and the terminal impedance is resistive. The capacitively phased reactive power peak of Fig. 3.7.1 has

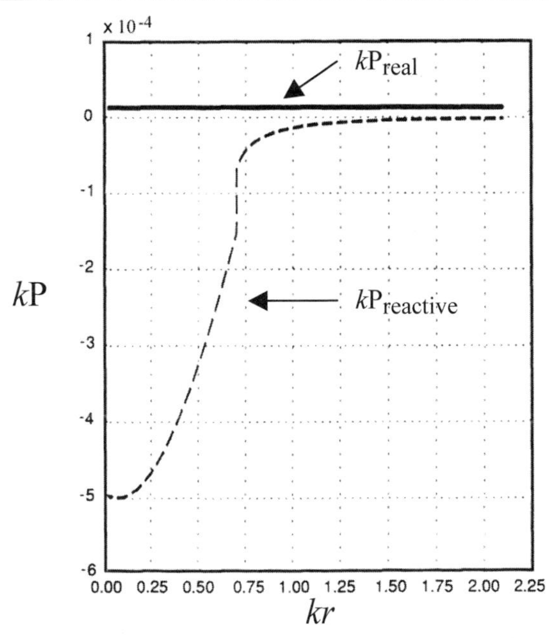

Figure 3.7.1. Normalized real and reactive powers versus *kr* for a biconical transmitting antenna; *applied voltage V(0) = a, cone angle ψ = 1°, and ka = 0.70.*

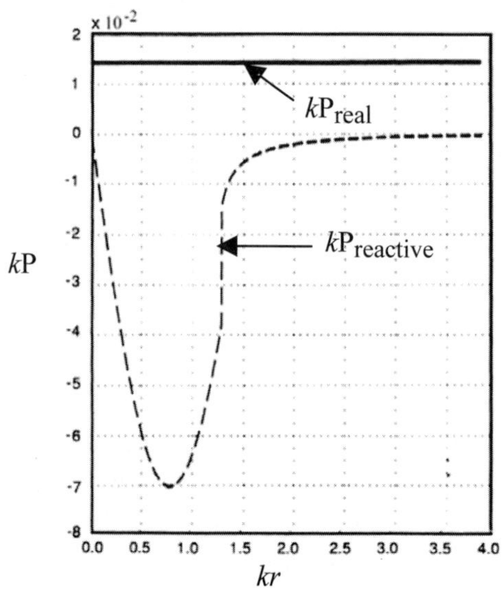

Figure 3.7.2. Real and reactive powers versus *kr* for a biconical transmitting antenna; *applied voltage V(0) = a, cone angle ψ = 1°, and ka = 1.28.*

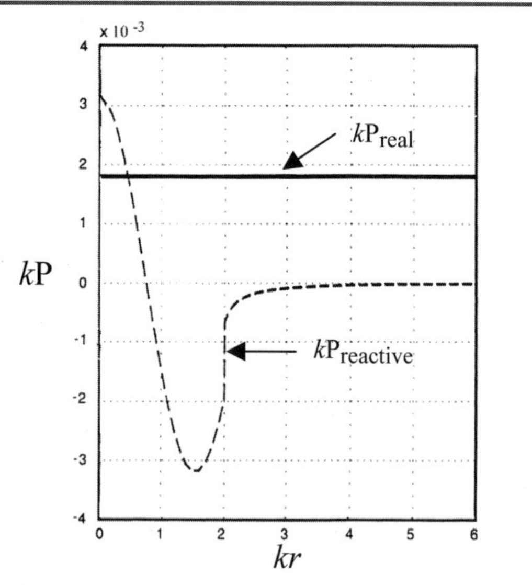

Figure 3.7.3. Real and reactive powers versus *kr* for a biconical transmitting antenna; *applied voltage V(0) = a, cone angle ψ = 1°, and ka = 2.00.*

moved outward to about $kr = 0.7$. From there it decreases slowly with decreasing radius to zero at the terminals. For increasing radius it behaves very similarly to Fig. 3.7.1.

The normalized complex power about an electrically long antenna, $ka = 2.00$, is shown in Fig. 3.7.3; the real power is less than that of Fig. 3.7.2 and the terminal impedance is inductive. The capacitively phased reactive power peak has moved outward to about $kr = 1.6$. From there it decreases slowly with decreasing radius, passes through zero and becomes inductively phased at the terminals.

References

V. Badii, "Numerical Analysis of the transmitting Biconical Antenna," Electrical Engineering doctoral dissertation, Pennsylvania State University, 1987. Available from University Microfilm, Ann Arbor MI

V. Badii, K. Tomiyama, D.M. Grimes, "Biconical Transmitting Antennas, A Numerical Analysis," *Appl. Comp. Electromag. Soc. J.*, vol. 5, pp. 62–93 (1990)

J.D. Jackson, *Classical Electrodynamics*, John Wiley & Sons, 2nd ed., (1975)

S.A. Schelkunoff, *Advanced Antenna Theory*, John Wiley (1952)

S.A. Schelkunoff, *Applied Mathematics for Engineers and Scientists*, 2nd ed., Van Nostrand (1965)

W. R. Smythe, *Static and Dynamic Electricity*, 3rd ed., Summa Press, New York, NY, p. 39 (1989)

Chapter 4

Receiving Biconical Antennas

$\mathcal{W}$e describe the analysis of a receiving biconical antenna in otherwise free space. Appropriately capped biconical antennas are of especial importance since they are the only antenna embodiment that approximates linear dipole antennas and for which exact expressions can be obtained. Expressions are obtained for the input impedance and for the full set of electric and magnetic fields at all points about a biconical antenna with perfectly conducting arms and caps. Surface currents, pattern directivity, field power density, momentum exchange with the far field, *etc.* follow from the fields. As with the transmitting biconical antenna, receiving biconical antennas may appear to have little direct connection with photon emission by an electron, which is the main concern of this book. However, analysis of biconical antennas serves as a quintessential example of the power of electromagnetic analysis techniques. Our analysis of a receiving biconical antenna in otherwise empty space obtains the full set of fields, near and far. The solution includes surface currents on the antenna and both the energy and the linear momentum transferred from the wave to the antenna as functions of antenna size and load impedance.

4.1 Receiving Biconical Antennas

For a receiving antenna to function it is necessary that a component of the applied electric field intensity be aligned parallel with the antenna axis; optimum operation is with full alignment. To analyze scattering by a sphere it is convenient for the wave to propagate in the z-direction. For analysis of scattering from a receiving antenna it is convenient for the antenna axis to lie along the z-axis and, therefore, the incoming wave propagates in the xy-plane. We make the arbitrary choice that it propagates in the *y*-direction and, therefore, we expand a *y*-directed wave in spherical coordinates. The technique is similar to that used in Sec. 2.8 for a *z*-directed plane wave. The desired exponential is:

$$e^{i(\omega t - ky)} = e^{i(\omega t - \sigma \sin\theta \sin\phi)}$$

(4.1.1)

Since Eq. 4.1.1 has no singularities the spherical coordinate expansion can contain no spherical Neumann functions, no fractional order Legendre functions, and no Legendre functions of the second kind. Only spherical Bessel functions and associated Legendre polynomials remain. The most general form for the expansion is:

$$e^{-i\sigma \sin\theta \sin\phi} = \sum_{\ell=0}^{\infty} \sum_{m=0}^{\ell} \left[{}^{c}G_\ell^m \cos m\phi - i\, {}^{s}G_\ell^m \sin m\phi \right] \frac{\ell(\ell+1)}{m} j_\ell(\sigma) P_\ell^m(\cos\theta)$$

(4.1.2)

It is necessary to evaluate the two infinite sets of coefficients ${}^{c}G_\ell^{\,m}$ and ${}^{s}G_\ell^{\,m}$. The reverse superscripts "c" and "s" indicate coefficients of cosine and sine, respectively. Multiplying factor $\ell(\ell+1)/m$ is chosen for later convenience. To evaluate the coefficients it is necessary to reformulate the equation as an infinite set of algebraic equations, each of which provides a value for one coefficient.

From the theory of spherical Bessel functions, Eq. A.18.6:

$$j_\ell(\sigma) = \frac{\sigma^\ell}{(2\ell+1)!!} + \text{higher order terms}$$

(4.1.3)

Substitute Eq. 4.1.3 into the right side of Eq. 4.1.2; take ℓ differentials of both sides with respect to σ and then go to the limit as σ goes to zero:

$$(-i)^\ell \sin^\ell\theta \sin^\ell\phi = \sum_{\ell=0}^{\infty} \sum_{m=0}^{\ell} \left[{}^{c}G_\ell^m \cos m\phi - i\, {}^{s}G_\ell^m \sin m\phi \right] \frac{\ell(\ell+1)}{m} \frac{\ell!}{(2\ell+1)!!} P_\ell^m(\cos\theta)$$

(4.1.4)

The left side of Eq. 4.1.4 contains powers of trigonometric functions. Identities that convert powers to multiples of the angles are:

n even

$$\sin^n \phi = \frac{2}{2^n} \left\{ \sum_{k=0}^{n/2-1} (-1)^{n/2-k} \frac{(n)!}{(n-k)!k!} \cos\left[(n-2k)\phi\right] + \frac{(n)!}{(n/2)!^2} \right\}$$

n odd

$$\sin^n \phi = \frac{2}{2^n} \left\{ \sum_{k=0}^{(n-1)/2} (-1)^{(n-1)/2-k} \frac{(n)!}{(n-k)!k!} \sin\left[(n-2k)\phi\right] \right\}$$

$$(4.1.5)$$

Substitute Eq. 4.1.5 into Eq. 4.1.4 then multiply by $\cos(q\phi)$, where q is an integer, and integrate over the azimuth angle. Next, multiply by $\sin(q\phi)$ and repeat the procedure. The results, where δ represents a Kronecker delta function and s represents an integer, are:

$$\frac{{}^cG_\ell^q}{q} \frac{\ell(\ell+1)}{(2\ell+1)!!} P_\ell^q(\cos\theta) = \frac{(-i)^{(\ell-q)}\,\delta(q,2s)}{(q/2)!\left[(\ell-q)/2\right]!}\sin^\ell\theta$$

$$\frac{{}^sG_\ell^q}{q} \frac{\ell(\ell+1)}{(2\ell+1)!} P_\ell^q(\cos\theta) = \frac{(-i)^{(\ell-q)}\,\delta(q,2s+1)}{(q/2)!\left[(\ell-q)/2\right]!}\sin^\ell\theta$$

$$(4.1.6)$$

By Eq. 4.1.6, ${}^cG_\ell^m$ is equal to zero if q is odd and ${}^sG_\ell^m$ is equal to zero if q is even; this reduces the total number of nonzero coefficients by half.

Next, multiply the top of Eq. 4.1.6 by $P_\ell^m(\cos\theta)$ and integrate over the zenith angle. Integral forms are Eq. 1 and 10 of Table A.16.1:

q even

$$\frac{2^{\ell+1}\ell!(\ell+q)!}{(q/2)!\left[(\ell-q)/2\right]!(2\ell+1)!}\delta(\ell+q,2s) = \frac{{}^cG_\ell^q}{q} \frac{\ell(\ell+1)}{(2\ell+1)!!} \frac{2}{(2\ell+1)} \frac{(\ell+q)!}{(\ell-q)!}$$

$$(4.1.7)$$

By Eq. 4.1.7, ${}^cG_\ell^m$ is equal to zero if $\ell+q$ is odd. Since q is even, it follows that ℓ is also even. Conducting the same operation on the bottom of Eq. 4.1.6 shows that ${}^sG_\ell^m$ is also equal to zero if $\ell+q$ is odd. Since, for this case, q is odd, it follows that ℓ is also odd. Therefore the two groups of coefficients form non-overlapping sets and the distinction may be dropped: G_ℓ^m represents both sets of functions. This reduces the set of non-zero coefficients to one-fourth of the original number in Eq. 4.1.2.

Simplifying results gives:

$$
G_\ell^m = \frac{2m(2\ell+1)(\ell-m)!\,\delta(\ell+m,2q)}{2^\ell \ell(\ell+1)\left(\dfrac{\ell+m}{2}\right)!\left(\dfrac{\ell-m}{2}\right)!} = \frac{2m(2\ell+1)(\ell-m)!\,\delta(\ell+m,2q)}{\ell(\ell+1)(\ell+m)!!(\ell-m)!!}
$$

$$(4.1.8)$$

This equation is correct for all modal combinations. Combining Eq. 4.1.2 and 4.1.8 shows that the spherical coordinate expansion for a y-directed, z-polarized plane wave is:

$$
e^{-i\sigma\sin\theta\sin\phi} = \left\{ \sum_{\ell=0;e}^{\infty} \sum_{me}^{\ell} \cos(m\phi) - i\sum_{\ell=1;o}^{\infty} \sum_{mo}^{\ell} \sin(m\phi) \right\} \frac{\ell(\ell+1)}{m} G_\ell^m j_\ell(\sigma) P_\ell^m(\cos\theta)
$$

$$(4.1.9)$$

4.2 Incoming TE Fields

To solve for the radial magnetic field component of a y-directed plane wave begin by noting that:

$$
\eta H_r = \sin\theta\cos\phi\, e^{-i\sigma\sin\theta\sin\phi} = \frac{i}{\sigma}\frac{\partial}{\partial\phi} e^{-i\sigma\sin\theta\sin\phi} \qquad (4.2.1)
$$

The first equality in Eq. 4.2.1 is by definition. Substituting Eq. 4.2.1 into Eq. 4.2.1 and differentiating gives:

$$
\eta H_r = \left[\sum_{\ell=1;o}^{\infty} \sum_{mo}^{\ell} \cos(m\phi) - i\sum_{\ell=2;e}^{\infty} \sum_{me}^{\ell} \sin(m\phi) \right] \ell(\ell+1)G_\ell^m \frac{j_\ell(\sigma)}{\sigma} P_\ell^m(\cos\theta)
$$

$$(4.2.2)$$

Combining Eq. 4.2.2 with the same field component of Eq. 1.13.9 determines the constant field coefficients. Knowledge of the constant field coefficients and the component forms of Eq. 1.13.9 are sufficient to obtain the full set of TE modes.

4.3 Incoming TM Fields

A longer procedure is necessary to obtain the coefficients for TM modes. It is convenient to define E_r in a way analogous with Eq. 4.2.2, using Eq. 4.3.1 to define coefficients ${}^cF_\ell{}^m$ and ${}^sF_\ell{}^m$ and then solve for the constants. The general form for the field is:

$$E_r = \sum_{\ell=0}^{\infty} \sum_{m=0}^{\ell} \left[{}^cF_\ell^m \cos(n\phi) - i\, {}^sF_\ell^m \sin(n\phi) \right] \ell(\ell+1) \frac{j_\ell(\sigma)}{\sigma} P_\ell^m(\cos\theta)$$

$$(4.3.1)$$

The radial component of the y-directed plane wave is:

$$E_r = \cos\theta e^{-i\sigma \sin\theta \sin\phi} = \frac{i}{\sigma \sin\phi} \frac{\partial}{\partial\theta} e^{-i\sigma \sin\theta \sin\phi}$$

$$(4.3.2)$$

Substituting Eq. 4.1.9 into Eq. 4.3.2 and taking the derivative gives:

$$E_r = \frac{i}{\sin\phi} \left[\sum_{\ell e;0}^{\infty} \sum_{me}^{\ell} \cos(m\phi) - i \sum_{\ell o;1}^{\infty} \sum_{mo}^{\ell} \sin(m\phi) \right]$$
$$\times \frac{\ell(\ell+1)}{m} G_\ell^m \frac{j_\ell(\sigma)}{\sigma} \frac{dP_\ell^m(\cos\theta)}{d\theta}$$

$$(4.3.3)$$

Symbols $\ell e;0$ and $\ell o;1$ indicate sums respectively over even integers starting at zero, and over odd integers starting at one. Symbols me and mo indicate sums respectively over even and odd integers only. A trigonometric identity that puts Eq. 4.3.3 in a more useful form is:

$$\frac{1}{\sin\phi} \equiv 2\sum_{s=0}^{\infty} \sin\left[(2s+1)\phi\right]$$

$$(4.3.4)$$

Combining Eq. 4.3.3 and 4.3.4 gives:

$$E_r = 2\sum_{s=0}^{\infty} \left[i\sum_{\ell e;0}^{\infty} \sum_{me}^{\ell} \cos m\phi + \sum_{\ell o;1}^{\infty} \sum_{mo}^{\ell} \sin m\phi \right] \sin\left[(2s+1)\phi\right]$$
$$\times \frac{\ell(\ell+1)}{m} G_\ell^m \frac{j_\ell(\sigma)}{\sigma} \frac{dP_\ell^m(\cos\theta)}{d\theta}$$

$$(4.3.5)$$

Other useful trigonometric identities are:

$$2\sin m\phi\sin(2s+1)\phi \equiv \cos\big[(m-2s-1)\phi\big]-\cos\big[(m+2s+1)\phi\big]$$
$$2\cos m\phi\sin(2s+1)\phi \equiv \sin\big[(m+2s+1)\phi\big]-\sin\big[(m-2s-1)\phi\big]$$

$$(4.3.6)$$

The procedure is to substitute Eq. 4.3.6 into Eq. 4.3.5, equate Eq. 4.3.1 and 4.3.5, differentiate both $(\ell-1)$ times by σ, and then go to the limit as σ becomes limitlessly small. The result, for ℓ odd, is:

$$\sum_{s=0}^{\infty}\left[\sum_{mo}^{\ell}\left\{\cos\big[(m-2s-1)\phi\big]-\cos\big[(m+2s+1)\phi\big]\right\}\right]\frac{G_{\ell}^{m}}{m}\frac{dP_{\ell}^{m}(\cos\theta)}{d\theta}$$
$$=\sum_{n=0}^{\ell}\left[{}^{c}F_{\ell}^{n}\cos(m\phi)-i\,{}^{s}F_{\ell}^{n}\sin(m\phi)\right]P_{\ell}^{n}(\cos\theta)\qquad(4.3.7)$$

Next multiply Eq. 4.3.7 by $\cos(q\phi)$, where q is an integer, and integrate from $\phi=0$ to 2π. This shows that ${}^{s}F_{\ell}^{m}$ is zero for odd values of ℓ; it follows similarly that ${}^{c}F_{\ell}^{m}$ is zero for even values of ℓ. Therefore the coefficients form non-overlapping sets and, again, the notation may be simplified by dropping the reverse superscript, with F_{ℓ}^{m} representing both sets of coefficients. This reduces the set of non-zero coefficients to one-fourth of the original number in Eq. 4.3.1. The resulting equality is:

$$\sum_{n=0}^{\ell}\delta(q,n)F_{\ell}^{n}P_{\ell}^{n}(\cos\theta)$$
$$=\sum_{s=0}^{\infty}\sum_{me}^{\ell}\left\{\delta(q,|m-2s-1|)-\delta(q,m+2s+1)\right\}\frac{G_{\ell}^{m}}{m}\frac{dP_{\ell}^{m}(\cos\theta)}{d\theta}\qquad(4.3.8)$$

Evaluating the delta functions and collecting terms gives, after some algebra and with $U(q)$ representing a step function of q:

$$F_{\ell}^{q}P_{\ell}^{q}(\cos\theta)=2U(q)\sum_{s=0}^{(\ell-q-1)/2}\frac{G_{\ell}^{q+2s+1}}{q+2s+1}\frac{d}{d\theta}P_{\ell}^{q+2s+1}(\cos\theta)\qquad(4.3.9)$$

With the aid of Table A.15.1 No. 1, Eq. 4.3.9 may be rewritten as:

$$F_\ell^q P_\ell^q\left(\cos\theta\right)=U\left(q\right)\sum_{s=0}^{(\ell-q-1)/2}\frac{G_\ell^{q+2s+1}}{q+2s+1}$$

$$\times\left[\left(\ell+q+2s+1\right)\left(\ell-q-2s\right)P_\ell^{q+2s}-P_\ell^{q+2s+2}\right] \tag{4.3.10}$$

Multiply Eq. 4.3.10 by $P_\ell^q(\cos\theta)$ and integrate over θ using the integrals of Table A.16.1 No. 2. After simplifying, the result is:

$$F_\ell^q=2\ell\left(\ell-q\right)!\sum_{s=0}^{(\ell-q-1)/2}\frac{G_\ell^{q+2s+1}}{\left(q+2s+1\right)}\frac{\left(-1\right)^s U\left(q\right)}{\left(\ell-q-2s-1\right)!} \tag{4.3.11}$$

Combining Eq. 4.1.9 and 4.3.11 with n equal to any positive integer gives:

$$F_\ell^q=\frac{4\left(2\ell+1\right)}{\left(\ell+1\right)}\left(\ell-q\right)!\sum_{s=0}^{(\ell-q-1)/2}\frac{\left(-1\right)^s U\left(q\right)\delta\left(\ell+q,2n+1\right)}{\left(\ell-q-2s-1\right)!!\left(\ell+q+2s+1\right)!!} \tag{4.3.12}$$

The sum of Eq. 4.3.12 is listed in Table A.15.1 No. 8. Incorporating the sum, replacing q by m to give the same dummy index as Eq. 4.1.9 and with n denoting any integer shows the two coefficient sets $F_\ell^{\,m}$ and $G_\ell^{\,m}$ are:

$$F_\ell^m=\frac{2\left(2\ell+1\right)}{\ell\left(\ell+1\right)}\frac{U\left(m\right)\left(\ell-m\right)!\delta\left(\ell+m,2n+1\right)}{\left(\ell+m-1\right)!!\left(\ell-m-1\right)!!}$$

$$\tag{4.3.13}$$

$$G_\ell^m\equiv\frac{2\left(2\ell+1\right)}{\ell\left(\ell+1\right)}\frac{m\left(\ell-m\right)!\delta\left(\ell+m,2n\right)}{\left(\ell+m\right)!!\left(\ell-m\right)!!}$$

Coefficients $F_\ell^{\,m}$ and $G_\ell^{\,m}$ have opposite parity in that $F_\ell^{\,m}$ is other than zero only if $\ell+m$ is odd and $G_\ell^{\,m}$ is other than zero only if $\ell+m$ is even. At degree $m=0$ coefficients $F_\ell^{\,m}$ have the maximum value and coefficients $G_\ell^{\,m}$ are equal to zero. Values through the first five orders are listed in Table 4.3.1.

Table 4.3.1. Values of field coefficients for a y-directed, z-polarized plane wave.

$$F_1^0 = \frac{3}{2} \quad G_1^1 = \frac{3}{2}$$

$$F_2^1 = \frac{5}{6} \quad G_2^2 = \frac{5}{12}$$

$$F_3^0 = \frac{7}{8} \quad G_3^1 = \frac{7}{72} \quad F_3^2 = \frac{7}{48} \quad G_3^3 = \frac{7}{96}$$

$$F_4^1 = \frac{9}{160} \quad G_4^2 = \frac{3}{160} \quad F_4^3 = \frac{3}{160} \quad G_4^4 = \frac{3}{320}$$

$$F_5^0 = \frac{11}{16} \quad G_5^1 = \frac{11}{240} \quad F_5^2 = \frac{11}{240} \quad G_5^3 = \frac{11}{1920} \quad F_5^4 = \frac{11}{5760} \quad G_5^5 = \frac{11}{11520}$$

4.4 Exterior Fields, Powers, and Forces

The radial field components of a y-directed, z-polarized plane wave are given by the combination of Eq. 4.2.2, 4.3.1, and 4.3.13:

$$E_r = \left\{ \sum_{lo;1}^{\infty} \sum_{me}^{\ell-1} \cos(m\phi) - i \sum_{le;2}^{\infty} \sum_{mo}^{\ell-1} \sin(m\phi) \right\} \ell(\ell+1) F_\ell^m \frac{j_\ell(\sigma)}{\sigma} P_\ell^m(\cos\theta)$$

$$\eta H_r = \left\{ \sum_{lo;1}^{\infty} \sum_{mo}^{\ell} \cos(m\phi) - i \sum_{le;2}^{\infty} \sum_{me}^{\ell} \sin(m\phi) \right\} \ell(\ell+1) G_\ell^m \frac{j_\ell(\sigma)}{\sigma} P_\ell^m(\cos\theta)$$

$$(4.4.1)$$

The angularly directed field components follow from Eq. 4.4.1.

Unlike the field forms of the incoming plane wave the scattered wave fields decrease with distance as $\exp[i(\omega t - \sigma)]/\sigma$ at limitlessly large ranges which, in turn, requires the radial dependent functions to be spherical Hankel functions of the second kind. It is, therefore, necessary to replace the spherical Bessel functions by spherical Hankel functions. As is the case for sphcrical scatterers, different modes scatter with different magnitudes and different phases. To account for these changes introduce two new infinite sets of field constants, α_ℓ^m and β_ℓ^m, as part of the scattered fields. Let α_ℓ^m be the coefficient of TE modes and β_ℓ^m be the coefficient of TM modes. Incorporating these results into the radial components of the scattered field gives:

$$E_r = \left\{ \sum_{\ell o;1}^{\infty} \sum_{me}^{\ell-1} \cos(m\phi) - i \sum_{\ell e;2}^{\infty} \sum_{mo}^{\ell-1} \sin(m\phi) \right\} \ell(\ell+1) \beta_\ell^m F_\ell^m \frac{h_\ell(\sigma)}{\sigma} P_\ell^m(\cos\theta)$$

$$\eta H_r = \left\{ \sum_{\ell o;1}^{\infty} \sum_{mo}^{\ell} \cos(m\phi) - i \sum_{\ell e;2}^{\infty} \sum_{me}^{\ell} \sin(m\phi) \right\} \ell(\ell+1) \alpha_\ell^m G_\ell^m \frac{h_\ell(\sigma)}{\sigma} P_\ell^m(\cos\theta)$$

$$(4.4.2)$$

Problem solution requires evaluation of the full parameter sets α_ℓ^m and β_ℓ^m. As with the spherical scatterer, the total field is the sum of the incoming plane wave fields and the outwardly directed scattered fields. Summing the radial field components of Eq. 4.4.1 and 4.4.2 gives the total radial field components. The components of the angular fields follow directly from the radial ones, see Eq. 1.13.9, and are:

$$E_\theta = \left[\sum_{\ell o;1}^{\infty} \sum_{me}^{\ell-1} \cos(m\phi) - i \sum_{\ell e;2}^{\infty} \sum_{mo}^{\ell-1} \sin(m\phi) \right] F_\ell^m \left(j_\ell^{\bullet} + \beta_\ell^m h_\ell^{\bullet} \right) \frac{dP_\ell^m}{d\theta}$$

$$- \left[\sum_{\ell e;2}^{\infty} \sum_{me}^{\ell} \cos(m\phi) - i \sum_{\ell o;1}^{\infty} \sum_{mo}^{\ell} \sin(m\phi) \right] G_\ell^m \left(j_\ell + \alpha_\ell^m h_\ell \right) \frac{mP_\ell^m}{\sin\theta}$$

$$\eta H_\phi = -i \left[\sum_{\ell o;1}^{\infty} \sum_{me}^{\ell-1} \cos(m\phi) - i \sum_{\ell e;2}^{\infty} \sum_{mo}^{\ell-1} \sin(m\phi) \right] F_\ell^m \left(j_\ell + \beta_\ell^m h_\ell \right) \frac{dP_\ell^m}{d\theta}$$

$$- i \left[\sum_{\ell e;2}^{\infty} \sum_{me}^{\ell} \cos(m\phi) - i \sum_{\ell o;1}^{\infty} \sum_{mo}^{\ell} \sin(m\phi) \right] G_\ell^m \left(j_\ell^{\bullet} + \alpha_\ell^m h_\ell^{\bullet} \right) \frac{mP_\ell^m}{\sin\theta}$$

$$-E_\phi = i \left[\sum_{\ell e;2}^{\infty} \sum_{mo}^{\ell-1} \cos(m\phi) - i \sum_{\ell o;3}^{\infty} \sum_{me}^{\ell-1} \sin(m\phi) \right] F_\ell^m \left(j_\ell^{\bullet} + \beta_\ell^m h_\ell^{\bullet} \right) \frac{mP_\ell^m}{\sin\theta}$$

$$(4.4.3)$$

$$- i \left[\sum_{\ell o;1}^{\infty} \sum_{mo}^{\ell} \cos(m\phi) - i \sum_{\ell e;2}^{\infty} \sum_{me}^{\ell} \sin(m\phi) \right] G_\ell^m \left(j_\ell + \alpha_\ell^m h_\ell \right) \frac{dP_\ell^m}{d\theta}$$

$$\eta H_\theta = \left[\sum_{\ell e;2}^{\infty} \sum_{mo}^{\ell-1} \cos(m\phi) - i \sum_{\ell o;3}^{\infty} \sum_{me}^{\ell-1} \sin(m\phi) \right] F_\ell^m \left(j_\ell + \beta_\ell^m h_\ell \right) \frac{mP_\ell^m}{\sin\theta}$$

$$+ \left[\sum_{\ell o;1}^{\infty} \sum_{mo}^{\ell} \cos(m\phi) - i \sum_{\ell e;2}^{\infty} \sum_{me}^{\ell} \sin(m\phi) \right] G_\ell^m \left(j_\ell^{\bullet} + \alpha_\ell^m h_\ell^{\bullet} \right) \frac{dP_\ell^m}{d\theta}$$

Power on a circumscribing virtual sphere of radius greater than a is obtained from the radial component of the complex Poynting vector. This equation leads to the four-part radial Poynting vector:

$$
N_{r11} = \frac{\mathrm{Re}}{2\eta}
\left\{
\begin{aligned}
&\left(\sum_{lo}^{\infty}\sum_{me}^{\ell-1}\cos(m\phi) - i\sum_{\ell e}^{\infty}\sum_{mo}^{\ell-1}\sin(m\phi) \right)
\left(\sum_{no}^{\infty}\sum_{pe}^{n-1}\cos(p\phi) + i\sum_{ne}^{\infty}\sum_{po}^{n-1}\sin(p\phi) \right) \\
&\times iF_{\ell}^{m}F_{n}^{p}\left(j_{\ell}^{\bullet} + \beta_{\ell}^{m}h_{\ell}^{\bullet} \right)\left(j_{n} + \beta_{n}^{p*}h_{n}^{*} \right)\left(\frac{dP_{\ell}^{m}}{d\theta}\frac{dP_{n}^{p}}{d\theta} \right) \\
&+\left(\sum_{\ell e}^{\infty}\sum_{mo}^{\ell-1}\cos(m\phi) - i\sum_{lo}^{\infty}\sum_{me}^{\ell-1}\sin(m\phi) \right)
\left(\sum_{ne}^{\infty}\sum_{po}^{n-1}\cos(p\phi) + i\sum_{no}^{\infty}\sum_{pe}^{n-1}\sin(p\phi) \right) \\
&\times iF_{\ell}^{m}F_{n}^{p}\left(j_{\ell}^{\bullet} + \beta_{\ell}^{m}h_{\ell}^{\bullet} \right)\left(j_{n} + \beta_{n}^{p*}h_{n}^{*} \right)\left(\frac{mP_{\ell}^{m}}{\sin\theta}\frac{pP_{n}^{p}}{\sin\theta} \right)
\end{aligned}
\right\}
\tag{4.4.4}
$$

$$
N_{r12} = -\frac{\mathrm{Re}}{2\eta}
\left\{
\begin{aligned}
&\left(\sum_{lo}^{\infty}\sum_{mo}^{\ell}\cos(m\phi) - i\sum_{\ell e}^{\infty}\sum_{me}^{\ell}\sin(m\phi) \right)
\left(\sum_{no}^{\infty}\sum_{po}^{n}\cos(p\phi) + i\sum_{ne}^{\infty}\sum_{pe}^{n}\sin(p\phi) \right) \\
&\times iG_{\ell}^{m}G_{n}^{p}(j_{\ell} + \alpha_{\ell}^{m}h_{\ell})(j_{n}^{\bullet} + \alpha_{n}^{p*}h_{n}^{\bullet*})\left(\frac{dP_{\ell}^{m}}{d\theta}\frac{dP_{n}^{p}}{d\theta} \right) \\
&+\left(\sum_{\ell e}^{\infty}\sum_{me}^{\ell}\cos(m\phi) - i\sum_{lo}^{\infty}\sum_{mo}^{\ell}\sin(m\phi) \right)
\left(\sum_{ne}^{\infty}\sum_{pe}^{n}\cos(p\phi) + i\sum_{no}^{\infty}\sum_{po}^{n}\sin(p\phi) \right) \\
&\times iG_{\ell}^{m}G_{n}^{p}(j_{\ell} + \alpha_{\ell}^{m}h_{\ell})(j_{n}^{\bullet} + \alpha_{n}^{p*}h_{n}^{\bullet*})\left(\frac{mP_{\ell}^{m}}{\sin\theta}\frac{pP_{n}^{p}}{\sin\theta} \right)
\end{aligned}
\right\}
\tag{4.4.5}
$$

$$
N_{r21} = \frac{\mathrm{Re}}{2\eta}
\left\{
\begin{aligned}
&\left(\sum_{lo}^{\infty}\sum_{me}^{\ell-1}\cos(m\phi) - i\sum_{\ell e}^{\infty}\sum_{mo}^{\ell-1}\sin(m\phi) \right)
\left(\sum_{ne}^{\infty}\sum_{pe}^{n}\cos(p\phi) + i\sum_{no}^{\infty}\sum_{po}^{n}\sin(p\phi) \right) \\
&\times iF_{\ell}^{m}G_{n}^{p}\left(j_{\ell}^{\bullet} + \beta_{\ell}^{m}h_{\ell}^{\bullet} \right)\left(j_{n}^{\bullet} + \alpha_{n}^{p*}h_{n}^{\bullet*} \right)\left(\frac{pP_{n}^{p}}{\sin\theta}\frac{dP_{\ell}^{m}}{d\theta} \right) \\
&+\left(\sum_{\ell e}^{\infty}\sum_{mo}^{\ell-1}\cos(m\phi) - i\sum_{lo}^{\infty}\sum_{me}^{\ell-1}\sin(m\phi) \right)
\left(\sum_{no}^{\infty}\sum_{po}^{n}\cos(p\phi) + i\sum_{ne}^{\infty}\sum_{pe}^{n}\sin(p\phi) \right) \\
&\times iF_{\ell}^{m}G_{n}^{p}\left(j_{\ell}^{\bullet} + \beta_{\ell}^{m}h_{\ell}^{\bullet} \right)\left(j_{n}^{\bullet} + \alpha_{n}^{p*}h_{n}^{\bullet*} \right)\left(\frac{mP_{\ell}^{m}}{\sin\theta}\frac{dP_{n}^{p}}{d\theta} \right)
\end{aligned}
\right\}
\tag{4.4.6}
$$

$$N_{r22} = -\frac{\mathrm{Re}}{2\eta} \left\{ \begin{array}{l} \left[\left(\sum_{\ell e}^{\infty}\sum_{me}^{\ell}\cos(m\phi) - i\sum_{\ell o}^{\infty}\sum_{mo}^{\ell}\sin(m\phi) \right)\left(\sum_{no}^{\infty}\sum_{pe}^{n-1}\cos(p\phi) + i\sum_{ne}^{\infty}\sum_{po}^{n-1}\sin(p\phi) \right) \right. \\[2ex] \times iF_\ell^m G_n^p \left(j_\ell + \alpha_\ell^m h_\ell \right)\left(j_n + \beta_n^{p*} h_n^* \right)\left(\frac{mP_\ell^m}{\sin\theta}\frac{dP_n^p}{d\theta} \right) \\[2ex] + \left(\sum_{\ell o}^{\infty}\sum_{mo}^{\ell}\cos(m\phi) - i\sum_{\ell e}^{\infty}\sum_{me}^{\ell}\sin(m\phi) \right)\left(\sum_{ne}^{\infty}\sum_{po}^{n-1}\cos(p\phi) + i\sum_{no}^{\infty}\sum_{pe}^{n-1}\sin(p\phi) \right) \\[2ex] \left. \times iF_\ell^m G_n^p \left(j_\ell + \alpha_\ell^m h_\ell \right)\left(j_n + \beta_n^{p*} h_n^* \right)\left(\frac{pP_n^p}{\sin\theta}\frac{dP_\ell^m}{d\theta} \right) \right\} \end{array} \right.$$

$$\tag{4.4.7}$$

The sum of the four surface integrals gives the total surface power. Integrating over the azimuth angle gives a Kronecker delta function of m and p, decreasing the number of sums by one. Results are shown in Eq. 4.4.8 and 4.4.9:

$$\int_0^{2\pi} d\phi \left(N_{r11} + N_{r12} \right) =$$

$$\mathrm{Re}\left[\begin{array}{l} \left(\sum_{\ell o}\sum_{no}\sum_{me} + \sum_{\ell e}\sum_{ne}\sum_{mo} \right) iF_\ell^m F_n^m \left(j_\ell^\bullet + \beta_\ell^m h_\ell^\bullet \right)\left(j_n + \beta_n^{m*} h_n^* \right) \\[2ex] - \left(\sum_{\ell o}\sum_{no}\sum_{mo} + \sum_{\ell e}\sum_{ne}\sum_{me} \right) iG_\ell^m G_n^m \left(j_\ell + \alpha_\ell^m h_\ell \right)\left(j_n^\bullet + \alpha_n^{m*} h_n^{\bullet*} \right) \end{array} \right]$$

$$\times \frac{\pi}{2\eta}\left[1 + \delta(m,0) \right]\left(\frac{dP_\ell^m}{d\theta}\frac{dP_n^m}{d\theta} + \frac{m^2 P_\ell^m P_n^m}{\sin^2\theta} \right) \tag{4.4.8}$$

$$\int_0^{2\pi} d\phi \left(N_{r21} + N_{r22} \right) =$$

$$\mathrm{Re}\left\{ \begin{array}{l} \left(\sum_{\ell e}\sum_{no}\sum_{mo} + \sum_{\ell o}\sum_{ne}\sum_{me} \right) iF_\ell^m G_n^m \left(j_\ell^\bullet + \beta_\ell^m h_\ell^\bullet \right)\left(j_n^\bullet + \alpha_n^{m*} h_n^{\bullet*} \right) \\[2ex] - \left(\sum_{\ell e}\sum_{no}\sum_{me} + \sum_{\ell o}\sum_{ne}\sum_{mo} \right) iF_\ell^m G_n^m \left(j_\ell + \alpha_\ell^m h_\ell \right)\left(j_n + \beta_n^{m*} h_n^* \right) \end{array} \right\} \frac{\pi}{\eta}\frac{m}{\sin\theta}\left(P_\ell^m P_n^m \right)$$

$$\tag{4.4.9}$$

To complete the evaluation it is necessary to integrate Eq. 4.4.8 and 4.4.9 over the zenith angle. The integral of Eq. 4.4.9 gives a null result. Evaluating the integral of Eq. 4.4.8 and replacing the coefficients by the values of Eq. 4.3.13 gives:

$$
\begin{aligned}
P_{av} = \frac{4\pi\sigma^2}{\eta k^2}\mathrm{Re}&\left\{\left[\sum_{lo}^{\infty}\sum_{me}^{\ell-1}+\sum_{le}^{\infty}\sum_{mo}^{\ell-1}\right]\frac{U(m)(2\ell+1)(\ell-m)!!(\ell+m)!!}{\ell(\ell+1)(\ell-m-1)!!(\ell+m-1)!!}\right. \\
&\left.\times i\left(\dot{j}_{\ell}+\beta_{\ell}^{m}\dot{h}_{\ell}\right)\left(j_{\ell}+\beta_{\ell}^{m*}h_{\ell}^{*}\right)\right\} \\
-\frac{4\pi\sigma^2}{\eta k^2}\mathrm{Re}&\left\{\left[\sum_{lo}^{\infty}\sum_{mo}^{\ell}+\sum_{le}^{\infty}\sum_{me}^{\ell}\right]\frac{m^2(2\ell+1)(\ell-m-1)!!(\ell+m-1)!!}{\ell(\ell+1)(\ell-m)!!(\ell+m)!!}\right. \\
&\left.\times i\left(j_{\ell}+\alpha_{\ell}^{m}h_{\ell}\right)\left(\dot{j}_{\ell}+\alpha_{\ell}^{m*}\dot{h}_{\ell}^{*}\right)\right\}
\end{aligned}
$$

$$(4.4.10)$$

In the limit of infinite radius, Eq. 4.4.10 goes to:

$$
\begin{aligned}
P_{av} = \frac{4\pi\sigma^2}{\eta k^2}\mathrm{Re}&\left\{\left[\sum_{lo}^{\infty}\sum_{me}^{\ell-1}+\sum_{le}^{\infty}\sum_{mo}^{\ell-1}\right]\frac{U(m)(2\ell+1)(\ell-m)!!(\ell+m)!!}{\ell(\ell+1)(\ell-m-1)!!(\ell+m-1)!!}\right. \\
&\left.\times i\left(\dot{j}_{\ell}+\beta_{\ell}^{m}\dot{h}_{\ell}\right)\left(j_{\ell}+\beta_{\ell}^{m*}h_{\ell}^{*}\right)\right\} \\
-\frac{4\pi\sigma^2}{\eta k^2}\mathrm{Re}&\left\{\left[\sum_{lo}^{\infty}\sum_{mo}^{\ell}+\sum_{le}^{\infty}\sum_{me}^{\ell}\right]\frac{m^2(2\ell+1)(\ell-m-1)!!(\ell+m-1)!!}{\ell(\ell+1)(\ell-m)!!(\ell+m)!!}\right. \\
&\left.\times i\left(j_{\ell}+\alpha_{\ell}^{m}h_{\ell}\right)\left(\dot{j}_{\ell}+\alpha_{\ell}^{m*}\dot{h}_{\ell}^{*}\right)\right\}
\end{aligned}
$$

$$(4.4.11)$$

The terms are interpreted similarly to those for scattering from a sphere: terms proportional to both $\alpha_n^m\,\alpha_n^{m*}$ and $\beta_n^m\,\beta_n^{m*}$ describe time-average power scattered away from the antenna; each term is positive. Terms proportional to $\mathrm{Re}\,\alpha_n^m$ and $\mathrm{Re}\,\beta_n^m$ are negative and describe inwardly directed power; the time integral of Eq. 4.4.11 is the negative of the extinction (absorbed plus scattered) energy. With an ideal, shorted antenna the two sets of terms have equal magnitudes, opposite signs, and sum to zero.

4.5 The Cross Sections

The purpose of this section is to compare and contrast the scattering properties of spheres with those of a receiving biconical antenna, and viewing a receiving antenna as a lossy scatterer. Cross sections were defined in Sec. 2.9 Analogously with Eq. 2.9.10, the scattering cross section C_{SC} is defined to equal the ratio of scattered power to the input power density. The geometric cross section is equal to the cross sectional area of the scatterer. Using Eq. 4.4.11, the scattering-to-geometric cross section ratio is:

$$\frac{C_{SC}}{C_{GE}} = \frac{8}{k^2a^2}\left\{\left[\sum_{lo}^{\infty}\sum_{me}^{\ell-1}+\sum_{le}^{\infty}\sum_{mo}^{\ell-1}\right]\frac{U(m)(2\ell+1)(\ell-m)!!(\ell+m)!!}{\ell(\ell+1)(\ell-m-1)!!(\ell+m-1)!!}\left(\beta_\ell^m\beta_\ell^{m*}\right)\right\}$$

$$+\frac{8}{k^2a^2}\left\{\left[\sum_{lo}^{\infty}\sum_{mo}^{\ell}+\sum_{le}^{\infty}\sum_{me}^{\ell}\right]\frac{m^2(2\ell+1)(\ell-m-1)!!(\ell+m-1)!!}{\ell(\ell+1)(\ell-m)!!(\ell+m)!!}\left(\alpha_\ell^m\alpha_\ell^{m*}\right)\right\} \quad (4.5.1)$$

The normalized extinction cross section C_{EX} is equal to the ratio of the total power extracted from the incoming plane wave to the geometric cross sectional area of the scatterer. Again using Eq. 4.4.11, the extinction-to-geometric cross section ratio is:

$$\frac{C_{EX}}{C_{GE}} = -\frac{4}{k^2a^2}\left\{\left[\sum_{lo}^{\infty}\sum_{me}^{\ell-1}+\sum_{le}^{\infty}\sum_{mo}^{\ell-1}\right]\frac{U(m)(2\ell+1)(\ell-m)!!(\ell+m)!!}{\ell(\ell+1)(\ell-m-1)!!(\ell+m-1)!!}\left(\mathrm{Re}\,\beta_\ell^m\right)\right\}$$

$$-\frac{8}{k^2a^2}\left\{\left[\sum_{lo}^{\infty}\sum_{mo}^{\ell}+\sum_{le}^{\infty}\sum_{me}^{\ell}\right]\frac{m^2(2\ell+1)(\ell-m-1)!!(\ell+m-1)!!}{\ell(\ell+1)(\ell-m)!!(\ell+m)!!}\left(\mathrm{Re}\,\alpha_\ell^m\right)\right\} \quad (4.5.2)$$

The absorption cross section, C_{AB}, is equal to the absorbed power-to-cross sectional area of the scatterer ratio and using Eq. 4.4.11, the value is:

$$\frac{C_{AB}}{C_{GE}} = -\frac{4}{k^2a^2}\left\{\left[\sum_{lo}^{\infty}\sum_{me}^{\ell-1}+\sum_{le}^{\infty}\sum_{mo}^{\ell-1}\right]\frac{(2\ell+1)(\ell-m)!!(\ell+m)!!}{\ell(\ell+1)(\ell-m-1)!!(\ell+m-1)!!}\left(\mathrm{Re}\,\beta_\ell^m-\beta_\ell^m\beta_\ell^{m*}\right)\right\}$$

$$-\frac{8}{k^2a^2}\left\{\left[\sum_{lo}^{\infty}\sum_{mo}^{\ell}+\sum_{le}^{\infty}\sum_{me}^{\ell}\right]\frac{m^2(2\ell+1)(\ell-m-1)!!(\ell+m-1)!!}{\ell(\ell+1)(v-m)!!(v+m)!!}\left(\mathrm{Re}\,\alpha_\ell^m-\alpha_\ell^m\alpha_\ell^{m*}\right)\right\}$$

$$(4.5.3)$$

The thrust on the antenna from the total power absorbed is c times the value of extinction power, Eq. 4.5.2. The thrust on the antenna from the scattered power is equal to the component of scattered wave in the direction of the incoming field integrated over a virtual surface:

$$F_{ySC} = -\frac{\sigma^2}{2\eta k^2} \int\limits_0^{2\pi} \sin\phi \, d\phi \int\limits_0^\pi \sin^2\theta \, d\theta \, \mathrm{Re}\left(N_r\right) \tag{4.5.4}$$

Inserting the scattered field terms of Eq. 4.4.2 into Eq. 4.5.4 gives:

$$\int\limits_0^{2\pi} N_r \sin\phi \, d\phi = \frac{\pi \mathrm{Re}}{4\eta}\left(\sum_{\ell o}\sum_{ne}\sum_{me} + \sum_{\ell e}\sum_{no}\sum_{mo}\right) \left\{ \begin{array}{l} \left(F_\ell^m F_n^{m+1} \beta_\ell^m \beta_n^{m+1} h_\ell^{\cdot} h_n^{*}\right)\left[1+\delta(m)\right] \\ \times\left(\dfrac{dP_\ell^m}{d\theta}\dfrac{dP_n^{m+1}}{d\theta} + \dfrac{m(m+1)}{\sin^2\theta}P_\ell^m P_n^{m+1}\right) \\ -\left(F_\ell^m F_n^{m-1}\beta_\ell^m \beta_n^{m-1} h_\ell^{\cdot} h_n^{*}\right)\left[1+\delta(m-1)\right] \\ \times\left(\dfrac{dP_\ell^m}{d\theta}\dfrac{dP_n^{m-1}}{d\theta} + \dfrac{m(m-1)}{\sin^2\theta}P_\ell^m P_n^{m-1}\right) \end{array} \right\}$$

$$-\frac{\pi \mathrm{Re}}{4\eta}\left(\sum_{\ell o}\sum_{ne}\sum_{mo} + \sum_{\ell e}\sum_{no}\sum_{me}\right) \left\{ \begin{array}{l} \left(G_\ell^m G_n^{m+1} \alpha_\ell^m \alpha_n^{m+1} h_\ell h_n^{\cdot *}\right)\left(\dfrac{dP_\ell^m}{d\theta}\dfrac{dP_n^{m+1}}{d\theta} + \dfrac{m(m+1)}{\sin^2\theta}P_\ell^m P_n^{m+1}\right) \\ -\left(G_\ell^m G_n^{m-1}\alpha_\ell^m \alpha_n^{m-1} h_\ell h_n^{\cdot *}\right)\left(\dfrac{dP_\ell^m}{d\theta}\dfrac{dP_n^{m-1}}{d\theta} + \dfrac{m(m-1)}{\sin^2\theta}P_\ell^m P_n^{m-1}\right) \end{array} \right\}$$

$$+\frac{\pi \mathrm{Re}}{4\eta}\left(\sum_{\ell o}\sum_{no}\sum_{me} + \sum_{\ell e}\sum_{ne}\sum_{mo}\right) \left\{ \begin{array}{l} \left(G_n^{m+1} F_\ell^m \alpha_n^{m+1*} \beta_\ell^m h_\ell^{\cdot} h_n^{\cdot *}\right)\left[1+\delta(m)\right] \\ \times\left(\dfrac{mP_\ell^m}{\sin\theta}\dfrac{dP_n^{m+1}}{d\theta} + \dfrac{(m+1)P_n^{m+1}}{\sin\theta}\dfrac{dP_\ell^m}{d\theta}\right) \\ -G_n^{m-1}F_\ell^m \alpha_n^{m-1*}\beta_\ell^m h_\ell^{\cdot} h_n^{\cdot *}\left(\dfrac{mP_\ell^m}{\sin\theta}\dfrac{dP_n^{m-1}}{d\theta} + \dfrac{(m-1)P_n^{m-1}}{\sin\theta}\dfrac{dP_\ell^m}{d\theta}\right) \end{array} \right\}$$

$$+\frac{\pi \mathrm{Re}}{4\eta}\left(\sum_{\ell o}\sum_{no}\sum_{mo} + \sum_{\ell e}\sum_{ne}\sum_{me}\right) \left\{ \begin{array}{l} \left(G_\ell^m F_n^{m-1}\alpha_\ell^m \beta_n^{m-1*} h_\ell h_n^{*}\right)\left[1+\delta(m-1)\right] \\ \times\left(\dfrac{mP_\ell^m}{\sin\theta}\dfrac{dP_n^{m-1}}{d\theta} + \dfrac{(m-1)P_n^{m-1}}{\sin\theta}\dfrac{dP_\ell^m}{d\theta}\right) \\ -G_\ell^m F_n^{m+1}\alpha_\ell^m \beta_n^{m+1*} h_\ell h_n^{*}\left(\dfrac{mP_\ell^m}{\sin\theta}\dfrac{dP_n^{m+1}}{d\theta} + \dfrac{(m+1)P_n^{m+1}}{\sin\theta}\dfrac{dP_\ell^m}{d\theta}\right) \end{array} \right\} \tag{4.5.5}$$

Next, with the aid of Table A.16, combine Eq. 4.5.4 and 4.5.5 to find the directed power through a virtual circumscribing sphere. In the limit where the scattered waves extend to infinite radius the normalized y-directed force due to the scattered field is:

$$
\frac{cf_{ySC}}{C_{GE}} = -\frac{4}{k^2 a^2} \left\{
\begin{aligned}
&\left[\left(\sum_{\ell o}\sum_{me} + \sum_{\ell e}\sum_{mo} \right)
\begin{pmatrix}
+\left(\beta_\ell^m \beta_{\ell+1}^{m+1*} + \beta_\ell^{m*} \beta_{\ell+1}^{m+1} \right) \dfrac{U(m)(\ell-m)!!(\ell+2)!!}{(\ell+1)^2(\ell-m-1)!!(\ell+m-1)!!} \\[2mm]
+\left(\beta_\ell^m \beta_{\ell-1}^{m+1*} + \beta_\ell^{m*} \beta_{\ell-1}^{m+1} \right) \dfrac{U(m)(\ell-m)!!(\ell+m)!!}{(\ell)^2(\ell-m-3)!!(\ell+m-1)!!}
\end{pmatrix} \right] \\[4mm]
&+\left(\sum_{\ell o}\sum_{mo} + \sum_{\ell e}\sum_{me} \right)
\begin{pmatrix}
+\left(\alpha_\ell^m \alpha_{\ell+1}^{m+1*} + \alpha_\ell^{m*} \alpha_{\ell+1}^{m+1} \right) \dfrac{m(m+1)(\ell-m-1)!!(\ell+m+1)!!}{(\ell+1)^2(\ell-m)!!(\ell+m)!!} \\[2mm]
+\left(\alpha_\ell^m \alpha_{\ell-1}^{m+1*} + \alpha_\ell^{m*} \alpha_{\ell-1}^{m+1} \right) \dfrac{m(m+1)(\ell-m-1)!!(\ell+m+1)!!}{(\ell)^2(\ell-m-2)!!(\ell+m)!!}
\end{pmatrix} \\[4mm]
&+\left(\sum_{\ell o}\sum_{me} + \sum_{\ell e}\sum_{mo} \right)\left(\alpha_\ell^{m+1*}\beta_\ell^m + \alpha_\ell^{m+1}\beta_\ell^{m*} \right)\left(\dfrac{(m+1)(2\ell+1)(\ell-m)!!(\ell+m)!!}{\ell^2(\ell+1)^2(\ell-m-1)!!(\ell+m-1)!!} \right) \\[4mm]
&-\left(\sum_{\ell o}\sum_{mo} + \sum_{\ell e}\sum_{me} \right)\left(\alpha_\ell^m\beta_\ell^{m+1*} + \alpha_\ell^{m*}\beta_\ell^{m+1} \right)\left(\dfrac{m(2\ell+1)(\ell-m-1)!!(\ell+m+1)!!}{\ell^2(\ell+1)^2(\ell-m-2)!!(\ell+m)!!} \right)
\end{aligned}
\right\}
$$

$$\text{(4.5.6)}$$

The normalized force due to the extinction power is:

$$
\frac{cf_{yEX}}{C_{GE}} = -\frac{8}{k^2 a^2} \left\{
\begin{aligned}
&\left[\sum_{\ell o}^{\infty}\sum_{me}^{\ell-1} + \sum_{\ell e}^{\infty}\sum_{mo}^{\ell-1} \right] \frac{U(m)(2\ell+1)(\ell-m)!!(\ell+m)!!}{\ell(\ell+1)(\ell-m-1)!!(\ell+m-1)!!} \operatorname{Re}\beta_\ell^m \\[3mm]
&+\left[\sum_{\ell o}^{\infty}\sum_{mo}^{\ell} + \sum_{\ell e}^{\infty}\sum_{me}^{\ell} \right] \frac{m^2(2\ell+1)(\ell-m-1)!!(\ell+m-1))!!}{\ell(\ell+1)(\ell-m)!!(\ell+m)!!} \operatorname{Re}\alpha_\ell^m
\end{aligned}
\right\}
$$

$$\text{(4.5.7)}$$

To compare results obtained using y- and z-directed incoming plane waves, consider scattering by an ideally conducting sphere. For a z-directed wave, the coefficients are given by Eq. 2.10.6; for a y-directed wave, the boundary conditions follow from Eq. 4.4.3, and are:

$$
j_\ell^{\cdot}(ka) + \beta_\ell^m h_\ell^{\cdot}(ka) = 0; \quad j_\ell(ka) + \alpha_\ell^m h_\ell(ka) = 0 \tag{4.5.8}
$$

These lead to:

$$\beta_\ell^m(ka) = -\frac{\ddot{j}_\ell(ka)}{\dot{h}_\ell(ka)}; \quad \alpha_\ell^m(ka) = -\frac{j_\ell(ka)}{h_\ell(ka)} \tag{4.5.9}$$

The coefficients are independent of degree. If the scatterer radius is much smaller than a wavelength the dipole coefficients are:

$$\beta_1^0(ka) = -\frac{2/3}{2/3 - i/(ka)^3} \cong \frac{2(ka)^3}{3i}; \quad \alpha_1^1(ka) = -\frac{(ka)/3}{(ka)/3 + i/(ka)^2} \cong \frac{i(ka)^3}{3} \tag{4.5.10}$$

The cross sections and normalized forces are:

$$\frac{C_{EX}}{C_{GE}} = \frac{C_{SC}}{C_{GE}} = \frac{cf_{EX}}{C_{GE}} = \frac{10(ka)^4}{3} \tag{4.5.11}$$

These results are equal to those of Eq. 2.10.8; comparative results emphasize that in the long wavelength limit the electrical size of a scatterer is significant but its shape is not.

4.6 General Comments

Biconical receiving antennas are of special significance for the same reasons biconical transmitting antennas are: they closely represent practical antennas and have mathematically complete solutions. We approach the receiving antenna as just another scatterer immersed in an otherwise steady state plane wave. Some incoming energy, the extinction energy, is transferred to the scatterer and the rest unperturbed; part of the extinction energy is absorbed and the rest radiated away as a scattered field. The solutions we seek are for arbitrary cone length, cone angle, and with an arbitrary value of impedance attached to the terminals. A full analysis, which requires knowing all fields at all points in space, is obtained by matching the full sets of possible field forms in the external and internal regions, see Fig. 4.1.1, to conducting boundary conditions at the antenna surfaces and virtual boundary conditions in the open aperture. From the fields one can determine the extinction energy and momentum, the scattered energy and momentum, the surface charge and current densities, the absorbed power, and the impedance at the antenna terminals.

For transmission, the calculation can begin with either voltage $V(0)$ from a constant voltage source or current $I(0)$ from a constant current source. Either is applied between the terminals of the cones at $r = b$ and, of course, there is no incident field. For reception the power sink at $r < b$ is a passive, isotropic energy absorber. Extinction power is extracted from an incident, y-directed plane wave, some of which is scattered and some is absorbed.

To make the analysis it is convenient to break space into regions similar to those of transmitting antennas:

$$\text{SINK: } r < b \,; \ 0 \leq \theta \leq \pi; \ 0 \leq \phi \leq 2\pi \tag{4.6.1}$$

INTERIOR

$$\textit{Arms}: \ b < r < a \,; \ 0 \leq \theta < \psi; \ \pi - \psi < \theta \leq \pi; \ 0 \leq \phi \leq 2\pi \tag{4.6.2}$$
$$\textit{Free space}: \ b < r < a \,; \ \psi < \theta < \pi - \psi; \ 0 \leq \phi \leq 2\pi$$

INTERFACE

$$\textit{Aperture}: \ r = a; \ \psi \leq \theta \leq \pi - \psi; \ 0 \leq \phi \leq 2\pi \tag{4.6.3}$$
$$\textit{Arms}: \ r = a; \ 0 \leq \theta \leq \psi; \ \pi - \psi < \theta \leq \pi; \ 0 \leq \phi \leq 2\pi$$

EXTERIOR

$$\textit{Free space}: \ r > a \,; \ 0 \leq \theta \leq \pi; \ 0 \leq \phi \leq 2\pi \tag{4.6.4}$$

The y-directed plane wave is the spherical Bessel function part of Eq. 4.4.2 and 4.4.3, and includes all integer orders from one to infinity and degree one. Since the z-axis is excluded from interior space interior modes have non-integer orders. Such modes are orthogonal but, of course, products of exterior and interior modes are not. Generally speaking, each exterior order contributes the most power to its interior numerical neighbors. Exterior modes are part and parcel of the surface charge and current densities on the caps and the interior modes are similarly on the cones. The zero degree plane wave modes excite external TM modes and both TM and TEM interior modes; these modes are similar to the transmitter ones and hence are transmitter modes. Higher degree exterior modes excite both TM and TE scattered and interior modes. With these modes and a lossless antenna the extinction and scattered energies are equal. They absorb momentum but not energy from the wave; these are receiver modes.

As an example of receiver modes, at low enough frequencies a surface current flows along the illuminated face of the antenna; it is largest near the caps. Detailed sketches are shown in Fig. 4.11.1. Going toward the conical apices at each differential length some current terminates on local electric charge densities. The remainder arrives at the terminals. At the cone-cap junction some charge is stored and some continues between the cap and arm. Since the currents into and out of the cone-cap junctions are not necessarily equal an oscillating ring of charge resides there. A similar current distribution is repeated, but oppositely directed, on the shadowed side of the antenna. The oppositely directed currents generate a magnetic dipole moment; the cross sectional area of the dipole is the geometrical cross section of the cones perpendicular both to the incoming wave and the antenna axis, in this case the x-direction. By Lenz's law, the phase of the generated magnetic moment is opposite that of the incoming magnetic field and results in a scattered wave.

In summary, the charge and current densities on the cones are functions of the interior fields and those on the caps are functions of the exterior fields. The signs of adjacent arm and cap surface and line charge densities may or may not be the same and the current that flows from the cone to the cone-cap junction is not necessarily equal to the current that flows from the junction to the cap. Differences result in a ring of charge at the junction.

4.7 Fields of Receiving Antennas

Combining Eq. 4.3.13 and 4.4.3 shows the TM and TE modes are respectively proportional to Kronecker's delta functions $\delta(\ell + m, 2n+1)$ and $\delta(\ell + m, 2n)$, and express the condition that the associated Legendre polynomials satisfy the symmetry conditions:

$$\text{TM modes;} \quad P_\ell^m\left(\cos\theta\right) = -P_\ell^m\left(-\cos\theta\right)$$
$$\text{TE modes;} \quad P_\ell^m\left(\cos\theta\right) = P_\ell^m\left(\cos\theta\right) \tag{4.7.1}$$

The total exterior field, *i.e.*, the modal fields of the plane and scattered waves, are equal to the sum of Eq. 4.4.1, 4.4.2, and 4.4.3. A complete field evaluation requires evaluation of the scattering field coefficients.

The interior modal structure depends upon the symmetries of the driving field and the antenna. For the antenna axis parallel with the direction of polarization, the antenna implementation retains the symmetry of the external fields in the internal region. As was the case for a transmitting antenna, non-infinite interior fields require the coefficients of all functions $y_\nu(\sigma)$ to equal zero for $\nu > 0$. Coefficients of $j_\nu(\sigma)$ are nonzero for both TM and TE modes. The symmetry of interior and exterior TM and TE modes are the same, with undetermined coefficients respectively defined to be $\Gamma_\nu{}^m$ and $\Lambda_\nu{}^m$. A full solution requires evaluation of the functional relationships between internal coefficients $\Gamma_\nu{}^m$ and $\Lambda_\nu{}^m$ and scattering coefficients $\alpha_\ell{}^m$ and $\beta_\ell{}^m$.

Combining all the above, the zero degree terms and the higher degree terms yield the expanded equation set:

$$E_r = \sum_{\nu>0}^{\infty} \left[\sum_{me}^{\infty} \cos m\phi - i \sum_{mo}^{\infty} \sin m\phi \right] \nu(\nu+1) \Gamma_\nu^m \frac{j_\nu(\sigma)}{\sigma} M_\nu^m(\cos\theta)$$

$$\eta H_r = \sum_{\nu>0}^{\infty} \left[\sum_{mo}^{\infty} \cos m\phi - i \sum_{me}^{\infty} \sin m\phi \right] \nu(\nu+1) \Lambda_\nu^m \frac{j_\nu(\sigma)}{\sigma} L_\nu^m(\cos\theta)$$

$$E_\theta = \frac{\eta G V(r)}{2\pi r \sin\theta} + \sum_{\nu>0}^{\infty} \left[\sum_{me}^{\infty} \cos m\phi - i \sum_{mo}^{\infty} \sin m\phi \right] \Gamma_\nu^m j_\nu^{\cdot}(\sigma) \frac{dM_\nu^m}{d\theta}$$

$$- \sum_{\nu>0}^{\infty} \left[\sum_{me}^{\infty} \cos m\phi - i \sum_{mo}^{\infty} \sin m\phi \right] \Lambda_\nu^m j_\nu(\sigma) \frac{mL_\nu^m}{\sin\theta}$$

$$\eta H_\phi = \frac{\eta I(r)}{2\pi r \sin\theta} - i \sum_{\nu>0}^{\infty} \left[\sum_{me}^{\infty} \cos m\phi - i \sum_{mo}^{\infty} \sin m\phi \right] \Gamma_\nu^m j_\nu(\sigma) \frac{dM_\nu^m}{d\theta}$$

$$- i \sum_{\nu>0}^{\infty} \left[\sum_{me}^{\infty} \cos m\phi - i \sum_{mo}^{\infty} \sin m\phi \right] \Lambda_\nu^m j_\nu(\sigma) \frac{mL_\nu^m}{\sin\theta} \tag{4.7.2}$$

$$E_\phi = -i \sum_{v>0}^{\infty} \left[\sum_{mo}^{\infty} \cos m\phi - i \sum_{me}^{\infty} \sin m\phi \right] \Gamma_v^m \, j_v^{\boldsymbol{\cdot}} (\sigma) \frac{m M_v^m}{\sin\theta}$$

$$+ \sum_{v>0}^{\infty} \left[\sum_{mo}^{\infty} \cos m\phi - i \sum_{me}^{\infty} \sin m\phi \right] \Lambda_v^m \, j_v (\sigma) \frac{dL_v^m}{d\theta}$$

$$\eta H_\theta = \sum_{v>0}^{\infty} \left[\sum_{mo}^{\infty} \cos m\phi - i \sum_{me}^{\infty} \sin m\phi \right] \Gamma_v^m \, j_v (\sigma) \frac{m M_v^m}{\sin\theta}$$

$$+ \sum_{v>0}^{\infty} \left[\sum_{mo}^{\infty} \cos m\phi - i \sum_{me}^{\infty} \sin m\phi \right] \Lambda_v^m \, j_v^{\boldsymbol{\cdot}} (\sigma) \frac{dL_v^m}{d\theta}$$

As was the case for transmission, in the limit as b goes to zero the only nonzero terms just off the $r = b$ surface are the TEM components of E_θ and H_ϕ. The TEM fields guide the energy through the interior region. Repeating the procedure of Eq. 3.6.6, evaluate the integral using Eq. 4.7.2 to obtain the algebraic equation:

$$\frac{V(a)}{a} = \frac{2iG}{Y_R(a)} \sum_{\ell} D_\ell^0 \left(j_\ell + \beta_\ell^0 h_\ell \right) \tag{4.7.3}$$

4.8 Boundary Conditions

Several boundary conditions have been built into field Eq. 4.4.2 and 4.4.3: rotational symmetry requires m to be an integer and noninfinite values on the exterior axes, $r > a$, assure only integer order Legendre functions of the first kind exist. The limiting condition as the radius becomes infinite requires spherical Hankel functions of the second kind. In the interior, noninfinite values at $r = b$ assure the coefficients of all negative-order Bessel functions are zero.

The boundary conditions still to be applied are:

- (1) Interior: $b < r < a$, $\theta = \psi$ and $\theta = \pi - \psi$. Tangential components of E_r and E_ϕ and normal component of H_θ, vanish.

- (2) Exterior: $r = a$, $\theta < \psi$ and $\theta > \pi - \psi$. Tangential components of E_θ and E_ϕ, and the normal component of H_r vanish.

- (3) Interface: $r = a$, $\psi < \theta < \pi - \psi$. All fields are continuous.

To satisfy the first condition, note the sums of Eq. 4.7.2 vanish for all interior radii only if:

$$M_\nu^m(\cos\psi) = 0; \quad dL_\nu^m(\cos\theta)/d\theta\big|_{\theta=\psi} = 0 \qquad (4.8.1)$$

For each degree, an infinite number of orders satisfy Eq. 4.8.1. Figure 3.4.2 includes plots of the first few values of ν and ψ for functions with $m = 1$ that satisfy these boundary conditions. In the limit as ψ approaches zero the first solutions for the odd and even functions occur respectively at $\nu = 2$ and 3. To satisfy the second condition it is necessary that:

$$\theta < \psi; \quad \theta > \pi - \psi$$

$$E_\theta(ka,\theta,\phi) = 0; \quad E_\phi(ka,\theta,\phi) = 0 \qquad (4.8.2)$$

To satisfy the third boundary condition, with δ a vanishingly small positive number, it is necessary that:

$$\psi < \theta < \pi - \psi$$

$$\mathbf{E}(ka-\delta,\theta,\phi) = \mathbf{E}(ka+\delta,\theta,\phi) \qquad (4.8.3)$$

$$\mathbf{H}(ka-\delta,\theta,\phi) = \mathbf{H}(ka+\delta,\theta,\phi)$$

The fields on the left side of Eq. 4.8.3 are those of Eq. 4.7.2. The fields on the right side of Eq. 4.8.3 are the sum of Eq. 4.4.1, 4.4.2, and 4.4.3. Desired algebraic equations are most easily obtained using the second and third boundary conditions to construct the four integral equalities of Eq. 4.8.4 through 4.8.7. In addition to these four equalities, the process is to be repeated with a similar set of integral equations after replacing $\sin(m\phi)$ by $\cos(m\phi)$ and $\cos(m\phi)$ by $-\sin(m\phi)$.

Although the zenith angle limits on both integrals are ψ to $\pi - \psi$ the second boundary condition permits changing the limits to 0 to π on the exterior electric field components.

$$\int_{\psi}^{\pi-\psi} \sin\theta d\theta \int_0^{2\pi} d\phi \left\{ E_\theta \frac{dP_\ell^m}{d\theta}\cos(m\phi) - E_\phi \frac{mP_\ell^m}{\sin\theta}\sin(m\phi) \right\}_{\sigma=ka-\delta}$$

$$= \int_0^{\pi} \sin\theta d\theta \int_0^{2\pi} d\phi \left\{ E_\theta \frac{dP_\ell^m}{d\theta}\cos(m\phi) - E_\phi \frac{mP_\ell^m}{\sin\theta}\sin(m\phi) \right\}_{\sigma=ka+\delta} \tag{4.8.4}$$

$$\int_{\psi}^{\pi-\psi} \sin\theta d\theta \int_0^{2\pi} d\phi \left\{ H_\phi \frac{dM_v^m}{d\theta}\cos(m\phi) + H_\theta \frac{mM_v^m}{\sin\theta}\sin(m\phi) \right\}_{\sigma=ka-\delta}$$

$$= \int_{\psi}^{\pi-\psi} \sin\theta d\theta \int_0^{2\pi} d\phi \left\{ H_\phi \frac{dM_v^m}{d\theta}\cos(m\phi) + H_\theta \frac{mM_v^m}{\sin\theta}\sin(m\phi) \right\}_{\sigma=ka+\delta} \tag{4.8.5}$$

$$\int_{\psi}^{\pi-\psi} \sin\theta d\theta \int_0^{2\pi} d\phi \left\{ E_\theta \frac{mP_\ell^m}{\sin\theta}\cos(m\phi) - E_\phi \frac{dP_\ell^m}{d\theta}\sin m\phi \right\}_{\sigma=ka-\delta}$$

$$= \int_0^{\pi} \sin\theta d\theta \int_0^{2\pi} d\phi \left\{ E_\theta \frac{mP_\ell^m}{\sin\theta}\cos(m\phi) - E_\phi \frac{dP_\ell^m}{d\theta}\sin m\phi \right\}_{\sigma=ka+\delta} \tag{4.8.6}$$

$$\int_{\psi}^{\pi-\psi} \sin\theta d\theta \int_0^{2\pi} d\theta \left\{ H_\phi \frac{mL_v^m}{\sin\theta}\cos(m\phi) + H_\theta \frac{dL_v^m}{d\theta}\sin(m\phi) \right\}_{\sigma=ka-\delta}$$

$$= \int_{\psi}^{\pi-\psi} \sin\theta d\theta \int_0^{2\pi} d\theta \left\{ H_\phi \frac{mL_v^m}{\sin\theta}\cos(m\phi) + H_\theta \frac{dL_v^m}{d\theta}\sin(m\phi) \right\}_{\sigma=ka+\delta} \tag{4.8.7}$$

Carrying out the integral operations of Eq. 4.8.4 through 4.8.7 with the similar set obtained by replacing sin($m\phi$) by cos($m\phi$) and cos($m\phi$) by $-$sin($m\phi$) results in the four linear equations:

$$\ell(\ell+1)\dot{j}_\ell F_\ell^m I_{\ell\ell} + \ell(\ell+1)\beta_\ell^m \dot{h}_\ell F_\ell^m I_{\ell\ell}$$

$$= \ell(\ell+1)\sum_{vo}^{\infty}\Gamma_v^m \dot{j}_v K_{\ell v} - \frac{\eta GV(a)}{\pi a}P_\ell\delta(m,0) + 2mP_\ell^m\sum_{\rho}^{\infty}\Lambda_\rho^m j_\rho L_\rho^m \tag{4.8.8}$$

$$v(v+1)\Gamma_v^m \dot{j}_v K_{vv} = \sum_n^{\infty} n(n+1)F_n^m \dot{j}_n K_{nv} + \sum_n^{\infty} n(n+1)F_n^m \beta_n^m \dot{h}_n K_{nv} \tag{4.8.9}$$

$$\ell(\ell+1)G_\ell^m j_\ell I_{\ell\ell} + \ell(\ell+1)G_\ell^m \alpha_\ell^m h_\ell I_{\ell\ell} = \sum_\rho^\infty \rho(\rho+1)\Lambda_\rho^m j_\rho I_{\ell\rho} \qquad (4.8.10)$$

$$\rho(\rho+1)\Lambda_\rho^m j_\rho^* I_{\rho\rho} = \rho(\rho+1)\sum_r^\infty G_r^m j_r^* I_{r\rho} + \rho(\rho+1)\sum_r^\infty G_r^m \alpha_r^m h_r^* I_{r\rho}$$

$$-2mL_\rho^m \sum_n^\infty F_n^m j_n P_n^m - 2mL_\rho^m \sum_n^\infty F_n^m \beta_n^m h_n P_n^m \qquad (4.8.11)$$

All but the following five terms are known in Eq. 4.8.8–4.8.11: $V(a)$, α_ℓ^m, β_ℓ^m, Λ_v^m, Γ_v^m. Problem solution requires evaluation of each of them; with a transmitter there are but three: $Y_T(a)$, β_ℓ, Γ_v.

4.9 Zero Degree Solution

We analyze zero degree modes first since only they carry absorbed power. The $m=0$ portion of Eq. 4.8.8 through 4.8.11 and Eq. 4.7.3 and are:

$$\ell(\ell+1)F_\ell^0 \beta_\ell^0 h_\ell^* I_{\ell\ell} = -\ell(\ell+1)F_\ell^0 j_\ell^* I_{\ell\ell} - \frac{\eta G V(a)}{\pi a} P_\ell + \ell(\ell+1)\sum_v^\infty \Gamma_v j_v^* K_{\ell v}$$

$$v(v+1)\Gamma_v j_v K_{vv} = \sum_{no}^\infty n(n+1)F_n^0 j_n K_{nv} + \sum_{no}^\infty n(n+1)F_n^0 \beta_n^0 h_n K_{nv} \qquad (4.9.1)$$

$$\frac{V(a)}{a} = \frac{2iG}{Y_R(a)}\sum_{no}^\infty F_n^0\left(j_n + \beta_n^0 h_n\right)$$

Transmitter coefficients β_ℓ and receiver products $F_\ell^0 \beta_\ell^0$ play similar roles: both sets of coefficients multiply TM fields that emanate from the antenna. Although Eq. 4.9.1 and Eq. 3.6.1–3.6.3 are similar in form a different approach to problem solution is helpful.

A case of special interest is an equated load. For this case, the receiver antenna load impedance equals the input impedance the transmitter antenna applies to incoming power. To analyze the case, adjust the driving field so $V(0)=a$. The antenna parameters given in Eq. 4.9.1 are then the same as those of the transmitter case of Eq. 3.6.1–3.6.3. Since identical equations give identical solutions:

$$\beta_\ell h_\ell / F_\ell^0 = j_\ell + \beta_\ell^0 h_\ell \qquad (4.9.2)$$

The presence of the j_ℓ term shows the relative phases and magnitudes of the transmitted and scattered fields per mode are not the same. Several coefficient values are tabulated in Table 4.9.1. Parameter values are then calculated using the numerical results of Table 4.9.1 and Eq. 4.9.2 for the special case $Y_R(a) = Y_T(a)$, $ka = 2$, $\psi = 5°$, and $F_\ell^0 (0) = a$.

Table 4.9.1. Values of β^0 for the special case of an equated load, $ka = 2$, $\psi = 5°$, and $V(0) = a$.

ℓ	real part, β_ℓ^0	imaginary part, β_ℓ^0	× order of magnitude
1	−1.0073	+0.15175	1
3	−1.3064	−0.14110	10^{-2}
5	−5.4652	−1.5762	10^{-3}
7	−1.1887	−3.4642	10^{-5}

Comparison of the transmitting and receiving equations shows that:

$$\Gamma_v^0 = \Gamma_v \qquad (4.9.3)$$

That is, the internal field coefficients for the two cases are the same.

The two equations contrast the relationships between the transmission and reception coefficients. Comparing Eq. 3.6.6 and 4.9.1 shows the termination admittances, $Y(a)$, are identical. The termination admittance is translated to the terminal admittance using Eq. 3.7.4 and confirms the terminal impedances of the antenna as a transmitter and as a receiver are identical.

$$Y_R(0) = Y_T(0) \qquad (4.9.4)$$

For an arbitrary but known load impedance, the solution procedure is to use Eq. 3.3.17 to solve for $Y(a)$ then combine with Eq. 4.9.1 to 4.9.3 to obtain the linear equation:

$$\beta_\ell^0 h_\ell = \frac{h_\ell}{F_\ell^0 h_\ell^{\bullet} I_{\ell\ell}} \left\{ \begin{array}{l} -F_\ell^0 j_\ell^{\bullet} I_{\ell\ell} - \dfrac{2\eta i G^2 P_\ell}{\pi\ell(\ell+1)Y_R(a)} \sum_{no}^{\infty} F_n^0 j_n + \sum_v^{\infty}\sum_{no}^{\infty} F_n^0 \dfrac{n(n+1)j_v^{\bullet} j_n K_{\ell v} K_{nv}}{v(v+1)j_v K_{vv}} \\[4mm] -\dfrac{2\eta i G^2 P_\ell}{\pi\ell(\ell+1)Y_R(a)} \sum_{no}^{\infty} F_n^0 \beta_n^0 h_n + \sum_v^{\infty}\sum_{no}^{\infty} F_n^0 \dfrac{n(n+1)j_v^{\bullet} K_{\ell v} K_{nv}}{v(v+1)j_v K_{vv}} \beta_n^0 h_n \end{array} \right\}$$

$$(4.9.5)$$

Symbols $I_{\ell\ell}$ and $K_{\ell v}$ *et al.* represent integrals listed in Tables A.16.1 and A.17.1. The equation form is the same as Eq. 3.6.5, and the solution technique is the same. Since all terms in Eq. 4.9.5 are known except β_ℓ^0 it may be solved first for $\beta_\ell^0 h_\ell$ and then for β_ℓ^0. Once the β_ℓ^0 are known, Eq. 4.9.1 may be used to solve for $V(a)$. The value of Γ_v^0 may be obtained using Eq. 4.9.3. The zero degree solution is then complete.

4.10 Non-Zero Degree Solutions

To find the solution for $m > 0$ solve for the exterior field parameters α_ℓ^m and β_ℓ^m using Eq. 4.8.8 through 4.8.11. The forms that most easily do this are:

$$\left(\beta_\ell^m h_\ell\right) = \frac{h_\ell}{\ell(\ell+1)h_\ell^{\bullet} F_\ell^m I_{\ell\ell}}$$

$$\times \left\{ \begin{array}{l} -\ell(\ell+1)F_\ell^m j_\ell^{\bullet} I_{\ell\ell} + \ell(\ell+1)\sum_v^{\infty}\sum_n^{\infty} F_n^m \dfrac{n(n+1)j_n j_v^{\bullet} K_{nv} K_{lv}}{v(v+1)j_v K_{vv}} \\[5mm] 2mP_\ell^m \sum_\rho^{\infty}\sum_r^{\infty} G_r^m \dfrac{j_\rho j_r^{\bullet} L_\rho^m I_{r\rho}}{j_\rho^{\bullet} I_{\rho\rho}} - 4m^2 P_\ell^m \sum_\rho^{\infty}\sum_n^{\infty} F_n^m \dfrac{j_\rho j_n \left(L_\rho^m\right)^2 P_n^m}{\rho(\rho+1)j_\rho^{\bullet} I_{\rho\rho}} \\[5mm] +2mP_\ell^m \sum_\rho^{\infty}\sum_r^{\infty} G_r^m \dfrac{j_\rho h_r^{\bullet} L_\rho^m I_{r\rho}}{j_\rho^{\bullet} h_r I_{\rho\rho}}\left(\alpha_r^m h_r\right) - 4m^2 P_\ell^m \sum_\rho^{\infty}\sum_n^{\infty} F_n^m \dfrac{j_\rho \left(L_\rho^m\right)^2 P_n^m}{\rho(\rho+1)j_\rho^{\bullet} I_{\rho\rho}}\left(\beta_n^m h_n\right) \\[5mm] +\ell(\ell+1)\sum_v^{\infty}\sum_n^{\infty} F_n^m \dfrac{n(n+1)j_v^{\bullet} K_{nv} K_{lv}}{v(v+1)j_v K_{vv}}\left(\beta_n^m h_n\right) \end{array} \right\}$$

$$(4.10.1)$$

$$\left(\alpha_\ell^m h_\ell\right) = \frac{1}{\ell(\ell+1)G_\ell^m I_{\ell\ell}}$$

$$\times \left\{ \begin{aligned} &-\ell(\ell+1)G_\ell^m j_\ell I_{\ell\ell} + \sum_\rho^\infty \sum_r^\infty G_r^m \frac{\rho(\rho+1)j_\rho \dot{j_r} I_{r\rho} I_{\ell\rho}}{\dot{j_\rho} I_{\rho\rho}} \\ &-2m \sum_\rho^\infty \sum_n^\infty F_n^m \frac{j_\rho j_n I_{\ell\rho} L_\rho^m P_n^m}{\dot{j_\rho} I_{\rho\rho}} \\ &\sum_\rho^\infty \sum_r^\infty G_r^m \frac{\rho(\rho+1)j_\rho \dot{h_r} I_{r\rho} I_{\ell\rho}}{\dot{j_\rho} h_r I_{\rho\rho}}\left(\alpha_r^m h_r\right) - 2m \sum_\rho^\infty \sum_n^\infty F_n^m \frac{j_\rho I_{\ell\rho} L_\rho^m P_n^m}{\dot{j_\rho} I_{\rho\rho}}\left(\beta_n^m h_n\right) \end{aligned} \right\}$$

$$(4.10.2)$$

These have the general algebraic form:

$$x_\ell + \sum_n^\infty N_{\ell n} x_n + \sum_r^\infty M_{\ell r} y_r = B_\ell$$

$$(4.10.3)$$

$$y_\ell + \sum_n^\infty S_{\ell n} x_n + \sum_r^\infty R_{\ell r} y_r + = A_\ell$$

All needed integrals are listed in Tables A.16.1 and A.17.1 and all other parameters are known. The sums of Eq. 4.10.1 and 4.10.2 may be truncated and solved concurrently for coefficients $\alpha_\ell^{\,m}$ and $\beta_\ell^{\,m}$, from which $\Gamma_v^{\,m}$ and $\Lambda_v^{\,m}$ follow directly. Knowledge of these four parameters provides the total solution of the fields around a receiving antenna.

Although the integral of products of Legendre functions of integer and noninteger orders, $M_v(\cos\theta)P_\ell(\cos\theta)$ for example, are largest if v is nearly equal to ℓ, none of them vanish. Therefore cross coupling exists between all modes of the same degree, m. Coupling between $\alpha_\ell^{\,m}$ and $\beta_\ell^{\,m}$ terms show that all modes, even and odd, of the same degree are coupled. That is, each TM mode and each TE mode interacts with all other modes, both TM and TE: each value of $\alpha_\ell^{\,m}$ and $\beta_\ell^{\,m}$ depends both upon all values of $\alpha_\lambda^{\,m}$ and $\beta_\lambda^{\,m}$ but are independent of $\alpha_\ell^{\,p}$ and $\beta_\ell^{\,p}$ for $m \neq p$.

Since for $m = 0$ and ψ approaching zero the order slowly approaches an integer value the integrals of cross modal terms remain significantly large and, therefore, cross modal coupling is significantly large even as ψ approaches zero.

4.11 Surface Current Densities

By definition, a conducting boundary condition is when the surface current density in amperes per meter is related to the magnetic field adjacent to it by the vector-phasor relationship

$$\tilde{\boldsymbol{I}} = n \times \tilde{\boldsymbol{H}} \tag{4.11.1}$$

Unit vector n is normal to and outbound from the conductor. Knowledge of the field coefficients permits the calculation of all antenna currents and Fig. 4.11.1 illustrates surface current patterns for the two lowest order exterior modes: $(\ell,m) = (1,0)$ and $(1,1)$; with δ representing a vanishingly small positive number, the figure also illustrates the three lowest order interior modes: $(\nu,m) = (0,0)$, $(1+\delta,0)$, and $(2+\delta,1)$. The exact interior modal number depends upon the value of ψ: with $5°$ cones the modal numbers $1+\delta$ and $2+\delta$ are respectively 1.444 484 and 2.022 029. The figure depicts current patterns for a small antenna, $a < \pi/4$. A z-polarized plane wave is incident from the left. In the interior

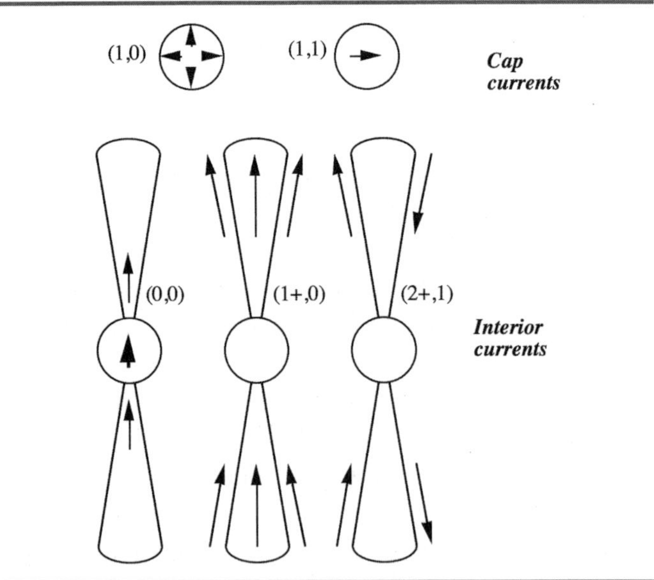

Figure 4.11.1. Receiving modal surface currents, wave incoming from left.

the TEM mode fields, see Eq. 4.3.1, are largest at $r = b$ and the peak value of current passes through the load. The current of mode $(1+\delta,0)$ has rotational symmetry around the cones. The current of mode $(2+\delta,1)$ is in phase with mode $(1+\delta,0)$ on the front face and out of phase on the back face; it is equal to zero in between. Both are TM driven modes, both are zero at the origin, the sink, and both are large near $r = a$. The current of mode $(1+\delta,0)$ produces a z-directed electric dipole moment. The current of mode $(2+\delta,1)$ produces a y-directed magnetic dipole mode, with a magnetic field phased according to Lenz's law. In the exterior, the currents of the two lowest modes, $(1,0)$ and $(1,1)$, produce respectively TM and TE fields. The cap current density of mode $(1,0)$ is θ-directed and zero at the center and the current density of mode $(1,1)$ is x-directed and maximum at the center.

In summary, the interior region at $r = b$: the $(0,0)$ current is maximum and both the $(1+,0)$ and $(2+,1)$ currents are zero. The $(1+,0)$ current is unidirectional and rotationally symmetric around the arms. The $(2+,1)$ current is bi-directional and it creates a magnetic moment directed in accordance with Lenz's law. In the exterior, the $(1,0)$ cap currents are directed along the zenith, rotationally symmetric, and zero at the midpoint. The $(2,1)$ currents are x-directed with a maximum at the midpoints.

4.12 Power

The time-average power on the surface of a virtual surface of radius σ/k that circumscribes the antenna is:

$$P = \frac{\sigma^2}{k^2} \int_0^{2\pi} d\phi \int_0^{\pi} \sin\theta d\theta \, \mathrm{Re}(N_r) \tag{4.12.1}$$

For receiving antennas, the time-average received power is equal to the negative of the real part of Eq. 4.12.1, after inserting the coefficients evaluated in Sec. 4.9 and 4.10. The power and the cross sections were calculated in Sec. 4.4 and 4.5. The normalized absorption cross section C_{AB}/C_{GE} is:

$$\frac{C_{AB}}{C_{GE}} = -\frac{4}{k^2 a^2}\left\{\left[\sum_{lo}^{\infty}\sum_{me}^{\ell-1}+\sum_{le}^{\infty}\sum_{mo}^{\ell-1}\right]\frac{(2\ell+1)(\ell-m)!!(\ell+m)!!}{\ell(\ell+1)(\ell-m-1)!!(\ell+m-1)!!}\left[\mathrm{Re}\left(\beta_\ell^m\right)+\beta_\ell^m\beta_\ell^{m*}\right]\right\}$$

$$-\frac{8}{k^2 a^2}\left\{\left[\sum_{lo}^{\infty}\sum_{mo}^{\ell}+\sum_{le}^{\infty}\sum_{me}^{\ell}\right]\frac{m^2(2\ell+1)(\ell-m-1)!!(\ell+m-1)!!}{\ell(\ell+1)(\ell-m)!!(\ell+m)!!}\left[\mathrm{Re}\left(\alpha_\ell^m\right)+\alpha_\ell^m\alpha_\ell^{m*}\right]\right\}$$

$$(4.12.2)$$

As we showed for a sphere, the portion of Eq. 4.12.2 proportional to the real part of the coefficients represents the extinction power extracted from the wave, and the portion proportional to sum of the square of the coefficients represents power scattered away from the antenna. The sum, expressed by Eq. 4.12.2, is the power absorbed by the antenna.

The incoming wave transfers both momentum and energy to the antenna. Since the incoming plane wave is y-directed, linear momentum is transferred to the antenna in that direction. The force on the scatterer is related to the momentum transferred as:

$$F_y = \frac{d}{dt}\left(linear\ momentum\right) \tag{4.12.3}$$

The net force applied to the antenna follows from the rate of momentum absorption and scattering, and is equal to:

$$F_y = \frac{\sigma^2}{ck^2}\int_0^{2\pi}\sin\phi\,d\phi\int_0^{\pi}N_r\sin^2\theta\,d\theta \tag{4.12.4}$$

By Eq. 4.5.7 the normalized force, f_y, due to the extinction power is:

$$\frac{cf_{yEX}}{C_{GE}} = -\frac{8\varepsilon}{k^2 a^2}\left\{\begin{array}{l}\left[\sum_{lo}^{\infty}\sum_{me}^{\ell-1}+\sum_{le}^{\infty}\sum_{mo}^{\ell-1}\right]\frac{U(m)(2\ell+1)(\ell-m)!!(\ell+m)!!}{\ell(\ell+1)(\ell-m-1)!!(\ell+m-1)!!}\mathrm{Re}\left(\beta_\ell^m\right)\\[2ex]+\left[\sum_{lo}^{\infty}\sum_{mo}^{\ell}+\sum_{le}^{\infty}\sum_{me}^{\ell}\right]\frac{m^2(2\ell+1)(\ell-m-1)!!(\ell+m-1))!!}{\ell(\ell+1)(\ell-m)!!(\ell+m)!!}\mathrm{Re}\left(\alpha_\ell^m\right)\end{array}\right\}$$

$$(4.12.5)$$

By Eq. 4.5.6 the normalized scattering force is:

$$\frac{cf_{ySC}}{C_{GE}} = -\frac{4}{k^2 a^2} \times$$

$$
\begin{aligned}
\Bigg[&\left(\sum_{\ell o} \sum_{me} + \sum_{\ell e} \sum_{mo} \right)
\begin{pmatrix}
\left(\beta_\ell^m \beta_{\ell+1}^{m+1*} + \beta_\ell^{m*} \beta_{\ell+1}^{m+1} \right) \dfrac{U(m)(\ell-m)!!(\ell+m+2)!!}{(\ell+1)^2 (\ell-m-1)!!(\ell+m-1)!!} \\[2mm]
+ \left(\beta_\ell^m \beta_{\ell-1}^{m+1*} + \beta_\ell^{m*} \beta_{\ell-1}^{m+1} \right) \dfrac{U(m)(\ell-m)!!(\ell+m)!!}{\ell^2 (\ell-m-3)!!(\ell+m-1)!!}
\end{pmatrix} \\[4mm]
+ &\left(\sum_{\ell o} \sum_{mo} + \sum_{\ell e} \sum_{me} \right)
\begin{pmatrix}
\left(\alpha_\ell^m \alpha_{\ell+1}^{m+1*} + \alpha_\ell^{m*} \alpha_{\ell+1}^{m+1} \right) \dfrac{m(m+1)(\ell-m-1)!!(\ell+m+1)!!}{(\ell+1)^2 (\ell-m)!!(\ell+m)!!} \\[2mm]
+ \left(\alpha_\ell^m \alpha_{\ell-1}^{m+1*} + \alpha_\ell^{m*} \alpha_{\ell-1}^{m+1} \right) \dfrac{m(m+1)(\ell-m-1)!!(\ell+m+1)!!}{\ell^2 (\ell-m-2)!!(\ell+m)!!}
\end{pmatrix} \\[4mm]
+ &\left(\sum_{\ell o} \sum_{me} + \sum_{\ell e} \sum_{mo} \right)
\left(\alpha_\ell^{m+1*} \beta_\ell^m + \alpha_\ell^{m+1} \beta_\ell^{m*} \right)
\left(\dfrac{(m+1)(2\ell+1)(\ell-m)!!(\ell+m)!!}{\ell^2 (\ell+1)^2 (\ell-m-1)!!(\ell+m-1)!!} \right) \\[4mm]
- &\left(\sum_{\ell o} \sum_{mo} + \sum_{\ell e} \sum_{me} \right)
\left(\alpha_\ell^m \beta_\ell^{m+1*} + \alpha_\ell^{m*} \beta_\ell^{m+1} \right)
\left(\dfrac{m(2\ell+1)(\ell-m-1)!!(\ell+m+1)!!}{\ell^2 (\ell+1)^2 (\ell-m-2)!!(\ell+m)!!} \right)
\Bigg]
\end{aligned}
$$

$$(4.12.6)$$

For the special case $m = 0$, the scattering cross section and normalized force terms are:

$$\frac{C_{SC}}{C_{GE}} = \frac{4}{k^2 a^2} \sum_{\ell o}^{\infty} \frac{(2\ell+1)(\ell!!)^2}{\ell(\ell+1)\left[(\ell-1)!!\right]^2} \beta_\ell^0 \beta_\ell^{0*} \tag{4.12.7}$$

$$\frac{cf_{ySC}}{C_{GE}} = -\frac{2}{k^2 a^2} \sum_{\ell o}
\begin{Bmatrix}
+ \left(\beta_\ell^0 \beta_{\ell+1}^{1*} + \beta_\ell^{0*} \beta_{\ell+1}^1 \right) \dfrac{\ell!!(\ell+2)!!}{(\ell+1)^2 \left[(\ell-1)!!\right]^2} \\[3mm]
+ \left(\beta_\ell^0 \beta_{\ell-1}^{1*} + \beta_\ell^{0*} \beta_{\ell-1}^1 \right) \dfrac{\ell!!^2}{\ell^2 (\ell-3)!!(\ell-1)!!} \\[3mm]
+ \left(\alpha_\ell^{1*} \beta_\ell^0 + \alpha_\ell^1 \beta_\ell^{0*} \right) \dfrac{2(2\ell+1)\ell!!^2}{\ell^2 (\ell+1)^2 \left[(\ell-1)!!\right]^2}
\end{Bmatrix}
\tag{4.12.8}$$

If the antenna is electrically small enough so only the dipole terms are significantly large, Eq. 4.12.8 shows that:

$$\frac{cf_{ySC}}{C_{GE}} = -\frac{3}{k^2 a^2}\left(\alpha_1^{1*}\beta_1^0 + \alpha_1^1\beta_1^{0*}\right) \tag{4.12.9}$$

Although the energy density-to-linear momentum density in the incoming plane wave is c, as discussed in the discussion of spherical scatterers the received energy-to-momentum ratio satisfies the relationship:

$$\frac{\text{Received Energy}}{\text{Received Momentum}} \le c \tag{4.12.10}$$

This is in contrast with a transmitting antenna. When transmitting power is radiated over a spread of angles and the average value of the cosine of the angle can never be greater than one and, therefore, the transmitted energy-to-momentum ratio obeys the relationship:

$$\frac{\text{Transmitted Energy}}{\text{Transmited Momentum}} \ge c \tag{4.12.11}$$

References

D.M. Grimes, "Biconical Receiving Antennas," *J. Math. Phys.*, vol. 23, pp. 897–914 (1982)

W.K.H. Panofsky, M. Phillips, *Classical Electricity and Magnetism*, 2nd ed., Addison-Wesley (1962)

S.A. Schelkunoff, *Advanced Antenna Theory*, John Wiley (1952)

S.A. Schelkunoff, *Applied Mathematics for Engineers and Scientists*, 2nd ed., Van Nostrand (1965)

W.R. Smythe, *Static and Dynamic Electricity*, 3rd ed., Summa Press, New York, NY, p. 39 (1989)

Chapter 5

Classical-Based Quantum Theory

Nearly a century after quantum mechanics became a fundamental branch of physics many basic questions remain unanswered. For example, is quantum mechanics the base on which the edifice of classical physics is built or is it an atomic level derivable result of classical physics? Although the correctness of quantum theory is beyond question the underlying physics on which it rests is not. Yet atomic level applications of the electromagnetic equations require information about objects that inhabit that level. In the early 20th century as the new ideas were being formed the physicists of the day, arguably led by Einstein and Bohr, struggled to incorporate the new facts into classical physics. Of course they were not privy to the many works amassed since. In particular, during the early 20th century the understanding of electrons was manifestly less than now, nearly a century later. Neither were they privy to the full range and complexity of in-depth analyses of electromagnetic radiation fields now available. For example, it wasn't until mid 20th century that Chu showed how quantitatively difficult it is for an electrically small antenna to emit or absorb radiant energy, and only then did Schelkunoff demonstrate how to detail characteristics of a transmitting antenna. It was another decade before Harrington detailed electromagnetic modal constructions that produce partially directed radiation. Harrington's result requires a transfer of linear momentum from a radiation field to its source. These are major steps beyond the ideas on which Planck based his lecture at the 1919 Nobel Prize ceremony where he explained his analysis of electromagnetic fields as a possible means of energy exchanges. These are also major steps beyond Einstein's comments in his historical and groundbreaking 1917 paper where he concluded his work made a quantum theory of fields almost unavoidable.

In this chapter we propose solutions to the underlying physics behind these issues; the solution includes a statistical basis for quantum theory that, in turn, leads to a modified interpretation of Schrödinger's equation. We begin with a derivation of that equation as a descriptor of electrons and, for that purpose, use a modified version of the presently accepted electron model. Our ultimate electron model violates no physical laws and maintains and supports all conservation theorems. For example, a satisfactory model is an electron composed of string-sized objects, which, when free, form into an object about the size of the Lorentz electron but, when trapped by an electrostatic potential, under the influence of radiation reaction forces expand throughout the eigenstate. The result, discussed in Sec. 5.1 through 5.7, is our development and interpretation of nonrelativistic quantum mechanics.

5.1 Electrons

Dirac noted nearly a century ago that an essential aspect of atoms is their extreme stability: nucleus-electron couplings remain stable over the lifetime of the universe. Since no explorer has diminished to the size of an electron, examined it, and returned to tell about it, it is necessary to use macroscopic scale physical laws and observations to interpret atomic scale events. Electrons possess fixed values of mass, charge, magnetic moment, and angular momentum, and they absorb and emit energy in units proportional to frequency. They act as matter waves that exhibit reflection and refraction. When trapped by a nuclear electrostatic potential, electrons form stable states with quantized values of energy and angular momentum. Since an orbiting charged object is both stable and consistent with Maxwell's equations only if it is a continuous current we infer an electron is distributed around a closed loop and it does not contain a charged kernel.

Natural objects are subject to the attractive and repulsive forces of four fundamental types of interactions: gravitation, electromagnetism, weak nuclear interactions, and strong nuclear interactions. The degree of dependence among these forces is unknown. Both gravitational and static electromagnetic forces decrease as the inverse square of the distance between interacting objects and the strong nuclear interaction is effective only locally within nuclei. Constituent nuclear particles are composed of two quite different structural blocks: leptons and hadrons. Strong interactions bind together hadrons; protons and neutrons are examples. Although both are complex structures on the nuclear scale of dimensions, on the atomic scale of dimensions both can be modeled as hard spheres approximately $8.5 \times 10^{-16} m$ in diameter.

Weak interactions bind together leptons; an electron is an example of a lepton and its structure is quite different from the hadrons. An historical survey of electron models is reminiscent of the Indian parable of the blind men and an elephant, see Sec. A.23: it depends upon the properties one chooses to emphasize. Although commonly regarded as a charged mass at a point, it is in many ways difficult to differentiate between point and spherically symmetric charges. The angular momentum and magnetic moment require that it spin about a central axis. If and only if electrons are sufficiently malleable to react to local force gradients by altering their shape and size is the model sufficient to support both stability and a classical description of radiative atomic properties.

An important electron parameter is its size. Lorentz made an early estimate; he related the electrostatic energy and mass using Eq. 1.3.13. By classical electrostatics the energy of a virtual shell of radius R_L that carries charge e is:

$$W = \frac{e^2}{4\pi\varepsilon R_L} = m_0 c^2 \tag{5.1.1}$$

This determined his radius R_L:

$$R_L = \left(\frac{e^2}{m_0}\right) \times 10^{-7} \cong 2.82 \times 10^{-15} m \tag{5.1.2}$$

R_L, the Lorentz radius of an electron, is the radius static charge e would have if it were somehow stabilized without introducing other forces. The idea, however, cannot be complete since Thomson's electromagnetic theorem states that an isolated, static array of electric charges cannot be held in equilibrium by electrostatic forces alone; since opposite charges attract and like charges repel charges either expand forever or contract upon themselves until something other than electrostatic forces are involved. Since an electron contains a negative electric charge and a magnetic dipole moment and since electrons are stable it follows that more than electrostatic forces are involved; that force is recognized as the weak nuclear force.

A quite different size estimate follows by including the magnetic moment and the speed of light in the analysis. An electron's magnetic moment is 0.928×10^{-23} *A* m^2 and, by definition, the magnetic moment of a current loop equals the product of peripheral current times the area enclosed by the loop. If distributed charge e rotates at the speed of light around a closed loop of radius R the radius of the loop needed to generate its magnetic moment is:

$$\mu_e = \frac{aec}{2}, \ R = 3.87 \times 10^{-13} \ m \qquad\qquad (5.1.3)$$

This is about 100 times larger than the Lorentz radius. It is larger still if, for example, the entire charge doesn't rotate at that speed. Therefore from this point of view Eq. 5.1.3 is the least expected size. As an aside, the ratio of this size to the measured size of a proton, about $8.5 \times 10^{-16} \ m$, is about 450; the size ratio is on the order of a football field to a football.

A possible scenario as a proton and an electron combine and form a hydrogen atom is: as the particles accelerate towards each other both emit energy until the system comes to rest with the proton at the electron's center. The electron retains its original symmetry but adjusts its form as needed to accommodate the inner charge while conserving its original angular momentum, magnetic moment, and radial oscillations; the expectation value of the first inverse radius, *i.e.*, Bohr orbit, is $\sim 5 \times 10^{-11}$ m.

There is no *a priori* reason to expect an electron in a stable state to be a circular current at a fixed position. If we regard intraelectron charge and current distributions as an ensemble we expect the detailed density distribution, created by changes in its internal structure, to be situation and time dependent within the constraints of the conservation laws. Such variations will surely carry over from the original formation of the atom, be excited by inter-atomic collisions, and, equally importantly, be excited by interactions with the Planck background radiation field. We seek to place such motion on a statistical basis. A necessary axiom is what Maxwell called 'the assumption of continuity of path': every non-radiative charge and current density distribution that satisfies the conservation laws is equally likely and furthermore, over time, the system traverses through all possibilities. According to Tolman although there is no proof of this assumption, even in non charge-bearing systems, there is extensive agreement between statistical mechanical results deduced from it and experimental results.

5.2 The Time-Independent Schrödinger Equation

The science of statistical mechanics consists of analyses of large numbers of identical, interacting particles grouped together as a single ensemble. The state of the ensemble is specified by the positions and velocities of the particles, and they, in turn, are sufficient to determine the kinetic and potential energies of the system. With particles modeled as realistically as possible there is little or no difficulty interpreting an experiment that measures the ensemble-average of a kinematic variable. Also the state of an isolated classical ensemble at any instant determines its future values. Since large ensembles contain too many degrees of freedom to detail, no attempt is made to obtain complete calculations. Instead, most probable values are obtained from properties of individual particles and using most-probable results as ensemble-average values.

In what follows we view a single electron as a statistical ensemble and, as with the statistical mechanics of electrically neutral objects, we use conservation laws to determine as much as possible about the properties of trapped electrons. Our derived results disagree with conventional quantum theory in only one significant way: Schrödinger's equation applies only to objects in equilibrium or near-equilibrium. His equation is not a descriptor of events during changes of state, for example changes of either state or entanglement.

We begin by noting that a dynamic charge distribution supports constant time-average values of concurrent charge and current densities that result from charge movements. Hence we define $\rho(r)$ and $\rho(p)$ to be, respectively, the charge and current density within differential spatial range r and $r+dr$ and differential momentum range p and $p+dp$. We also note that an ensemble of current eddies create local magnetic fields and such fields act on each other in a way that causes the orbiting array to remain in continual motion.

Since an electron's charge is negative and, when trapped, orbits in a single direction we require the sign of the charge densities to be positive real. This constraint requires the densities to be expressible as a complex function times its own complex conjugate. Therefore we define complex functions $U(r)$ and $\Gamma(p)$ as:

$$eU^*(r)U(r) = \rho(r); \quad e\Gamma^*(p)\Gamma(p) = \rho(p) \tag{5.2.1}$$

By definition $U(r)$ and $\Gamma(p)$ are normalized, complex wave functions. With dV and dV_p representing, respectively, differential volume in space and momentum coordinates the wave functions obey the normalizing integral relationships:

$$\int U^*(r)U(r)dV = \int \Gamma^*(p)\Gamma(p)dV_p = 1 \tag{5.2.2}$$

Functions $U(r)$ and $\Gamma(p)$ are relatable since they describe the same dynamic charge distribution. Each position in coordinate space receives contributions from the full range of momentum in proportion to the value of $\Gamma(p)$ and *vice versa*. Therefore we seek a linear transformation between the two coordinate systems that satisfies the conditions:

$$U(r) = L\{\Gamma(p)\}; \quad \Gamma(p) = L^{-1}\{U(r)\} \tag{5.2.3}$$

L is a linear operator and L^{-1} is its inverse. A widely applicable linear transformation that meets these requirements is the Fourier integral transform pair:

$$U(r) = \left[\frac{1}{2\pi\hbar}\right]^{3/2} \int \Gamma(p)\exp\left(\frac{ir \cdot p}{\hbar}\right)dV_p$$

$$\Gamma(p) = \left[\frac{1}{2\pi\hbar}\right]^{3/2} \int U(r)\exp\left(\frac{r \cdot p}{i\hbar}\right)dV \tag{5.2.4}$$

Constant $\hbar$ determines dimensions and it, in turn, must be determined by experiment. Dropping to one dimension for simplicity, Eq. 5.2.4 takes the form:

$$U(x) = \left[\frac{1}{2\pi\hbar}\right]^{1/2} \int \Gamma(p)\exp\left(\frac{ixp}{\hbar}\right)dp$$

$$\Gamma(p) = \left[\frac{1}{2\pi\hbar}\right]^{1/2} \int U(x)\exp\left(\frac{xp}{i\hbar}\right)dx$$

$$(5.2.5)$$

This procedure for going from charge and current densities to wave functions and the Fourier transforms of Eq. 5.2.4 parallels the method of going from electric energy density to electric field intensity, except the field intensities form a vector field and the wave functions form a scalar field.

The average, *i.e.* expectation, value of momentum $\langle p \rangle$ follows from the above and is given in statistical form by the equation:

$$\langle p \rangle = \int p\Gamma^{*}(p)\Gamma(p)dp \tag{5.2.6}$$

The same value may be calculated using $U(x)$; to do so substitute $\Gamma(p)$ from the second of Eq. 5.2.5 into Eq. 5.2.6:

$$\langle p \rangle = \left(\frac{1}{2\pi\hbar}\right)^{1/2} \int_{-\infty}^{\infty} p\Gamma^{*}(p)\int_{-\infty}^{\infty} U(x)\exp\left(\frac{xp}{i\hbar}\right)dx \tag{5.2.7}$$

Integrating the second integral by parts gives:

$$\langle p \rangle = \left(\frac{1}{2\pi\hbar}\right)^{1/2} \int_{-\infty}^{\infty} \Gamma^{*}(p)\left\{i\hbar U(x)\Big|_{-\infty}^{\infty} - i\hbar\int_{-\infty}^{\infty} \frac{\partial U(x)}{\partial x}\exp\left(\frac{xp}{i\hbar}\right)dx\right\}$$

Since an acceptable wave function vanishes at infinity the first term within the brackets vanishes. Substituting the complex conjugate of the first of Eq. 5.2.5 into the second term just above and reversing the order of integration gives:

$$\langle p \rangle = \int_{-\infty}^{\infty} U^*(x) \left[\frac{\hbar}{i} \frac{\partial U(x)}{\partial x} \right] dx \tag{5.2.8}$$

This is an example of a general case: a dynamic variable in momentum space may be replaced by an operation in dimensional space, and *vice versa*. Letting O indicate the operator form of the variable, in three dimensions the momentum operator is:

$$O(p) = \frac{\hbar}{i} \nabla \tag{5.2.9}$$

It is understood that the operator acts on wave function $U(r)$. Repeating the above procedure for p^n shows, after 'n' partial integrations, that the result generalizes to:

$$O(p^n) = \left(\frac{\hbar}{i} \right)^n \nabla^n \tag{5.2.10}$$

This result makes it unnecessary to know both $U(r)$ and $\Gamma(p)$ to solve a kinematic problem. One functional type is adequate, typically $U(r)$, if the conjugate variable is expressed in operator form.

We seek to impose conservation of energy on a trapped electron as it moves at slow enough speeds for relativistic effects to be safely ignored. With W defined as the time-average total energy, the kinetic energy within each differential volume in momentum space is kinetic energy $p^2/2m$ times the fraction of the ensemble within the momentum range p to $p+dp$, and the integrand of first integral on the right side of Eq. 5.2.11. The potential energy within each differential volume in coordinate space is potential energy $\Lambda(r)$ times the fraction of the ensemble within the coordinate range r to $r+dr$, the second integral on the right side of Eq. 5.2.11. Both terms are integrated over the full spatial extent of the ensemble and Eq. 5.2.11 is a statement of conservation of the energy of an electron viewed as a statistical ensemble:

$$W = \frac{1}{2m}\int dV_{\mathrm{p}}\left[p^2\Gamma^*(\mathbf{p})\Gamma(\mathbf{p})\right] + \int dV\left[\Lambda(\mathbf{r})U^*(\mathbf{r})U(\mathbf{r})\right] \qquad (5.2.11)$$

An arbitrary constant, such as the self-energy of the electron, may be added to the potential without affecting results to follow.

Applying Eq. 5.2.10 to 5.2.11 and using 5.2.3 gives:

$$\int dV U^*(r)\left\{-\frac{\hbar^2}{2m}\nabla^2 U(r) + \left[\Lambda(r) - W\right]U(r)\right\} = 0 \qquad (5.2.12)$$

Although only the integral equation is required the more restrictive condition that the integrand be positive real may be applied. Doing so yields the differential equation:

$$-\frac{\hbar^2}{2m}\nabla^2 U(r) + \Lambda(r)U(r) = WU(r) \qquad (5.2.13)$$

This is the Schrödinger time independent differential equation and was first published by him in 1926. Although Schrödinger found that solutions of his equation correctly describe electrons, he was not led to the result by first principles. The equation is invaluable for describing atomic level phenomena but solutions are also statistical in character and there has been no unique physical basis for it. The equation *per se* is no help in determining whether $eU(r)U^*(r)$ represents an actual static charge density, the fraction of the time a point electron occupies a particular differential volume, or something in between. It is only known that solving Eq. 5.2.13 for $U(r)$ then evaluating $eU(r)U^*(r)$ gives the correct time-average spatial charge distribution for nonrelativistic electron systems.

It has been shown that several quite different postulate sets yield Schrödinger's equation as a derived result and the accompanying interpretation depends upon the nature of the postulates used to derive it. Of course, because a specific postulate set produces the necessary equation does not constitute a proof the model is correct. This section derives the Schrödinger equation using the above statistical approach and we interpret results accordingly. It is adequate to obtain time-average values of kinematic properties and it is adequate to correctly predict parameters that vary linearly with applied forces. However, and most importantly, since this derivation requires either equilibrium or near-equilibrium conditions it follows that Schrödinger's equation cannot describe events that occur away from equilibrium. Such conditions occur during transitions between eigenstates.

5.3 The Uncertainty Principle

By the uncertainty principle it is not possible to simultaneously determine exact values of conjugate variables, for example position and momentum. The more accurately the position of an electron is known the less accurately the momentum can be known, and *vice versa*. As a simple example, consider an electron described by a one dimensional Gaussian wave function in which $U(x)$ is proportional to $\exp(-x^2/B)$, where $B \geq 0$. The electron is confined to position zero if B is limitlessly large; the smaller B's magnitude the more extensive the extended charge distribution.

The system is normalized with a charge density at each point of:

$$U^*(x)U(x) = \sqrt{\frac{2}{\pi B}}\exp\left(-\frac{2x^2}{B}\right) \tag{5.3.1}$$

The expectation value of x^2 follows using the integrals of Table 5.3.1:

$$\langle x^2 \rangle = \sqrt{\frac{2}{\pi B}}\int_{-\infty}^{\infty}\exp\left(-\frac{2x^2}{B}\right)x^2 dx = \frac{B}{4} \tag{5.3.2}$$

Substituting $U(x)$ into the second of Eq. 5.2.4 results in the wave function in momentum space:

$$\Gamma(p) = \left(\frac{B}{2\pi\hbar^2}\right)^{1/4}\exp\left(-\frac{B}{4\hbar^2}p^2\right) \tag{5.3.3}$$

Table 5.3.1. Gaussian integrals.

$$\int_0^\infty \exp\left(-a^2 x^2\right) dx = \frac{\sqrt{\pi}}{2a}$$

$$\int_0^\infty x^2 \exp\left(-x^2\right) dx = \frac{\sqrt{\pi}}{4}$$

$$\int_0^\infty \exp\left(-a^2 x^2\right) \cos(bx)\, dx = \frac{\sqrt{\pi}}{2a} \exp\left(-\frac{b^2}{4a^2}\right)$$

Using Eq. 5.3.3 to calculate the root-mean-square, rms, value of momentum gives:

$$\langle p^2 \rangle = \left(\frac{B}{2\pi\hbar^2}\right)^{1/2} \int_{-\infty}^{\infty} \exp\left(-\frac{Bp^2}{2\hbar^2}\right) p^2 dp = \frac{\hbar^2}{B} \tag{5.3.4}$$

Recalculating $\langle p^2 \rangle$ in coordinate space using operator notation gives the rms value:

$$\langle p^2 \rangle = -\hbar^2 \left(\frac{2}{\pi B}\right)^{1/2} \int_{-\infty}^{\infty} \exp\left(-\frac{x^2}{B}\right) \frac{d^2}{dx^2}\left[\exp\left(-\frac{x^2}{B}\right)\right] dx = \frac{\hbar^2}{B} \tag{5.3.5}$$

Inspection shows the rms values of position and momentum satisfy the parabolic relationship:

$$\sqrt{\langle x^2 \rangle \langle p^2 \rangle} = \hbar/2 \tag{5.3.6}$$

This equation confirms with a Gaussian wave function it is not possible to know both position and momentum more accurately than $\Delta x \Delta p \approx \hbar$, where Δx and Δp are, respectively, uncertainty in the measurement of position and momentum. This is a quantitative statement of the uncertainty principle. It results from the properties of the Fourier integral transform relationships between the momentum and coordinate space wave functions. The same is true for all conjugate pairs, *i.e.*, pairs related by Fourier transforms; they satisfy the parabolic uncertainty relationship of Eq. 5.3.6 or larger. It may be shown that a Gaussian wave function provides the least possible uncertainty with all other functions providing greater uncertainty.

5.4 The Time-Dependent Schrödinger Equation

Solutions of the time-independent Schrödinger equation describe time-average values of the kinematic parameters and, of course, time-average values are constant. The derivation also requires equilibrium or near-equilibrium conditions; these will not be met if changes are large enough or fast enough. In this section we examine changes over periods that are short on the time scale of macroscopic events but long on the time scale required to establish electronic equilibrium within an eigenstate. The result is a shift of expectation values away from equilibrium positions and permits obtaining event probabilities during an initial slowly varying change of time-average values.

To find expectation values that are time dependent on the time scale of macroscopic events, begin by defining time-dependent wave function $\psi(r,t)$ as:

$$\psi(r,t) = U(r)e^{i\omega t} \tag{5.4.1}$$

Restricting to exponential time functions places no restrictions on well-behaved functions since the actual time variation follows by integrating over time-dependent coefficients. With this notation the time dependence of the charge density in coordinate space is:

$$\rho(r,t) = e\psi^*(r,t)\psi(r,t) \tag{5.4.2}$$

Time-average momentum density is also shown in Eq. 5.2.8 and 5.2.9 and, if the charge-mass density ratio is constant throughout the eigenstate, the current density is:

$$J(r,t) = \frac{e}{m}p(r,t) \tag{5.4.3}$$

Combining Eq. 5.2.9 with 5.4.3 and requiring the current density to be real give the expression:

$$J(r,t) = \frac{\hbar e}{2im}\left[\psi^*(r,t)\nabla\psi(r,t) - \psi(r,t)\nabla\psi^*(r,t)\right] \tag{5.4.4}$$

The time-dependent Schrödinger equation follows by combining Eq. 5.2.13, the time-independent Schrödinger equation, with Eq. 1.5.3, the continuity equation:

$$\nabla\cdot J(r,t) + \frac{\partial\rho(r,t)}{\partial t} = 0 \tag{5.4.5}$$

Using Eq. 5.4.2 the rate of change of charge density is:

$$\frac{\partial\rho(r,t)}{\partial t} = e\left[\psi^*(r,t)\frac{\partial}{\partial t}\psi(r,t) + \psi(r,t)\frac{\partial}{\partial t}\psi^*(r,t)\right] \tag{5.4.6}$$

Combining equations and then combining the result with the divergence of Eq. 5.4.4 gives:

$$\nabla\cdot J(r,t) = \frac{\hbar e}{2im}\left[\psi^*(r,t)\nabla^2\psi(r,t) - \psi(r,t)\nabla^2\psi^*(r,t)\right] \tag{5.4.7}$$

Substituting Eq. 5.4.5 and 5.4.6 into Eq. 5.4.7, multiplying by ($\hbar/ie$), and then adding and subtracting potential $\Lambda(r)$ gives:

$$\psi^*(r,t)\left[-\frac{\hbar^2}{2m}\nabla^2\psi(r,t) + \Lambda(r)\psi(r,t) + \frac{\hbar}{i}\frac{\partial}{\partial t}\psi(r,t)\right]$$
$$-\psi(r,t)\left[-\frac{\hbar^2}{2m}\nabla^2\psi^*(r,t) + \Lambda(r)\psi^*(r,t) - \frac{\hbar}{i}\frac{\partial}{\partial t}\psi^*(r,t)\right] = 0 \tag{5.4.8}$$

We seek wave function $\psi(r,t)$ that in the limit as the time dependence becomes vanishingly slow goes to:

$$U^*(r)\left[-\frac{\hbar^2}{2m}\nabla^2 U(r)+\Lambda U(r)-WU(r)\right]$$

$$-U(r)\left[-\frac{\hbar^2}{2m}\nabla^2 U^*(r)+\Lambda U^*(r)-WU^*(r)\right]=0 \qquad (5.4.9)$$

Integrating Eq. 5.4.9 over all space and comparing with Eq. 5.2.13 shows that each line of Eq. 5.4.9 vanishes. Since Eq. 5.4.8 holds for all nonrelativistic speeds an acceptable time-dependent wave equation is:

$$\int \psi^*(r,t)\left[-\frac{\hbar^2}{2m}\nabla^2\psi(r,t)+\Lambda(r)\psi(r,t)+\frac{\hbar}{i}\frac{\partial}{\partial t}\psi(r,t)\right]dV=0 \quad (5.4.10)$$

Separately requiring that the integrand be positive real gives the Schrödinger time-dependent equation:

$$-\frac{\hbar}{i}\frac{\partial}{\partial t}\psi(\mathbf{r},t)=H\psi(\mathbf{r},t)=W\psi(\mathbf{r},t) \qquad (5.4.11)$$

By definition H is the Hamiltonian operator. It is the operator that acts on the wave function and yields both the time dependence and the state energy.

From the first equality of Eq. 5.4.11:

$$H\psi(\mathbf{r},t)=-\frac{\hbar^2}{2m}\nabla^2\psi(\mathbf{r},t)+\Lambda(\mathbf{r})\psi(\mathbf{r},t) \qquad (5.4.12)$$

From the second equality of Eq. 5.4.11:

$$\psi(r,t)=\psi(r,t_0)\exp\left(\frac{iW}{\hbar}t\right) \qquad (5.4.13)$$

The initial value of the wave function is the equilibrium value:

$$\psi(r,t_0)=U(r) \qquad (5.4.14)$$

A significant result of Eq. 5.4.13 is that the equilibrium frequency of an eigenstate is related to its energy by proportionality factor $\hbar$:

$$\omega = W/\hbar \tag{5.4.15}$$

This equation is a basis for the experimental determination of $\hbar$.

The time-independent Schrödinger equation follows from applying the conservations of energy and charge to a statistical model of an electron. It is thermodynamic-like in the sense that time-average values are obtained only over times long enough to establish intra-state equilibrium. Although Eq. 5.4.11 provides the correct changes in time-average values but, with our electron model, applies only to energy changes that are small enough for the system to remain in a near-equilibrium condition and does not imply an actual time-line of events during non-equilibrium conditions.

5.5 Quantum Operators

The logic that supports using operators in spatial coordinates to calculate expectation values of momentum generalizes to functions of momentum. To do so consider the integral:

$$I = \int_{-\infty}^{\infty} \left[\Gamma_R^*(\boldsymbol{p}) \, \boldsymbol{p} \, \Gamma_T(\boldsymbol{p}) \right] dV_p \tag{5.5.1}$$

$\Gamma_R(\boldsymbol{p})$ and $\Gamma_T(\boldsymbol{p})$ represent different eigenfunctions that are solutions of the same differential equation. For each solution in momentum space, $\Gamma_R(\boldsymbol{p})$, there exists a unique Fourier integral transform to a function in coordinate space, $U_R(\boldsymbol{r})$. To rewrite the integral of Eq. 5.5.1 in terms of spatial functions, repeat the procedure used to develop Eq. 5.2.8. Taking the gradient in the direction of the momentum and working with the Γ_T-state function, the integral of Eq. 5.5.1 becomes:

$$I = \int_{-\infty}^{\infty} U_R^*(\boldsymbol{r}) \left[\frac{\hbar}{i} \nabla U_T(\boldsymbol{r}) \right] dV \tag{5.5.2}$$

Similarly, working with the Γ_R-state function gives:

$$I = \int_{-\infty}^{\infty} U_T(r)\left[\frac{\hbar}{i}\nabla U_R(r)\right]^* dV \qquad (5.5.3)$$

Since physical events are real and since Eq. 5.5.3 represents something that is physically real it is equal its own complex conjugate:

$$\int_{-\infty}^{\infty} U_T^*(r)\left[\frac{\hbar}{i}\nabla U_R(r)\right]dV = \int_{-\infty}^{\infty} U_T(r)\left[-\frac{\hbar}{i}\nabla U_R^*(r)\right]dV \qquad (5.5.4)$$

Combining Eq. 5.5.2 and 5.5.4 gives:

$$\int_{-\infty}^{\infty} U_R^*(r)\left[\frac{\hbar}{i}\nabla U_T(r)\right]dV = \int_{-\infty}^{\infty} U_T(r)\left[\frac{\hbar}{i}\nabla U_R(r)\right]^* dV \qquad (5.5.5)$$

The result generalizes to:

$$\int_{-\infty}^{\infty} U_R^*(r)O\left[U_T(r)\right]dV = \int_{-\infty}^{\infty} U_T(r)\left\langle O\left[U_R(r)\right]\right\rangle^* dV \qquad (5.5.6)$$

The symbol "O" indicates any quantum mechanical operator; equality is widely used in quantum theory analyses. An operator that satisfies Eq. 5.5.6 is by definition a Hermitian operator.

5.6 Wave Function Orthogonality

To examine orthogonality properties of wave function $\psi(r,t)$, let O be a quantum theory operator, $\psi_R(r,t)$ and $\psi_T(r,t)$ be time-dependent eigenfunctions, and I_R and I_T be the corresponding state values:

$$O\psi_R(r,t) = I_R\psi_R(r,t); \quad O\psi_T(r,t) = I_T\psi_T(r,t) \qquad (5.6.1)$$

All functions $\psi_R(r,t)$ that satisfy this equation are eigenfunctions and constants I_R are state values. Multiplying the left equation by $\psi_T^*(r,t)$, the right equation by $\psi_R{}^*(r,t)$, subtracting one from the other, and then integrating over the volume gives:

$$\int \left[\psi_R^* I_T \psi_T - \psi_T^* I_R \psi_R \right] dV = \int \left[\psi_R^* O \psi_T - \psi_T \left(O \psi_R \right)^* \right] dV$$

$$= \left(I_T - I_R \right) \int \psi_R^* \psi_T dV \tag{5.6.2}$$

Since 'O' is an Hermitian operator, from Eq. 5.5.6:

$$\int \left[\psi_R^* \left(O \psi_T \right) - \psi_T \left(O \psi_R \right)^* \right] dV = 0 \tag{5.6.3}$$

Combining Eq. 5.6.2 and 5.6.3 gives:

$$\left(I_T - I_R \right) \int \psi_R^* \psi_T dV = 0 \tag{5.6.4}$$

If a system has more than one eigenfunction with the same state energy the system is degenerate; the number of solutions that produce the same state energy is the degree of degeneracy. A conclusion from Eq. 5.6.4 is that if the states are not degenerate the functions are orthogonal; if the state energies are equal the functions are degenerate and the function may or may not be orthogonal.

If solutions of a single operator result in many eigenfunctions, $\psi_T(r,t)$, the physical result is a sum, weighted by constants a_T, over all possible eigenfunctions:

$$\Psi(r,t) = \sum_{T=1}^{\infty} a_T \psi_T(r,t) \tag{5.6.5}$$

Functions $\psi_T(r,t)$ are normalized wave functions. Requiring that the total wave function be normalized gives:

$$\int \Psi^* \Psi dV = \sum_{R=1}^{\infty} \sum_{T=1}^{\infty} a_R a_T^* \int \psi_R \psi_T^* dV = \sum_{R=1}^{\infty} a_R a_R^* = 1 \tag{5.6.6}$$

This shows that the sum over the magnitudes of all coefficients is one, and leads to the conclusion:

$$\langle O \rangle = \int \psi^* O \psi \, dV = \sum_{R=1}^{\infty} \sum_{S=1}^{\infty} a_R a_T^* \int \psi_T^* O \psi_R \, dV$$

$$= \sum_{R=1}^{\infty} I_R a_R a_R^* \qquad (5.6.7)$$

In words, the expectation value of any dynamic function "O" is the sum over the probabilities that the electron occupies a particular state multiplied by the state value. For any particular measurement, the use of operator $\langle O \rangle$ produces only a particular value $a_R a_R^*$. Within a linear system, a single electron enters a statistically weighted fraction of all available eigenstates.

The initial rate of change of an expectation value follows by differentiating the first equality of Eq. 5.6.7:

$$\frac{d}{dt}\langle O \rangle = \int \left(\frac{\partial \psi^*}{\partial t} O \psi + \psi^* O \frac{\partial \psi}{\partial t} \right) dV \qquad (5.6.8)$$

Using Eq. 5.5.11, this may be written as:

$$\frac{d}{dt}\langle O \rangle = -\frac{i}{\hbar} \int \left(\psi^* O (H\psi) - (H\psi)^* (O\psi) \right) dV \qquad (5.6.9)$$

Incorporating Eq. 5.5.6:

$$\frac{d}{dt}\langle O \rangle = -\frac{i}{\hbar} \int \psi^* (OH - HO)\psi \, dV \qquad (5.6.10)$$

For the special case where $O = r$:

$$\langle p \rangle = m \frac{d}{dt}\langle r \rangle = \frac{im}{\hbar}(Hr - rH) \qquad (5.6.11)$$

The bracket on the right side of Eq. 5.6.11 is, by definition, the commutator of the indicated variable; the bracket shown is the commutator of position.

5.7 Electron Spin

Electron spin does not directly enter the Schrödinger equation; it is necessary to separately add it as a separate entity. Dirac's equations, in contrast, are relativistically correct, apply in all inertial systems, and spin is an integral part of the whole. Although Dirac's work is of singular importance to quantum theory it does not assist in resolving basic issues of photon exchanges considered here and therefore is not detailed here.

Well before Schrödinger's equation was recognized as correctly describing quantum phenomena it was known that a complete description of optical events required four quantum numbers. An integral part of Dirac's equations is that electron characteristics include more than charge and mass: there is a permanent angular momentum and a permanent magnetic dipole moment.

By definition the angular momentum of a particle in terms of its mass and velocity is:

$$l = m\,\boldsymbol{r} \times \boldsymbol{v} \tag{5.7.1}$$

Consider a point mass with angular momentum l, mass m, at position $\boldsymbol{r}$ from a fixed point that moves with velocity $\boldsymbol{v}$. If the particle also supports charge q, the magnetic moment is:

$$\Omega = q\mathbf{r} \times \mathbf{v}/2 \tag{5.7.2}$$

Comparison of Eq. 5.7.1 and 5.7.2 shows that

$$\Omega = ql/2m \tag{5.7.3}$$

These equations show the ideal relationship between an electron's magnetic and mechanical moments is:

$$\Omega = -e\mathbf{l}/2m \qquad (5.7.4)$$

The measured value for electrons, however, is approximately twice this ratio.

The angular momentum of an electron obeys rules similar to those of orbital motion about a central force field except an electron's intrinsic spin is magnitude one half. That is, if s represents the electron angular momentum, the square of its magnitude is:

$$\mathbf{s} \cdot \mathbf{s} = s(s+1)\hbar^2 \qquad (5.7.5)$$

The component along a particular axis is:

$$s_z = m_s \hbar \qquad (5.7.6)$$

Quantum numbers m_s are equal to $\pm 1/2$. Combining terms shows that the magnetic moment can have either of the two values:

$$\Omega_z = \pm e\hbar/2m \qquad (5.7.7)$$

The absolute value of the magnetic moment, by definition the Bohr magneton, is:

$$\Omega_z = 9.274 \times 10^{-24} J/T \qquad (5.7.8)$$

Considering S to be an eigenfunction associated with operator s_z gives:

$$\hat{s}_z S = \pm \hbar S/2 \qquad (5.7.9)$$

The total spin wave function is:

$$S(s_z) = c_+ S_+(s_z) + c_- S_-(s_z) \qquad (5.7.10)$$

These wave functions are orthogonal and normalized.

5.8 Harmonic Oscillators

Physical phenomena that satisfy the harmonic oscillator equation appear in many manifestations. Examples include RC circuits, violin strings, and atoms forming a molecule. Consider a two-atom molecule as an example of a one-dimensional problem. To make a quantum mechanical analysis of such a harmonic oscillator let Hamiltonian operator H represent its total energy; the numeric value is W.

$$H = \frac{p^2}{2m} + \frac{1}{2}\alpha x^2 \tag{5.8.1}$$

Constant α is specific to the particular molecule. The time independent and dependent Schrödinger equations are:

$$\int U^*(x)\left(-\frac{\hbar^2}{2m} + \frac{1}{2}\alpha x^2\right)U(x)\,dx = W$$

$$W\psi(x,t) = -\frac{\hbar}{i}\frac{\partial\psi(x,t)}{\partial t} \tag{5.8.2}$$

Since energy W is constant it follows from the second of Eq. 5.8.2 that:

$$\psi(x,t) = U(x)e^{-iWt/\hbar} \tag{5.8.3}$$

The first of Eq. 5.8.2 is satisfied if:

$$-\frac{\hbar^2}{2m}\frac{d^2U(x)}{dx^2} + \left(\frac{1}{2}\alpha x^2 - W\right)U(x) = 0 \tag{5.8.4}$$

It is convenient to define new variable ρ as:

$$\rho = \gamma x \tag{5.8.5}$$

Combining Eq. 5.8.4 and 5.8.5 gives:

$$\frac{d^2U(\rho)}{d\rho^2} + \left(\frac{2m\gamma^2}{\hbar^2}W - \frac{m\alpha}{\hbar^2}\gamma^4\rho^2\right)U(\rho) = 0 \tag{5.8.6}$$

It is also convenient to make the substitutions:

$$\gamma^4 = \frac{m\alpha}{\hbar^2} \quad \text{and} \quad \lambda = \frac{2W}{\hbar\sqrt{\alpha/m}} \tag{5.8.7}$$

With these substitutions the dimensionless differential equation is:

$$U''(\rho) + \left(\lambda - \rho^2\right)U(\rho) = 0 \tag{5.8.8}$$

Primes indicate derivatives with respect to ρ. In the limit as ρ becomes large the term $\lambda U(\rho)$ becomes negligibly small. In that case the equation and its solution have the forms:

$$U''(\rho) - \rho^2 U(\rho) = 0$$
$$U(\rho) \approx e^{\pm \rho^2/2} \tag{5.8.9}$$

The positive sign is inconsistent with a localized charge and physical reality requires the negative sign. Therefore, with $H(\rho)$ a slowly varying function of ρ at large radii Eq. 5.8.9 may be written:

$$U(\rho) = AH(\rho)e^{-\rho^2/2} \tag{5.8.10}$$

Substituting Eq. 5.8.10 into 5.8.8 gives:

$$H''(\rho) - 2\rho H'(\rho) + \left(\lambda - 1\right)H(\rho) = 0 \tag{5.8.11}$$

The solution is most conveniently found using a series expansion:

$$H(\rho) = \sum_{s=0}^{\infty} a_s \rho^s \tag{5.8.12}$$

Substituting Eq. 5.8.12 into 5.8.11 gives:

$$\frac{a_{s+2}}{a_s} = \frac{2s - \left(\lambda - 1\right)}{(s+2)(s+1)} \tag{5.8.13}$$

In the limit as 's' becomes limitlessly large the ratio of Eq. 5.8.13 goes to:

$$\frac{a_{s+2}}{a_s} \Rightarrow \frac{2}{s} \tag{5.8.14}$$

The value of Eq. 5.8.14 also equals the ratio of the equivalent terms in the expansion of $\exp(\rho^2)$. Combining it with Eq. 5.8.10 gives the unacceptable result that it becomes infinite as ρ becomes limitlessly large. Therefore an acceptable function requires the numerator of Eq. 5.8.13 to vanish. This occurs if, with n an integer:

$$\lambda = 2n+1 \tag{5.8.15}$$

With Eq. 5.8.15 as a constraint the series terminates into a polynomial with largest power λ, making $H_n(\rho)$ a polynomial with largest power λ. The resulting polynomials are defined as Hermite polynomials, $H_n(\rho)$, and follow from the recursion relationship Eq. 5.8.13. Combining Eq. 5.8.7 and 5.8.15 shows the energy to be:

$$W_n = (2n+1)\hbar\omega_0/2 \quad \text{where} \quad \omega_0 = \sqrt{\alpha/m} \tag{5.8.16}$$

We note eigenstate energies are equally spaced energy $\hbar\omega_0$ apart and the lowest possible eigenstate energy is $\hbar\omega/2$. Since the lowest energy state is occupied at temperature absolute zero even at that temperature the molecule still oscillates with energy $\hbar\omega/2$.

5.9 Angular Momentum, Central Force Fields

By definition, the angular momentum, l, in kinematic and operator forms is:

$$l \equiv r \times p = \frac{\hbar}{i} r \times \nabla \tag{5.9.1}$$

The first equality follows from classical mechanics and the second includes operator notation for momentum. The angular momentum operator about each of the three rectangular axes takes the specific forms:

$$l_x = \frac{\hbar}{i}\left(y\frac{\partial}{\partial z} - z\frac{\partial}{\partial y}\right) = -\frac{\hbar}{i}\left(\sin\phi\frac{\partial}{\partial\theta} + \cot\theta\cos\phi\frac{\partial}{\partial\phi}\right)$$

$$l_y = \frac{\hbar}{i}\left(z\frac{\partial}{\partial x} - x\frac{\partial}{\partial z}\right) = \frac{\hbar}{i}\left(\cos\phi\frac{\partial}{\partial\theta} - \cot\theta\sin\phi\frac{\partial}{\partial\phi}\right) \qquad (5.9.2)$$

$$l_z = \frac{\hbar}{i}\left(x\frac{\partial}{\partial y} - y\frac{\partial}{\partial x}\right) = \frac{\hbar}{i}\left(\frac{\partial}{\partial\phi}\right)$$

The first column of equalities in Eq. 5.9.2 follows directly from Eq. 5.9.1 and the second after transforming to spherical coordinates.

To obtain the magnitude of the angular momentum we calculate its square. Its operator form follows from Eq. 5.9.2:

$$l_x^2 + l_y^2 = -\hbar^2\left(\frac{\partial^2}{\partial\theta^2} + \cot^2\theta\frac{\partial^2}{\partial\phi^2} + \cot\theta\frac{\partial}{\partial\theta}\right); \quad l_z^2 = -\hbar^2\left(\frac{\partial^2}{\partial\phi^2}\right)$$

The sum is:

$$l^2 = l_x^2 + l_y^2 + l_z^2 = -\hbar^2\left(\frac{1}{\sin\theta}\frac{\partial}{\partial\theta}\left[\sin\theta\frac{\partial}{\partial\theta}\right] + \frac{1}{\sin^2\theta}\frac{\partial^2}{\partial\phi^2}\right) \qquad (5.9.3)$$

Electrostatic force fields about atomic nuclei have spherical symmetry and hence, with spherical symmetry, the potential depends upon the magnitude of the radius only; there is no angular dependence. For this case the Schrödinger equation has the form:

$$-\frac{\hbar^2}{2m}\nabla^2 U(r) + \Lambda(r)U(r) = WU(r) \qquad (5.9.4)$$

Inserting the spherical Laplacian operator into Eq. 5.9.4 and rearranging terms gives:

$$\frac{1}{\sin\theta}\frac{\partial}{\partial\theta}\left[\sin\theta\frac{\partial U(r,\theta,\phi)}{\partial\theta}\right]+\frac{1}{\sin^2\theta}\left[\frac{\partial^2 U(r,\theta,\phi)}{\partial\phi^2}\right]$$

$$=-\frac{\partial}{\partial r}\left[r^2\frac{\partial U(r,\theta,\phi)}{\partial r}\right]+\frac{2m}{\hbar^2}\left[W+\Lambda(r)\right]r^2 U(r,\theta,\phi) \tag{5.9.5}$$

To integrate this equation, break function $U(r,\theta,\phi)$ into functions of a single variable; that is, put $U(r,\theta,\phi)$ equal to product function $R(r)\Theta(\theta)\Phi(\phi)$ then sum over all possible solutions:

$$U(r,\theta,\phi)=\sum R(r)\Theta(\theta)\Phi(\phi) \tag{5.9.6}$$

Substituting the product into Eq. 5.9.5 then multiplying by the inverse gives the equality:

$$\frac{1}{\Theta\sin\theta}\frac{d}{d\theta}\left(\sin\theta\frac{d\Theta}{d\theta}\right)+\frac{1}{\Phi\sin^2\theta}\left(\frac{d^2\Phi}{d\phi^2}\right)=-\frac{d}{Rdr}\left(r^2\frac{dR}{dr}\right)+\frac{2m}{\hbar^2}\left[W+\Lambda(r)\right]r^2$$

$$\tag{5.9.7}$$

Since the left side of the equation is a function of angles only and the right side a function of radius only, each side is constant. It is most convenient to put the separation constant equal to $-\ell(\ell+1)$. In a similar way, with separation constant m, the terms on the left side of Eq. 5.9.7 break into functions of θ alone and ϕ alone. The result is the two complete differential equations:

$$\frac{d^2\Phi}{d\phi^2}+m^2\Phi=0$$

$$\tag{5.9.8}$$

$$\frac{1}{\sin\theta}\frac{d}{d\theta}\left[\sin\theta\frac{d\Theta}{d\theta}\right]+\left[\ell(\ell+1)-\frac{m^2}{\sin^2\theta}\right]\Theta=0$$

The ϕ solutions may be written as one of the following:

$$\Phi(\phi)=A_m\cos(m\phi)+B_m\sin(m\phi)=C_m e^{im\phi}+D_m e^{-im\phi} \tag{5.9.9}$$

Since physically real solutions cannot be multivalued, m must be an integer. If the solution is proportional to either A_m or B_m, by the third of Eq. 5.9.2 the z-component of angular momentum is zero; if the solution is proportional to C_m or D_m the z-component of angular momentum, l_z, is:

$$l_z = \pm m\hbar \tag{5.9.10}$$

Combining Eq. 5.9.3 with the θ-dependent part of Eq. 5.9.8 shows the angular momentum satisfies the equation:

$$l^2 = \ell(\ell+1)\hbar^2 \tag{5.9.11}$$

The θ-equation provides a physically real solution only if ℓ is an integer and solutions with integer values of ℓ are associated Legendre functions.

Therefore both separation constants ℓ and m are integers and in quantum theory are called quantum numbers. Solutions exist for quantum numbers in the ranges:

$$0 \le \ell < \infty; \ -\ell \le m \le \ell \tag{5.9.12}$$

Combining the above shows that under all circumstances:

$$l^2 > l_z^2 \tag{5.9.13}$$

It follows that orbital angular momentum is never entirely about a single axis.

References

G. Auletta, *Foundations and Interpretation of Quantum Mechanics*, World Scientific (2001)

D. Bohm, *Quantum Theory*, Prentice Hall, Princeton NJ (1951)

P.A.M. Dirac, *The Principles of Quantum Mechanics*, 4th ed., Oxford (1958)

G. Greenstein, A.G. Zajonc, *The Quantum Challenge: Modern Research on the Foundations of Quantum Mechanics*, Jones and Bartlett Publishers (1997)

L.D. Landau, E.M. Lifshitz, *Quantum Mechanics: Non-Relativistic Theory*, Addison-Wesley, Reading MA (1958)

A. Messiah, *Quantum Mechanics*, vol. 1, trans. by G. M. Temmer, John Wiley, New York (1966)

R. Omnès, *Understanding Quantum Mechanics*, Princeton University Press (1999)

L.I. Schiff, *Quantum Mechanics*, McGraw-Hill Book Co., New York (1949)

F. Schwabl, *Quantum Mechanics*, 2nd ed., Springer, New York (1995)

R.C. Tolman, *The Principles of Statistical Mechanics* (1938), reprinted by Dover Publications (1979)

Chapter 6
Quantized Energy Exchanges

$\mathcal{T}$his chapter reviews the current understanding of photon exchanges between the radiation field and quantized energy states. In subsequent chapters we detail our view and contrast it with this one. The purpose is to place the two quite different explanations in appropriate technical and historical perspectives.

The first four sections of this chapter are devoted to Planck's black body radiation law, its implications, and the background electromagnetic radiation field. The quantum theory of radiation incorporates Planck's radiation law; Planck used Hertz' photoelectric discovery in the development of his law. Both his black body radiation law and Hertz' measurement of the photoelectric effect were later explained by Einstein in a pragmatic way. The explanation included a pragmatic description of the absorption and emission of radiation by eigenstate electrons. It is based upon the conservation laws, first of energy and then of linear momentum; both used the relativistic transformations between moving systems discussed in Sec. 1.3. Since Einstein's energy-based derivation is more succinct than Planck's and moves smoothly into the momentum based argument, we follow Einstein.

Einstein presented his pragmatic approach as an expedient to help understand the onrush of experimental data about atomic phenomena. He wrote as if he expected the viewpoint to be a temporary one that would later yield to a more general theory. This type of wide-open pragmatism is quite different from developments in quantum theory a few years later. Then quantum theory pragmatists argued that quantum theory explanations were complete and no deeper understanding is possible than the relationship presented at that time between the classical and quantum theories.

Sections 6.5 and 6.6 apply Schrödinger's equation first to a Coulomb potential and then a hydrogen atom. Perturbation analysis, described in Sec. 6.7, is a powerful method of problem analysis. However, as the name suggests, its use is restricted to perturbative changes from otherwise known problem solutions. Sections 6.8 and 6.9 repeat the accepted analyses of electron transitions between eigenstates using perturbation theory. However, results of that theory are valid only to small changes from known solutions and electron transitions between eigenstates are certainly not small. Yet the correctness of predictions based upon perturbation analyses are beyond dispute. (For reasons that will become obvious our explanation of why perturbation theory works for eigenstate transitions is placed in Chap. 9.) Section 6.9 describes photon absorption and emission and Sec. 6.10 develops optical selection rules. Section 6.11 examines the Exclusion Principle, first formulated by Pauli, which states that only antisymmetric wave functions exist. From it and our electron model we conclude that on some dimensional scale an electron's charge distribution is granular.

6.1 Blackbody Radiation, Long Wavelength Limit

Atoms undergo thermal oscillations and the higher the temperature the larger the oscillatory energy. Oscillating charges undergo accelerations that, in turn, result in the absorption and emission of energy in the form of electromagnetic waves. The number of absorptive and emissive energy exchanges increase with temperature. We seek to use conservation laws to obtain a thermodynamically quantitative understanding of the general properties of such phenomena; that is, we seek to achieve maximum field characterization with minimum reliance upon details of specific emission and absorption processes.

Consider a walled cavity with a contained equilibrated internal thermal radiation field; equal amounts of energy and momentum are continuously exchanged among all parts. To examine the radiation field let its energy density be represented by $w(\omega,T)$ and given by:

$$w(\omega,T) = \frac{\varepsilon}{2}E^2 + \frac{\mu}{2}H^2 \qquad (6.1.1)$$

Since the energy density is a function of frequency, define energy function $w_\omega(\omega,T)d\omega$ as the energy between frequencies ω and $\omega + d\omega$. It follows that:

$$w(\omega) = \int_0^\infty w_\omega(\omega)d\omega \qquad (6.1.2)$$

Next consider two objects, A and B, which are composed of different materials. The objects contain cavities in which the energy densities are $w_{\omega A}$ and $w_{\omega B}$. The objects are placed in thermal contact and achieve equilibrium at temperature T. Next thermally isolate the objects and construct an optical system that exchanges radiation between them through an optical filter that passes only radiation between frequencies ω and $\omega + d\omega$. If placing the exchange mechanism into position produces a net energy exchange each cavity will achieve a new equilibrium with the temperature of one cavity increased and the other decreased. That temperature difference can be used to run a heat engine. Since this violates the second law of thermodynamics, it follows that there is no net energy exchange between the chambers at any frequency and that the equilibrium energy density is independent of the material of which the cavities are formed. Since this is true for all materials, it is independent of the color of the walls, the fraction of the incident radiation the walls reflect or absorb, or any other wall or cavity parameter. Kirchhoff first pointed out this remarkable result in 1859; it is expressed by the equation:

$$w_{\omega A}(\omega,T) = w_{\omega B}(\omega,T) \qquad (6.1.3)$$

The result is exact for a cavity with a negligibly small aperture, and the cavity-contained radiation field is, by definition, the isothermal cavity radiation field. Next suppose that one cavity wall is ideally black. Since ideally black objects absorb all incident radiation, this energy field is also that of an ideal black object. Hence the radiation field is also the blackbody radiation field.

Since the equilibrium electromagnetic energy density is not a function of the bounding material, without loss of generality we can construct the enclosure of the material most convenient to analyze and apply the result to all materials. The enclosure of choice has walls of harmonic oscillators. Radiation is reflected or absorbed and re-emitted by the walls, thereby maintaining temperature T. The number of available frequency states per unit volume available is given by Eq. 2.3.14 and repeated here:

$$\frac{1}{V} dN = \frac{\omega^2}{\pi^2 c^3} d\omega \qquad (6.1.4)$$

By the equipartition theorem of statistical mechanics each oscillator supports energy $kT/2$ per degree of freedom, where k is the Boltzman constant. Since the walls are two-dimensional, each oscillator supports energy kT per mode. Substituting this into Eq. 6.1.4 gives:

$$w_\omega(\omega, T) = \frac{\omega^2}{\pi^2 c^3} kT \qquad (6.1.5)$$

This is the long wavelength limit of the contained radiation field; it is the Rayleigh-Jeans formula for blackbody radiation. It is valid at low enough frequencies and high enough temperatures so $\hbar\omega << kT$.

6.2 Blackbody Radiation Law Using Energy

Let a cavity in thermal equilibrium at temperature T contain N identical and large gas molecules, where N is a very large number. Each molecule supports two non-degenerate energy levels, states Z_n and Z_s. The eigenstate energies of the levels are, respectively, W_n and W_s with $W_n > W_s$. The molecular mass is large enough so the molecular speeds are much less than the speed of light. Einstein postulated that Maxwell statistics apply for the special case of large molecules in equilibrium. Therefore the number of large molecules in state Z_s is N_s where:

$$N_s = \frac{N \exp(-W_s/kT)}{\exp(-W_n/kT) + \exp(-W_s/kT)} \qquad (6.2.1)$$

In conformance with the theory of photoelectricity and with ω_n^s equal the frequency, he required that molecules emit and absorb radiation in energy units:

$$\hbar\omega_n^s = W_n - W_s \tag{6.2.2}$$

Since harmonic oscillators satisfy this condition, see Eq. 5.8.16, walls of harmonic oscillators satisfy the requirement. He also required the Rayleigh-Jeans formula, Eq. 6.1.5, to apply in the limit of low frequencies. Although that equation is based upon classical electrodynamics with continuous energy changes, in the low frequency limit the energy differences are vanishingly small and the Rayleigh-Jeans limit is satisfied.

Let the transition rate $d\Gamma/dt$ from state Z_n to state Z_s, with the accompanying emission of energy $\hbar\omega$, be:

$$\frac{d\Gamma_n^s}{dt} = N_n \left[A_n^s + B_n^s w\left(\omega_n^s\right) \right] \tag{6.2.3}$$

This transition rate is the sum of two parts: (1) spontaneous emission A_n^s is due to internal processes within the molecules and independent of the radiation field and (2) induced emission B_n^s is driven by the radiation field and therefore proportional to the applied field intensity. In the absence of spontaneous absorption the complete transition rate from state Z_s to Z_n is:

$$\frac{d\Gamma_s^n}{dt} = N_s B_s^n w\left(\omega_n^s\right) \tag{6.2.4}$$

Equal rates of absorption and emission are required at equilibrium. Equating them and collecting terms gives:

$$w\left(\omega_n^s\right) = \frac{A_n^s}{B_n^s \exp\left(\hbar\omega/kT\right) - B_s^n} \tag{6.2.5}$$

In the limit of small values of the exponential, Eq. 6.2.5 goes to:

$$w\left(\omega_n^s\right) = \frac{A_n^s}{B_n^s - B_s^n - B_n^s\left(\hbar\omega/kT\right)} \tag{6.2.6}$$

Equating Eq. 6.1.5 and 6.2.6 gives:

$$B_n^s = B_s^n; \quad A_n^s = \frac{\left(\omega_n^s\right)^2}{\pi^2 c^3}\left(\hbar\omega_n^s\right)B_n^s \tag{6.2.7}$$

Combining Eq. 6.2.5 and 6.2.7 yields Planck's radiation law:

$$w\left(\omega,t\right) = \frac{\omega^2}{\pi^2 c^3}\left(\frac{\hbar\omega}{\exp\left(\hbar\omega/kT\right)-1}\right) \tag{6.2.8}$$

This law describes the energy density in an equilibrated electromagnetic radiation field as a function of radiation frequency and the temperature of the surroundings. In the low frequency limit it satisfies the Rayleigh-Jeans formula:

$$\lim_{\hbar\omega\to 0} w_\omega\left(\omega,T\right) = \frac{\omega^2}{\pi^2 c^3}kT$$

In the high frequency limit the distribution obeys the Wien formula, a form of the Maxwell distribution law:

$$\lim_{\hbar\omega\to\infty} w\left(\omega\right) = \frac{\hbar\omega^3}{\pi^2 c^3}\exp\left(-\hbar\omega/kT\right) \tag{6.2.9}$$

Actual values of $A_n^{\ s}$ and $B_n^{\ s}$ can be calculated using Eq. 6.8.10 for molecules with known eigenstates.

6.3 Blackbody Radiation Law Using Momentum

In the second part of his 1917 paper Einstein explained that thermodynamic problems are typically analyzed using energy exchanges; momentum exchanges, which are less by a factor of c, are ignored. For theoretical considerations, however, energy and momentum are on a par. A theory is only correct if the momentum transferred during energy exchanges leads to the same statistical results as energy considerations.

He then presented a development of the radiation equation using momentum exchanges, and found it results from momentum exchanges only if the radiation is fully directed. That is, as a molecule emits or absorbs energy $\hbar\omega$ the exchange is accompanied by an exchange of linear momentum $\hbar\omega/c$: the molecule undergoes a push of $\hbar\omega/c$. Although laser beams may approach this condition, in the case of atomic radiation the diameter of the emitter may be less than a hundredth of an optical wavelength. Einstein viewed absorption from a plane wave as containing the energy and momentum in the same ratio as the wave. The analyses in Sec. 2.9 and 2.10 show this can occur only if there is no scattering. He viewed emission as being carried by spherical Hertzian waves, *i.e.*, a multimodal expansion, with rotational symmetry about the radiation axis; for such waves the source suffers compression but not a push. This argument led him to conclude that a quantum theory of radiation is almost unavoidable. He also explained that his theory has two weaknesses: it does nothing to illuminate the connection between quantized energy exchanges on the one hand and Maxwell's wave theory on the other, and it leaves the time and direction of elementary processes to chance.

Following Einstein, we imagine a cavity containing gaseous molecules in thermal equilibrium at temperature T, but no radiation. The mass M of the molecules is large enough so molecular speeds are much less than c so relativistic speed corrections are not needed. Next introduce electromagnetic radiation as an isotropic distribution of plane waves that interact with the molecules in a way that retains the original temperature, T. Also, as in Sec. 6.2, only eigenstates Z_s and Z_n undergo energy exchanges.

Since at equilibrium the average molecular speed is zero, on the average all radiation-molecule interactions occur with stationary molecules. During events both energy and impulse are exchanged and produce an acquired molecular speed v. Although on the average a molecule sits in a uniform radiation field, a post-event one does not and for moving molecules the radiation field is not isotropic. There is a net field unbalance in the direction of motion and a frequency alteration. This unbalance produces an altered event rate that, in turn, damps the molecular motion.

Let each event transfer momentum Δ between the field and the molecule. Molecular damping is denoted as Rv where R is a field-dependent constant to be determined and acts in the direction opposite to the motion. Equilibrium requires, on the average, the speed to return to zero before the next event. It follows that during the period between events the momentum of the molecule is:

$$Mv - Rv\tau + \Delta$$

τ is time since the last event and mass $M \gg R\tau$.

For a system to remain in equilibrium the average velocity v at the time of an event remains equal to zero. This may be stated as:

$$\left\langle \left(Mv - Rv\tau + \Delta \right)^2 \right\rangle = \left\langle \left(Mv \right)^2 \right\rangle \tag{6.3.1}$$

Combining the above gives:

$$\left\langle \Delta^2 \right\rangle = 2RvM \left\langle v^2 \right\rangle \tag{6.3.2}$$

A system analysis using momentum exchanges can be correct only if Eq. 6.3.2 is satisfied.

The mean-square velocity may be expressed in terms of temperature using Boltzman's one-dimensional law:

$$\frac{1}{2} M \left\langle v^2 \right\rangle = \frac{1}{2} kT \tag{6.3.3}$$

Damping Product Rv

To examine if Eq. 6.3.2 is satisfied, using the results of Sec. 6.2 we calculate the rate of momentum transfer from a moving molecule to an otherwise uniform radiation field that is unbalanced by the motion. Let the molecule be moving along the z-axis with speed v in stationary coordinate system K. To calculate radiation damping we analyze conditions in coordinate system K' in which the molecule is at rest. Since radiation in system K is isotropic the field intensity and the radiation energy within differential solid angle $d\kappa$ and frequency range $d\omega$ is:

$$w(\omega)d\omega\frac{d\kappa}{4\pi} \tag{6.3.4}$$

The frequencies in systems K and K' differ by the Doppler frequency; in K' the intensity is a function of angle with respect to the z-axis:

$$w(\omega',\theta')d\omega'\frac{d\kappa'}{4\pi} \tag{6.3.5}$$

The relationship between Eq. 6.3.4 and 6.3.5 follows by using Eq. 1.3.2 to act on the fields of Eq. 6.3.4 arrayed in the form of Eq. 1.6.4. For v/c much less than one the energy density, frequency, and angle with the z-axis transform as:

$$w(\omega',\theta') = w(\omega)\frac{d\omega}{d\omega'}\frac{d\kappa}{d\kappa'}\left(1-2\frac{v}{c}\cos\theta\right) \tag{6.3.6}$$

$$\omega' = \omega\left(1-\frac{v}{c}\cos\theta\right) \tag{6.3.7}$$

$$\cos\theta' = \cos\theta - \frac{v}{c}\left(1-\cos^2\theta\right) \tag{6.3.8}$$

Since $v \ll c$ it follows from Eq. 6.3.7 and 6.3.8 that:

$$\omega = \omega'\left(1+\frac{v}{c}\cos\theta'\right) \tag{6.3.9}$$

$$w(\omega) = w\left(\omega'+\frac{v}{c}\omega'\cos\theta'\right) \cong w(\omega')+\left(\frac{v}{c}\omega'\cos\theta'\right)\frac{\partial w(\omega)}{\partial\omega}\Big|_{\omega=\omega'} \tag{6.3.10}$$

From Eq. 6.3.9:

$$\frac{d\omega}{d\omega'} = \left(1 + \frac{v}{c}\cos\theta'\right)$$

(6.3.11)

Combining the definitions of $d\kappa$ and $d\kappa'$ with Eq. 6.3.8 gives:

$$\frac{d\kappa}{d\kappa'} = \frac{\sin\theta d\theta}{\sin\theta' d\theta'} = \frac{d(\cos\theta)}{d(\cos\theta')} = 1 - 2\frac{v}{c}\cos\theta'$$

(6.3.12)

The radiation intensity in system K' follows by combining the above and gives, through first order:

$$w(\omega',\theta') = \left[w(\omega') + \frac{v}{c}\omega'\cos\theta'\frac{\partial w(\omega)}{\partial\omega}\Big|_{\omega=\omega'}\right]\left(1 - 3\frac{v}{c}\cos\theta'\right)$$

(6.3.13)

For systems in equilibrium the rate of absorption events is:

$$B_s^n w(\omega',\theta') d\omega' \frac{d\kappa'}{4\pi}$$

(6.3.14)

Since this transfer rate is correct for a system in equilibrium, for a moving molecule it is also the rate at which momentum transfers would occur if the subsequent decay were instantaneous. The moving molecule, however, is not in equilibrium. By the ergodic theorem of statistics, for a constraint-free system in equilibrium the spatial average population of states is equal to the fraction of time an individual molecule spends in each state. For this case the average fraction of equilibrated molecules in state Z_s, see Eq. 6.2.1, is equal to the fraction of time a specific molecule spends in that state. Therefore multiplying Eq. 6.3.14 by the fraction of the time the molecule spends in that state, Eq. 6.2.1, gives the number of absorption transitions due to radiation within differential solid angle $d\kappa'$ per second:

$$\frac{e^{-W_s/kT}}{e^{-W_n/kT} + e^{-W_s/kT}} B_s^n w(\omega',\theta') \frac{d\kappa'}{4\pi}$$

(6.3.15)

Next make the basic postulate that all radiation events are fully directed in that for either absorption or emission each event transfers momentum Δ to the molecule where:

$$\Delta = \pm \frac{W_n - W_s}{c} \cos\theta'$$

(6.3.16)

Applying the above arguments to emission, combining the absorption and emission results, and incorporating the first of Eq. 6.2.8 gives the net momentum transferred from the molecule to the radiation:

$$\left(\frac{\exp(-W_s/kT) - \exp(-W_n/kT)}{\exp(-W_s/kT) + \exp(-W_n/kT)} \right) \frac{\hbar\omega}{4\pi c} B_s^n \int w(\omega',\theta') \cos\theta' d\kappa'$$

(6.3.17)

Substituting Eq. 6.3.13 into Eq. 6.3.17 and integrating over the solid angle gives the momentum transferred from the molecules to the radiation:

$$-\frac{\hbar\omega}{c^2} B_s^n \left(\frac{\exp(-W_s/kT) - \exp(-W_n/kT)}{\exp(-W_s/kT) + \exp(-W_n/kT)} \right) \left(w(\omega') - \frac{1}{3}\omega' \frac{\partial w(\omega)}{\partial\omega} \bigg|_{\omega=\omega'} \right) v$$

(6.3.18)

For v much less than c the difference between frequencies ω and ω' is small. Putting them equal and equating the expression to R gives:

$$R = \frac{\hbar\omega}{c^2} B_s^n \left(\frac{\exp(-W_s/kT)}{\exp(-W_s/kT) + \exp(-W_n/kT)} \right) \left(1 - e^{-\hbar\omega/kT} \right) \left(w(\omega) - \frac{1}{3}\omega \frac{\partial w(\omega)}{\partial\omega} \right)$$

(6.3.19)

It follows from Planck's radiation law that:

$$\left(w(\omega) - \frac{1}{3}\omega \frac{\partial w(\omega)}{\partial\omega} \right) \left(1 - e^{-\hbar\omega/kT} \right) = \frac{\hbar\omega}{3kT} w(\omega)$$

(6.3.20)

Combining:

$$R = \frac{\hbar^2\omega^2}{c^2} \left(\frac{\exp(-W_s/kT)}{\exp(-W_s/kT) + \exp(-W_n/kT)} \right) NB_s^n \frac{w(\omega)}{3kT}$$

(6.3.21)

Since spontaneous emission is postulated to be both fully directed and randomly oriented it does not affect the exchanged momentum and therefore is not considered here.

Momentum Transfer $\langle \Delta^2 \rangle$

The effect of random processes on the mechanical behavior of molecules is much easier to derive. Let z-directed linear momentum λ be transferred to a molecule with each energy exchange and let it be of varying magnitude and direction. The average value, however, is along the z-axis and is equal to zero. If the mean value of the momentum changes is zero and if l is the number of such events it follows that:

$$\langle \Delta^2 \rangle = \langle l\, \lambda^2 \rangle \tag{6.3.22}$$

With each energy exchange, and consistent with the radiation results, the momentum transferred to the molecule is:

$$\lambda = \pm \frac{\hbar \omega}{c} \cos\theta \tag{6.3.23}$$

With this notation the limits on angle θ are $-\pi/2$ to $\pi/2$, from which it follows that:

$$\langle \lambda^2 \rangle = \frac{1}{3}\left(\frac{\hbar \omega}{c}\right)^2 \tag{6.3.24}$$

The number of events that occur in time τ is just twice the number of absorption processes:

$$l = \left(\frac{2\exp\left(-W_s/kT\right)}{\exp\left(-W_s/kT\right)+\exp\left(-W_n/kT\right)}\right) B_s^n w(\omega)\tau \tag{6.3.25}$$

Combining Eq. 6.3.22, 6.3.24, and 6.3.25 gives:

$$\frac{\langle \Delta^2 \rangle}{\tau} = \frac{2}{3}\left(\frac{\hbar \omega}{c}\right)^2 \left(\frac{\exp\left(-W_s/kT\right)}{\exp\left(-W_s/kT\right)+\exp\left(-W_n/kT\right)}\right) B_s^n w(\omega) \tag{6.3.26}$$

Comparing Eq. 6.3.26 and 6.3.20 after including Eq. 6.3.3 gives:

$$\frac{\langle \Delta^2 \rangle}{\tau} = 2RkT \tag{6.3.27}$$

This result confirms that fully directed radiation exchanges is a sufficient basis for Planck's radiation law. The requirement is only consistent with classical electromagnetism, Eq. 4.12.10 and 4.12.11, if absorbed and emitted energy-to-momentum ratios are equal; this, in turn, requires that both equal c.

6.4 The Zero-Point Field

According to the equipartition theorem of statistical mechanics the energy per degree of freedom in a statistical system of equilibrated particles is:

$$\text{Energy} \, / \, \text{degree of freedom} = kT/2 \tag{6.4.1}$$

Within an equilibrated electromagnetic energy field and at high temperatures the energy density at frequency ω is given by Planck's equation, Eq. 6.2.8. In the limit as the temperature goes to absolute zero that expression may be expressed as Eq. 6.2.9. The limiting value at very large temperatures is also of interest. For that case Eq. 6.2.8 goes to:

$$\lim_{T \to \infty} w(\omega,t) = \frac{\omega^2}{\pi^2 c^3} \left(\frac{\hbar\omega}{\hbar\omega/kT + \left(\hbar\omega/kT\right)^2 / 2} \right) \approx \frac{\omega^2}{\pi^2 c^3} \left(kT - \frac{\hbar\omega}{2} \right) \tag{6.4.2}$$

It follows from Eq. 2.3.14 that the spatial density of available energy states is:

$$N = \frac{\omega^2}{\pi^2 c^3} \tag{6.4.3}$$

Combining Eq. 6.4.2 and 6.4.3 shows that the energy per state is:

$$kT = \frac{\hbar\omega}{2} \tag{6.4.4}$$

Since Eq. 6.4.1 is not consistent with 6.4.4 it follows that the Planck expression should be amended to:

$$w_\omega = \frac{\omega^2}{\pi^2 c^3}\left(\frac{\hbar\omega}{e^{\hbar\omega/kT}-1} + \frac{\hbar\omega}{2}\right) \tag{6.4.5}$$

In the limit as T approaches zero Planck's original radiation energy distribution term vanishes leaving only the zero-point energy. That energy, in turn, is in equilibrium with the zero-point energy of the enclosing cavity walls. Therefore both the harmonic oscillators and the radiation field maintain an irreducible energy density at absolute zero temperature.

6.5 Coulomb Potential Well

Arguably the most ubiquitous electron trap, and upon which the entire science of chemistry is based, is the coulomb potential of positive nuclei. Therefore we seek detailed results. The potential energy of an electron with charge $-e$ in the vicinity of a nucleus with charge $+Ze$ is:

$$\Lambda(r) = -\frac{Ze^2}{4\pi\varepsilon r} \tag{6.5.1}$$

Combining Eq. 6.5.1 and the radial portion of Eq. 5.9.7 gives:

$$-\frac{h^2}{2m}\frac{1}{r^2}\frac{d}{dr}\left(r^2\frac{dR}{dr}\right) - \frac{Ze^2}{4\pi er}R + \frac{l(l+1)h^2}{2mr^2}R = WR \tag{6.5.2}$$

Total energy W is respectively less than or greater than zero for bound or free electrons.

The equation is most easily solved by introducing parameter α and variable ρ where, by definition:

$$\rho = \alpha r \tag{6.5.3}$$

We also introduce the definitions:

$$\alpha^2 = \frac{8m|W|}{\hbar^2}; \quad n = \frac{Ze^2}{4\pi\varepsilon\hbar}\sqrt{\frac{m}{2|W|}} = \frac{Ze^2 m}{2\pi\varepsilon\alpha\hbar^2} \tag{6.5.4}$$

If the electron energy is negative, substituting ρ back into Eq. 6.5.2 and using Eq. 6.5.4 gives:

$$\frac{1}{\rho^2}\frac{d}{d\rho}\left(\rho^2\frac{dR}{d\rho}\right) - \frac{\ell(\ell+1)}{\rho^2}R + \frac{nR}{\rho} - \frac{R}{4} = 0 \tag{6.5.5}$$

To solve Eq. 6.5.5 begin with the asymptotic solution at limitlessly large distances. As ρ increases without limit the asymptotic differential equation is:

$$\frac{d^2R}{d\sigma^2} - \frac{R}{4} = 0 \tag{6.5.6}$$

The solution of Eq. 6.5.6 is:

$$R(\rho) = F(\rho)e^{-\rho/2} + G(\rho)e^{\rho/2} \tag{6.5.7}$$

At limitlessly large values of radius, a condition on Eq. 6.5.6 is that the exponential terms vary much more rapidly with radius than either $F(\rho)$ or $G(\rho)$. Also in that limit Eq. 6.5.7 shows, *prima facie*, that $G(\rho) = 0$. Substituting $F(\rho)$ into Eq. 6.5.5 gives the differential equation:

$$\frac{d^2F}{d\rho^2} + \left(\frac{2}{\rho} - 1\right)\frac{dF}{d\rho} + \left(\frac{n-1}{\rho} - \frac{\ell(\ell+1)}{\rho^2}\right)F = 0 \tag{6.5.8}$$

A convenient solution method is a power series expansion. The solution procedure begins with the summation:

$$F(\rho) = \rho^s \sum_{j=0}^{\infty} a_j \rho^j \tag{6.5.9}$$

Solution requires that $a_0 \neq 0$ and $a_j \geq 0$. Substituting Eq. 6.5.9 into Eq. 6.5.8 results in the sum:

$$\left[s(s+1)-\ell(\ell+1)\right]\rho^{-2}$$

$$+\sum_{j=0}^{\infty}\left\{\left[(s+j+1)(s+j+2)-\ell(\ell+1)\right]a_{j+1}-(s+j+1-n)a_{j}\right\}\rho^{j-1}=0$$

$$(6.5.10)$$

Since a_0 is not equal to zero, the first term of Eq. 6.5.10 requires either $s=\ell$ or $s=-(\ell+1)$; since the latter is singular at the origin it cannot represent physical reality. Substituting $s=\ell$ into Eq. 6.5.10 gives the coefficient ratio:

$$\frac{a_{j+1}}{a_{j}}=\frac{j+1+\ell-n}{(j+1)(j+2\ell+2)}$$

$$(6.5.11)$$

As index 'j' increases without limit Eq. 6.5.11 goes asymptotically to the index of the expansion for the exponential e^{ρ}. Combining this with Eq. 6.5.7 shows that the radial function is proportional to $e^{\rho/2}$ and that, in the limit of very large ρ, is an unacceptable result. Hence a nontrivial solution of $F(\rho)$ exists if and only if the series terminates and the series terminates only if n is an integer. For that case, Eq. 6.5.4 shows that state energy W_n is equal to:

$$W_{n}=-\frac{mZ^{2}e^{4}}{32\pi^{2}\varepsilon^{2}n^{2}\hbar^{2}}$$

$$(6.5.12)$$

There are an infinite number of energy values, energy is independent of quantum numbers ℓ and m, and the state energy varies as the inverse square of quantum number n.

It is helpful to define radius r_0 as:

$$r_{0}=\frac{2Z}{n\alpha}; \text{ hence } r_{0}=\frac{4\pi\varepsilon\hbar^{2}}{me^{2}}$$

$$(6.5.13)$$

Evaluating Eq. 6.5.13 shows that:

$$r_{0}=5.29172\times10^{-11}m$$

$$(6.5.14)$$

By definition r_0 is the first Bohr radius.

The electrostatic energy may be written as:

$$W_n = \frac{Ze^2}{8\pi\varepsilon}\langle 1/r \rangle \tag{6.5.15}$$

Combining Eq. 6.5.12, 6.5.13, and 6.5.15 gives:

$$\langle 1/r \rangle = Z/\left(n^2 r_0\right) \tag{6.5.16}$$

In the limit of large values of n, the energy goes to zero and the expectation value of the radius of the electron state becomes infinite: the electron is distributed over all space.

To solve for the radial function, rewrite Eq. 6.5.11 as:

$$a_{j+1} = \frac{(j+\ell+1-n)}{(j+2\ell+2)(j+1)} a_j \tag{6.5.17}$$

In terms of the initial constant, Eq. 6.5.17 takes the form:

$$a_j = (-1)^j \left(\frac{(n-\ell-1)!(2\ell+1)!}{(n-j-\ell-1)!\,j!\,(2\ell+j+1)!} \right) a_0$$

To bring this into agreement with common usage, define a_0 as:

$$a_0 = -\frac{\left[(n+\ell)!\right]^2}{(n-\ell-1)!(2\ell+1)!}$$

Combining gives:

$$a_j = (-1)^{j+1} \left\{ \frac{\left[(n+\ell)!\right]^2}{(n-j-\ell-1)!\,j!\,(2\ell+j+1)!} \right\} \tag{6.5.18}$$

It follows that:

$$\ell < n \tag{6.5.19}$$

Combining all results shows that $R(\rho)$ depends upon both quantum numbers n and ℓ, and is equal to:

$$R_n^\ell(\rho) = \rho^\ell e^{-\rho/2} \sum_{j=0}^{n-\ell-1} (-1)^{j+1} \frac{\left[(n+\ell)!\right]^2 \rho^j}{(n-j-\ell-1)!\,j!\,(2\ell+j+1)!} \qquad (6.5.20)$$

Functional values for $n = 1$ through 4 are listed in Table 6.5.1.

Table 6.5.1. Values of $R_n^\ell(\sigma)$ for n from one to four.

	$\ell = 1$	$\ell = 2,\ \times e^{-\rho/2}$	$\ell = 3,\ \times e^{-\rho/2}$	$\ell = 4,\ \times e^{-\rho/2}$
$R_n^0(\sigma)$	$-e^{-\rho/2}$	$-(2)!(2-\rho)$	$-3\left(6-6\rho+\rho^2\right)$	$-24\left(\begin{array}{l}4-6\rho\\+2\rho^2-\rho^3\end{array}\right)$
$R_n^1(\sigma)$		$-(3)!\rho$	$-(4)!\rho(4-\rho)$	$-60\rho\left(\begin{array}{l}20-10\rho\\+3\rho^2\end{array}\right)$
$R_n^2(\sigma)$			$-(5)!\rho^2$	$-(6)!\rho^2(6-\rho)$
$R_n^3(\sigma)$				$-(7)!\rho^3$

Wave function normalization follows from Eq. 6.5.20; the resulting radial integral is:

$$I_n^\ell = \int_0^\infty \rho^2 d\rho\, R_n^\ell(\rho) R_q^\ell(\rho) = \frac{2n\left[(n+\ell)!\right]^3}{(n-\ell-1)!}\,\delta(n,q) \qquad (6.5.21)$$

The first row of Table 6.5.1 shows the function is largest at the origin. Including the origin, there are a total of ℓ maxima as a function of radius. The remaining rows show that the radius raised to power ℓ multiplies each function and therefore the function vanishes at the origin for all values of ℓ except zero. The radius of the region with a negligibly small value of charge increases with increasing values of ℓ.

6.6 Hydrogen Atom Eigenfunctions

The product of radial and angular functions with the central force field of Sec. 5.9 gives the full expression for an eigenfunction:

$$U_{n\ell m}\left(r,\theta,\phi\right)=A_{n\ell m}R_n^{\ell}\left(\alpha_n r\right)P_{\ell}^m\left(\cos\theta\right)e^{-im\phi} \tag{6.6.1}$$

In free space since the solution exists for the full azimuth and zenith angular ranges ℓ and m are both integers. From the theory of Legendre polynomials:

$$\int_0^{\pi}\sin\theta d\theta\left[P_{\ell}^m\left(\cos\theta\right)\right]^2=\frac{4\pi}{\left(2\ell+1\right)}\frac{\left(\ell+m\right)!}{\left(\ell-m\right)!} \tag{6.6.2}$$

Using Eq. 6.6.1 and 6.6.2 to solve for the probability coefficient of each state, normalizing that value to unity, and solving for the constant coefficients of Eq. 6.6.1 gives:

$$\left|A_{n\ell m}\right|=\left\{\frac{\alpha_n^3}{4\pi n}\frac{\left(n-\ell-1\right)!}{\left(n+\ell\right)!^3}\frac{\left(2\ell+1\right)\left(\ell-m\right)!}{\left(\ell+m\right)!}\right\}^{1/2} \tag{6.6.3}$$

Each wave function has $(n-\ell)$ zeros, including infinity, and undergoes $(n-\ell)$ functional maxima, *i.e.*, nodes, as a function of radius. Several complete eigenfunctions are listed in Table 6.6.1.

The energy levels of Eq. 6.5.12 show that the energy depends upon quantum number n but not upon quantum numbers ℓ and m. In common with other boundary value problems only eigenfunction solutions can exist. Parameter r_0 of Eq. 6.5.13 is a normalizing radial factor that shows atomic radii to be on the order of 0.1 nm.

Since wave functions with $m=0$ have spherical symmetry the charge density associated with them produces monopolar electrostatic fields. There is a charge density node at the origin and $(n-1)$ others at increasing values of radius. Wave functions with $m=1$ have bilateral symmetry: there is a null in the charge density at the origin and $(n-1)$ nodes. Wave functions with $m=2$ have quadrilateral symmetry: there is a charge density null at the origin and $(n-2)$ nodes, *etc.*

Table 6.6.1. Hydrogen atom eigenfunctions for $n = 1$ through 3, effective radius ($1/<r>$) depends upon quantum number n only.

$$U_{100} = \sqrt{\frac{1}{\pi}}\left(\frac{Z}{r_0}\right)^{3/2} e^{-Zr/r_0} \qquad\qquad \frac{1}{\langle 1/r\rangle} = \frac{r_0}{Z}$$

$$U_{200} = \frac{1}{\sqrt{32\pi}}\left(\frac{Z}{r_0}\right)^{3/2}\left(2 - \frac{Zr}{r_0}\right)e^{-Zr/2r_0} \qquad\qquad \frac{1}{\langle 1/r\rangle} = \frac{4r_0}{Z}$$

$$U_{210} = \frac{1}{\sqrt{32\pi}}\left(\frac{Z}{r_0}\right)^{3/2}\left(\frac{Zr}{r_0}\right)e^{-Zr/2r_0}\cos\theta$$

$$U_{21\pm1} = \frac{1}{8\sqrt{\pi}}\left(\frac{Z}{r_0}\right)^{3/2}\left(\frac{Zr}{r_0}\right)e^{-Zr/2r_0}\sin\theta e^{\pm i\phi}$$

$$U_{300} = \frac{1}{81\sqrt{3\pi}}\left(\frac{Z}{r_0}\right)^{3/2}\left(27 - 18\frac{Zr}{r_0} + 2\frac{Z^2r^2}{r_0^2}\right)e^{-Zr/3r_0} \qquad \frac{1}{\langle 1/r\rangle} = \frac{9r_0}{Z}$$

$$U_{310} = \frac{\sqrt{2}}{81\sqrt{\pi}}\left(\frac{Z}{r_0}\right)^{3/2}\left(6 - \frac{Zr}{r_0}\right)\left(\frac{Zr}{r_0}\right)e^{-Zr/3r_0}\cos\theta$$

$$U_{31\pm1} = \frac{1}{81\sqrt{\pi}}\left(\frac{Z}{r_0}\right)^{3/2}\left(6 - \frac{Zr}{r_0}\right)\left(\frac{Zr}{r_0}\right)e^{-Zr/3r_0}\sin\theta e^{\pm i\phi}$$

$$U_{320} = \frac{1}{81\sqrt{6\pi}}\left(\frac{Z}{r_0}\right)^{3/2}\left(\frac{Z^2r^2}{r_0^2}\right)e^{-Zr/3r_0}\left(3\cos^2\theta - 1\right)$$

$$U_{32\pm1} = \frac{1}{81\sqrt{\pi}}\left(\frac{Z}{r_0}\right)^{3/2}\left(\frac{Z^2r^2}{r_0^2}\right)e^{-Zr/3r_0}\sin\theta\cos\theta e^{\pm i\phi}$$

$$U_{32\pm2} = \frac{1}{162\sqrt{\pi}}\left(\frac{Z}{r_0}\right)^{3/2}\left(\frac{Z^2r^2}{r_0^2}\right)e^{-Zr/3r_0}\sin^2\theta e^{\pm i2\phi}$$

For $n = 1$, both ℓ and m are equal to zero and there is but one eigenfunction and no degeneracy. For $n = 2$ there are two types of solutions: one is $\ell = 1$ with an accompanying triplet of state values of m: $m = -1$, 0, $+1$. The other is the singlet $\ell = 0$ with the accompanying singlet state $m = 0$. Since energy depends only upon n, and since for each value of ℓ there are $(2\ell + 1)$ values of m, the result is $(2\ell + 1)$ fold energy degeneracy. For each value of n there are $n-1$ values of ℓ. Hence the total energy degeneracy is:

$$\sum_{\ell=1}^{n-1}(2\ell+1)=n^2 \tag{6.6.4}$$

There are n^2 possible solutions for each value of energy.

The degeneracy is lifted if the electron system is immersed in a static electric or magnetic field: the results being, respectively, the Stark or Zeeman effect. The $m=\pm1$ states extend further outward from the nucleus than the more tightly bound $m=0$ states, and hence a static electric field affects different states differently. The $m=0$ state supports neither angular momentum nor magnetic moment and the $m=\pm1$ states support both angular momentum and magnetic moment. Both Stark and Zeeman effects remove the degeneracy of affected states.

6.7 Perturbation Analysis

Consider an atom immersed in an applied force field with a magnitude much less than that of the trapping potential; the wave functions retain their original character and solutions entail a re-scrambling of the occupied states. Consider, as an important example, the special case of an atom immersed in a static applied electric field that is small compared with the Coulomb field.

Let the Hamiltonian operator H_0 characterize the energy of the isolated system. The total eigenfunction is a sum over wave functions that are solutions of the Schrödinger equation with operator H_0. The resulting eigenfunctions are U_{n0} and energies W_{n0} are known. Each satisfies the relationship:

$$\int\left(U_{n0}^*H_0U_{n0}\right)dV=W_{n0}\int U_{n0}^*U_{n0}dV \tag{6.7.1}$$

The applied field modifies the Hamiltonian to the operational form:

$$H=H_0+H_1 \tag{6.7.2}$$

Since the external field is controllable by external means, for example the intensity of an applied laser beam, the actual operational form may be written as:

$$H = H_0 + \alpha H_1 \tag{6.7.3}$$

An experimenter may vary the value of α from zero to one, and, with it, change both eigenfunctions and energies. If the applied force is small enough a power series in powers of α converges, giving:

$$U_n = U_{n0} + \alpha U_{n1} + \alpha^2 U_{n2} + \ldots = \sum_r \alpha^r U_{nr}$$

$$W_n = W_{n0} + \alpha W_{n1} + \alpha^2 W_{n2} + \ldots = \sum_s \alpha^r W_{nr} \tag{6.7.4}$$

The "0" subscripts form the solution in the absence of the external field, the "1" subscripts describe the first order correction, the "2" subscripts describe the second order correction, *etc.* For small fields, only correction terms of first order in α are large enough to be of interest and terms proportional to α^2 are ignored. The first order terms are:

$$\int \left\{ \left(U_{n0}^* + \alpha U_{n1}^* \right) \left(H_0 + \alpha H_1 \right) \left(U_{n0} + \alpha U_{n1} \right) \right\} dV$$

$$= \int \left\{ \left(U_{n0}^* + \alpha U_{n1}^* \right) \left(W_{n0} + \alpha W_{n1} \right) \left(U_{n0} + \alpha U_{n1} \right) \right\} dV$$

$$\cong \int \left\{ U_{n0}^* H_0 U_{n0} + \alpha \left[U_{n1}^* H_0 U_{n0} + U_{n0}^* H_1 U_{n0} + U_{n0}^* H_0 U_{n1} \right] \right\} dV$$

$$= \int \left\{ W_{n0} U_{n0}^* U_{n0} + \alpha \left[W_{n0} U_{n1}^* U_{n0} + W_{n1} U_{n0}^* U_{n0} + W_{n0} U_{n0}^* U_{n1} \right] \right\} dV \tag{6.7.5}$$

In the last equality, the first terms in Eq. 6.7.1 are separately equal and may be subtracted from the equation; so are the first terms within the square brackets. Applying the Hermitian operator property, Eq. 5.5.6, to the last terms shows they too are equal. Eliminating these three term pairs leaves the equality:

$$\int U_{n0}^* H_1 U_{n0} dV = W_{n1} \int U_{n0}^* U_{n0} dV = W_{n1} \tag{6.7.6}$$

This is the first order correction term. It shows first order energy changes can be calculated without knowing the corrected wave function; it is only necessary to know the first order correction to the Hamiltonian operator. The result is the basis for predicting many eigenstate energy transitions and hence is significant.

6.8 Non-Ionizing Transitions

If Hamiltonian operator H_0, eigenfunctions ψ_n, and energies W_n apply to an electron in an unperturbed atom, for every value of n the Schrödinger equation is:

$$H_0 \psi_n (r,t) = W_n \psi_n (r,t) \tag{6.8.1}$$

The total wave function is a weighted sum over all possible wave functions:

$$\Psi(r,t) = \sum_n a_n \psi_n (r,t) \tag{6.8.2}$$

If a second electron is attached to the same atom it will have the same Hamiltonian operator, the same energies, and the same set of eigenfunctions but a different set of multiplying coefficients:

$$\Psi(r,t) = \sum_n b_n \psi_n (r,t) \tag{6.8.3}$$

Assign subscript '1' and '2' to the different electrons and make the following definitions:

$$\psi_{mn} (r,t) = \psi_m (r_1,t) \psi_n (r_2,t)$$

$$U_{mn} (r) = u_m (r_1) u_n (r_2)$$

$$W_{mn} = W_m + W_n$$

The equation that describes both electrons is:

$$\Psi(r,t) = \sum_m \sum_n a_m b_n U_{mn}(r_1,r_2)\exp\left(\frac{iW_{mn}t}{\hbar}\right) = \sum_m \sum_n c_{mn}\psi_{mn}(r,t) \quad (6.8.4)$$

After applying a perturbing field, the Hamiltonian operator changes from H_0 to $H_0 + H_1$ but by Eq. 6.7.6 the wave functions are unchanged. However, the probability coefficients, c_{mn}, become time dependent. To show this, write Eq. 6.8.4 in the form:

$$(H_0 + H_1)\sum_m \sum_n c_{mn}(t)\psi_{mn}(r,t) = \frac{\hbar}{i}\frac{\partial}{\partial t}\sum_m \sum_n c_{mn}(t)\psi_{mn}(r,t) \quad (6.8.5)$$

Writing out the equation term by term gives:

$$\sum_m \sum_n \left[c_{mn}(t)H_0\psi_{mn} + c_{mn}(t)H_1\psi_{mn}\right]$$
$$= \sum_m \sum_n \left[\frac{\hbar}{i}c_{mn}(t)\frac{\partial\psi_{mn}(t)}{\partial t} + \frac{\hbar}{i}\frac{\partial c_{mn}(t)}{\partial t}\psi_{mn}\right] \quad (6.8.6)$$

Combining Eq. 6.7.6 and 6.8.6 gives the equality:

$$\sum_m \sum_n c_{mn}(t)H_1\psi_{mn}(r,t) = \frac{\hbar}{i}\sum_m \sum_n \frac{\partial c_{mn}(t)}{\partial t}\psi_{mn}(r,t) \quad (6.8.7)$$

Multiplying through by $\psi^*_{pq}(r,t)$ and integrating over all space gives:

$$\frac{d}{dt}c_{pq}(t) = \frac{i}{\hbar}\sum_m \sum_n c_{mn}(t)\int \psi^*_{pq}H_1\psi_{mn}dV \quad (6.8.8)$$

The integral is over the volume occupied by both electrons. Make the definition:

$$\langle pq|H_1|mn\rangle = \int U^*_{pq}H_1 U_{mn}dV \quad (6.8.9)$$

The terms of Eq. 6.8.9 are defined to be the matrix elements of interaction energy H_1. With the aid of Eq. 5.4.13, Eq. 6.8.5 may be written as:

$$\frac{dc_{pq}(t)}{dt} = \frac{i}{\hbar} \sum_m \sum_n c_{mn} \langle pq|H_1|mn \rangle \exp\left[\frac{i\left(W_m + W_n - W_p - W_q\right)t}{\hbar}\right] \qquad (6.8.10)$$

A significant exercise is to let the electrons be in states m and n at time $t = 0$ and find the probability as a function of time they will occupy states p and q. At time $t = 0$ coefficient c_{mn} is equal to one and all others are zero:

$$\frac{dc_{pq}(t)}{dt} = \frac{i}{\hbar} \langle pq|H_1|mn \rangle \exp\left[\frac{i\left(W_p + W_q - W_m - W_n\right)t}{\hbar}\right] \qquad (6.8.11)$$

Make the definition that:

$$\Delta W = W_p + W_q - W_m - W_n$$

While restricting the analysis to times short enough so c_{mn} remains nearly equal to one, the time integral shows the initial time dependence of the coefficient is:

$$c_{pq}(t) = \langle pq|H_1|mn \rangle \frac{\exp(i\Delta W t / \hbar) - 1}{\Delta W} \qquad (6.8.12)$$

The probability of state (p,q) being occupied is:

$$c_{pq}^* c_{pq} = 4\langle pq|H_1|mn \rangle^2 \frac{\sin^2\left(\Delta W t / 2\hbar\right)}{\left(\Delta W\right)^2} \qquad (6.8.13)$$

This shows the probability a particular transition will occur is proportional to the square of the matrix element. The magnitude of the matrix element depends upon both sets of quantum numbers, *pq* and *mn*. Transitions are "forbidden" if the matrix element is equal to zero. Since the ratio $\sin^2(x)/x^2$ has maximum magnitude at $x = 0$, it follows from energy conservation the most probable value of ΔW is zero.

6.9 Absorption and Emission of Radiation

The purpose of this section is to describe the absorption and emission of radiation by an atom with a full compliment of electrons. The atom is immersed in an externally applied plane wave of radian frequency ω and the wavelength is much larger than the atom.

The calculation begins with the relationships between the Hamiltonian, H, and the time dependence of the applied field. The relationships are, see Eq. 5.4.11:

$$\frac{\hbar}{i}\frac{\partial}{\partial t}\psi_n(r,t) = H\psi_n(r,t) = W_n\psi_n(r,t) \tag{6.9.1}$$

By Eq. 5.4.13 and 5.4.14 the time-dependent eigenfunctions are related to the time-independent ones as:

$$\psi_n(r,t) = U_n(r)e^{iW_n t/\hbar} \tag{6.9.2}$$

The multiplying coefficients are time dependent and the complete wave function $\Psi(r,t)$ is a weighted sum over all possible eigenfunctions:

$$\Psi(r,t) = \sum_n c_n(t)U_n(r)e^{i\omega_n t}; \quad \omega_n = W_n/\hbar \tag{6.9.3}$$

It follows from first order perturbation equation, Eq. 6.7.6, and Eq. 6.9.1 to 6.9.3 that the rate of change of coefficients as a function of the coefficients themselves is:

$$\frac{\hbar}{i}\sum_n \frac{\partial c_n(t)}{\partial t}U_n(r)e^{i\omega_n t} = \sum_n c_n(t)H_1 U_n(r)e^{i\omega_n t} \tag{6.9.4}$$

As before, symbol H_1 represents the modification to the Hamiltonian due to the perturbation. This equation shows the affect of the applied field is to make the state coefficients time dependent.

The electric field intensity in the perturbing plane wave is $E(r,t)$ and it varies with time as $\cos(\omega t)$. It is convenient to rewrite it in complex terms as the sum of complex conjugate functions:

$$E(r,t) = E_0 e^{i(\omega t - k \cdot r)} + c.c. \tag{6.9.5}$$

For atoms of diameter much less than a wavelength, the perturbing energy is approximately equal to:

$$H_1 = -eE(t) \cdot r(t) \tag{6.9.6}$$

Combining Eq. 6.9.6 with 6.9.4 gives:

$$\sum_n \frac{\partial c_n(t)}{\partial t} e^{i\omega_n t} U_n(r)$$

$$= -\frac{ie}{\hbar} \sum_n \left\{ e^{i(\omega_n t + \omega t - k \cdot r)} + e^{i(\omega_n t - \omega t + k \cdot r)} \right\} c_n(t) E_0 \cdot r U_n(r) \tag{6.9.7}$$

To determine the effect of interaction between states n and k, multiply Eq. 6.9.7 through by $U_k(r)$ and integrate over all space. The result is:

$$\frac{\partial c_k(t)}{\partial t} = -\frac{ie}{\hbar} \sum_n c_n(t) \left\{ e^{ik \cdot r i(\omega_n - \omega_k - \omega)t} + e^{-ik \cdot r} e^{i(\omega_n - \omega_k + \omega)t} \right\} E_0 \cdot \langle U_k | r | U_n \rangle$$

$$\tag{6.9.8}$$

A boundary condition is that at initial time $t = 0$ only the single state 'n' is occupied and, therefore, $c_n = 1$ and all other coefficients equal zero. With this condition the time integral of Eq. 6.9.8 is:

$$c_k(t) = -\frac{ie}{\hbar} \left\{ e^{ik \cdot r} \left[\frac{e^{i(\omega_n - \omega_k - \omega)t} - 1}{(\omega_n - \omega_k - \omega)} \right] + e^{-ik \cdot r} \left[\frac{e^{i(\omega_n - \omega_k + \omega)t} - 1}{(\omega_n - \omega_k + \omega)} \right] \right\} E_0 \cdot \langle U_k | r | U_n \rangle$$

$$\tag{6.9.9}$$

Expanding the exponential gives $[e^{-ik \cdot r} \approx 1 - ik \cdot r + ...]$. Since atomic sizes are much less than a wavelength only the initial term is retained and the square of the magnitude is:

$$\left|c_k(t)\right|^2 = \left(2eE_0\right)^2 \langle U_k|\mathbf{r}|U_n\rangle^2 \left\{ \frac{\sin^2\left[\dfrac{1}{2\hbar}\left(W_n - W_k \pm \hbar\omega\right)t\right]}{\left(W_n - W_k \pm \hbar\omega\right)^2} \right\} \tag{6.9.10}$$

The term within the curly brackets of Eq. 6.9.10 is significantly different from zero only if the argument of the sine vanishes, or:

$$\omega_k = \omega_n \pm \omega \tag{6.9.11}$$

Multiplying Eq. 6.9.11 through by $\hbar$ shows that $(W_k - W_n) = \pm\,\hbar\omega$. If $(W_k > W_n)$, energy $\hbar\omega$ is added to the system and the transition is associated with energy absorption. If $(W_k < W_n)$, energy $\hbar\omega$ is removed from the system and the transition is associated with energy emission. Therefore, it seems reasonable to ascribe the upper or lower sign of Eq. 6.9.11 respectively to an electron's energy absorption or emission; the transition probabilities for emission and absorption are the same. The equation is valid only over times so small that $c_n(t)$ remains nearly equal to one and all other $c_k(t)$ remain much less than one and hence only during the initial changes of the electron from its equilibrium state.

If the radius of the atom equals the first Bohr orbit and if the wavelength is 500 nm, the center of the optical band:

$$2kr = 6.65 \times 10^{-4}\left(n^2/Z\right) \tag{6.9.12}$$

If n is small enough so the magnitude is much less than one, the perturbation expansion of Eq. 6.9.4 converges rapidly and calculated results are limited to the first correction term only.

Although results calculated in this section are correct, a significant question remains. Perturbation theory is applicable only to small changes from equilibrated values and quantum jumps can hardly be viewed as a small change. Still, pragmatically, the equations correctly predict the probability of electronic transitions with an associated creation or annihilation of a photon. We return to the subject in Chap. 9.

6.10 Dipole Radiation Selection Rules

The probability of an energy exchange between two states is, by Eq. 6.9.10, proportional to the square of the joint matrix element. For a single-electron atom the wave functions are given by Eq. 6.6.1 and the components of the matrix elements directed along the three rectangular coordinate axes are:

$$X = \int_0^\infty r^2 dr \int_0^\pi \sin\theta d\theta \int_0^{2\pi} d\phi\, r\sin\theta\cos\phi\, U^*_{n'\ell'm'} U_{n\ell m}$$

$$Y = \int_0^\infty r^2 dr \int_0^\pi \sin\theta d\theta \int_0^{2\pi} d\phi\, r\sin\theta\sin\phi\, U^*_{n'\ell'm'} U_{n\ell m} \tag{6.10.1}$$

$$Z = \int_0^\infty r^2 dr \int_0^\pi \sin\theta d\theta \int_0^{2\pi} d\phi\, r\cos\theta\, U^*_{n'\ell'm'} U_{n\ell m}$$

The onset of a field-induced transition between eigenfunctions is proportional to the corresponding integral of Eq. 6.10.1. To calculate onset probabilities note from Eq. 6.6.1 the wave function is expressible as:

$$U_{n\ell m}(r,\theta,\phi) = A_{n\ell m} R_n^\ell(\alpha_n r) P_\ell^m(\cos\theta) e^{-im\phi} \tag{6.10.2}$$

Inserting Eq. 6.10.2 into Eq. 6.10.1 and evaluating the integral for all possible values of ℓ,m and ℓ',m' gives the initial transition probabilities.

As an example, the angular portion of integral X is:

$$X_{ang} = \int_0^\pi P_{\ell'}^{m'} P_\ell^m \sin^2\theta d\theta \int_0^{2\pi} e^{i(m'-m)\phi} \cos\phi d\phi \tag{6.10.3}$$

To evaluate the azimuth integral we note:

$$\int_0^{2\pi} d\phi \cos\phi \left[\cos(m'-m)\phi + i\sin(m'-m)\phi \right] = \pi\delta(m', m\pm 1) \tag{6.10.4}$$

To evaluate the zenith integral note from the identities of Table A.15.1:

$$\sin^2\theta\, P_{\ell'}^{m+1} P_{\ell}^{m} = \frac{1}{2\ell+1}\sin\theta\left(P_{\ell+1}^{m+1} - P_{\ell-1}^{m+1}\right)P_{\ell'}^{m+1}$$

$$\sin^2\theta\, P_{\ell'}^{m-1} P_{\ell}^{m} = \frac{1}{2\ell'+1}\sin\theta\left(P_{\ell'+1}^{m} - P_{\ell'-1}^{m}\right)P_{\ell}^{m}$$

(6.10.5)

Combining Eq. 6.10.3 and 6.10.5 gives:

$$X_{ang} = \int_0^{\pi}\left[\begin{array}{l}\dfrac{\pi}{2\ell+1}\left(P_{\ell+1}^{m+1}P_{\ell'}^{m+1} - P_{\ell-1}^{m+1}P_{\ell'}^{m+1}\right)\delta(m',m+1) \\[2ex] +\dfrac{\pi}{2\ell'+1}\sin\theta\left(P_{\ell'+1}^{m}P_{\ell}^{m} - P_{\ell'-1}^{m}P_{\ell}^{m}\right)\delta(m',m-1)\end{array}\right]\sin\theta d\theta \qquad (6.10.6)$$

Integrating and combining gives:

$$X_{ang} = \frac{2\pi}{(2\ell+1)}\left\{\begin{array}{l}\dfrac{1}{(2\ell+3)}\dfrac{(\ell+m+2)!}{(\ell-m)!}\delta(\ell',\ell+1)\delta(m',m+1) \\[2ex] -\dfrac{1}{(2\ell+3)}\dfrac{(\ell+m)!}{(\ell-m)!}\delta(\ell',\ell+1)\delta(m',m-1) \\[2ex] +\dfrac{1}{(2\ell-1)}\dfrac{(\ell+m)!}{(\ell-m)!}\delta(\ell',\ell+1)\delta(m',m-1) \\[2ex] -\dfrac{1}{(2\ell-1)}\dfrac{(\ell+m)!}{(\ell-m-2)!}\delta(\ell',\ell-1)\delta(m',m+1)\end{array}\right\} \qquad (6.10.7)$$

Function X_{ang} is different from zero only if:

$$m' = m \pm 1 \text{ and } \ell' = \ell \pm 1 \qquad (6.10.8)$$

The corresponding integral over Y_{ang} gives the same result.
   The angular portion of Z is:

$$Z_{ang} = \int_0^{\pi} P_{\ell'}^{m'} P_{\ell}^{m}\sin\theta\cos\theta d\theta \int_0^{2\pi} e^{i(m'-m)\phi}d\phi \qquad (6.10.9)$$

The integral of the azimuth angle is:

$$\int_0^{2\pi} e^{i(m'-m)\phi}\, d\phi = 2\pi\delta(m',m) \tag{6.10.10}$$

Table A.15.1 contains the identity:

$$\cos\theta P_\ell^m = \frac{1}{2\ell+1}\left[(\ell-m+1)P_{\ell+1}^m - (\ell+m)P_{\ell-1}^m\right] \tag{6.10.11}$$

Substituting Eq. 6.10.11 into Eq. 6.10.9 and integrating gives:

$$\frac{1}{(2\ell+1)}\int_0^\pi \sin\theta\left[(\ell-m+1)P_{\ell+1}^m - (\ell+m)P_{\ell-1}^m\right]P_{\ell'}^m\, d\theta =$$

$$= \frac{2}{(2\ell+1)}\left\{\frac{(\ell-m+1)}{(2\ell+3)}\frac{(\ell+m+1)!}{(\ell-m+1)!}\delta(\ell',\ell+1) - \frac{(\ell+m)}{(2\ell+3)}\frac{(\ell+m-1)!}{(\ell-m-1)!}\delta(\ell',\ell-1)\right\} \tag{6.10.12}$$

Combining Eq. 6.10.10 and 6.10.12 gives:

$$Z_{ang} = \frac{4\pi}{(2\ell+1)}\delta(m',m)\left\{\begin{array}{l}\dfrac{(\ell-m+1)}{(2\ell+3)}\dfrac{(\ell+m+1)!}{(\ell-m+1)!}\delta(\ell',\ell+1)\\[2ex] -\dfrac{(\ell+m)}{(2\ell+3)}\dfrac{(\ell+m-1)!}{(\ell-m-1)!}\delta(\ell',\ell-1)\end{array}\right\} \tag{6.10.13}$$

Function Z_{ang} is different from zero if:

$$m' = m \text{ and } \ell' = \ell \pm 1 \tag{6.10.14}$$

Since the radial function places no restrictions on n, a nonzero probability of dipolar radiation exists if:

$$\Delta\ell = \pm 1 \text{ and } \Delta m = \pm 1 \text{ or } 0 \tag{6.10.15}$$

The interpretation of Eq. 6.10.8 and 6.10.14 is that the z-component of radiated angular momentum is equal to zero or $\pm\hbar$. Since the angular momentum of the source changes by the amount shown by Eq. 6.10.15, by Newton's law the radiation supports an equal, oppositely directed angular momentum.

6.11 Many-Electron Systems

Although the full Hamiltonian of a system of n electrons may depend in a complicated way on the both locations and internal structure of individual electrons, independently of such complications is the critical property that electrons are indistinguishable. An interchange of two electrons leaves all results invariant and hence all energies are symmetrical with respect to electronic interchanges:

$$\psi^{Sy}(r,t) = \psi^{Sy}(-r,t) \quad \text{and} \quad \psi^{As}(r,t) = -\psi^{As}(-r,t) \tag{6.11.1}$$

In general, any physically real function of time can be expressed as a wave function that is a superposition of symmetric and antisymmetric parts:

$$\psi(r,t) = \psi^{Sy}(r,t) + \psi^{As}(r,t) \tag{6.11.2}$$

The probability density due to this particular wave function is:

$$\psi(r,t)^{*}\psi(r,t) = \left[\psi^{Sy*}(r,t)\psi^{Sy}(r,t) + \psi^{As*}(r,t)\psi^{As}(r,t)\right]$$
$$+ \left[\psi^{Sy*}(r,t)\psi^{As}(r,t) + \psi^{As*}(r,t)\psi^{Sy}(r,t)\right] \tag{6.11.3}$$

The first line of Eq. 6.11.3 is invariant with respect to electron interchange but the second one is not. Since the second line is zero only if either the symmetric or antisymmetric part is zero we conclude wave functions may be either symmetric or antisymmetric with respect to electron interchange but not a mixture thereof.

The simplest multi-electron system has two electrons. We designate them as electron "a" and electron "b" and define P_{ab} as a permutation operator that interchanges the electrons. It follows that:

$$P_{ab}\psi(a,b)=\psi(b,a) \text{ and } P_{ab}P_{ab}\psi(a,b)=\psi(a,b) \tag{6.11.4}$$

It follows from Eq. 6.11.4 that:

$$P_{ab}=\pm 1 \text{ and hence } P_{ab}P_{ab}=1 \tag{6.11.5}$$

It follows that

$$P_{ab}\psi^{As}(r,t)=-\psi^{As}(r,t);\ \ P_{ab}\psi^{Sy}(r,t)=\psi^{Sy}(r,t) \tag{6.11.6}$$

Taking operations with respect to time has no effect upon positional symmetry. A state that is initially symmetric or antisymmetric before a quantum mechanical operation has the same symmetry afterward and hence the initial symmetry is retained for all time. The argument generalizes to include an arbitrary number of electrons and the conclusion is that wave functions are either symmetric or antisymmetric. The conclusion is either one type is always zero or there are two separate realities: one has symmetric wave functions and the other has antisymmetric wave functions.

The difference between the two symmetrics follows by considering a two-electron atom. Both total and individual wave functions obey quantum theory relationships and r' and r'' represent separate spatial locations:

$$\psi(r',r'')=\psi_a(r')\psi_b(r'')=\frac{1}{2}\Big[\psi_a(r')\psi_b(r'')+\psi_b(r')\psi_a(r'')\Big]$$
$$+\frac{1}{2}\Big[\psi_a(r')\psi_b(r'')-\psi_b(r')\psi_a(r'')\Big] \tag{6.11.7}$$

The first and second terms on the right side are, respectively, symmetric and antisymmetric. Introduce separate notion for the two parts:

$$\Gamma_{ab}(r',r'')=\Big[\psi_a(r')\psi_b(r'')+\psi_b(r')\psi_a(r'')\Big]$$
$$\Lambda_{ab}(r',r'')=\Big[\psi_a(r')\psi_b(r'')-\psi_b(r')\psi_a(r'')\Big] \tag{6.11.8}$$

Both electrons occupy all points, and we examine conditions at $r' = r''$:

$$\Gamma_{ab}(r',r') = 2\psi_a(r')\psi_b(r')$$
$$\Lambda_{ab}(r',r'') = 0$$

(6.11.9)

The electrostatic interaction energy between the two electrons is:

$$W_{ab} = \frac{e^2}{4\pi\varepsilon}\int\left(\frac{1}{|r'-r''|}\right)\psi_{ab}(r',r'')\psi_{ab}^{*}(r',r'')dV$$

(6.11.10)

If the electron charge density is a constant function of position, Eq. 6.11.10 gives a physically acceptable result with either electron symmetry although, as expected from Eq. 6.11.9, the energy with symmetric functions is larger than with antisymmetric functions. However, the definite integral of Eq. 6.11.10 correctly represents the system energy if and only if the wave functions are continuous functions of position. If there is a dimensional scale below which the charge density is granular, on that scale of dimensions it is necessary to replace the integration by a sum over interaction energies. With Eq. 6.11.9 granular charges are adjacent or overlapping and the sum is singular; there is no parallel with antisymmetric wave functions since the overlapping densities vanish. It follows that if there is a dimensional scale below which the charge density is granular only antisymmetric wave functions exist. The Exclusion Principle, first formulated by Pauli, states that only antisymmetric wave functions exist in nature. Therefore, it also suggests that, on an appropriate dimensional scale, at least a portion of an electron charge distribution is granular.

If the same Hamiltonian affects two electrons they, in turn, are described by the antisymmetric wave function of Eq. 6.11.6. By definition, such coupled electrons are entangled and each contributes to wave function $\Psi(r',r'')$ over the full range of variables. The total wave function remains valid even if the electrons are later separated an arbitrary distance, in a way that does not destroy entanglement. Upon the breakup of entanglement the two electrons become independent entities. Such a transition is not an equilibrium event and hence not subject to analysis on the basis of the Schrödinger equation.

References

G. Auletta, *Foundations and Interpretation of Quantum Mechanics*, World Scientific (2001)

M. Born, "Obituary, Max Karl Ernst Ludwig Planck," *Obituary Notices of Fellows of the Royal Society*, vol. 6, (1948) pp. 161–180, also in *The World of the Atom*, H. A. Boorse and L. Motz, eds., Basic Books (1966) p. 462–484

A.H. Compton, *A Quantum Theory of the Scattering of X-Rays by Light Elements*, Pys. Rev., vol. 21, pp. 483–522 (1923)

A. Einstein, "The Quantum Theory of Radiation," *Phys. Z.*, vol. 18, 121–128 (1917); also in *The World of the Atom*, H. A. Boorse and L. Motz, eds., Basic Books (1966) p. 888–901

H. Hertz, *Electric Waves: Researches on the Propagation of Electric Action with Finite Velocity Through Space* (1893) Translated by D. E. Jones, Dover Publications (1962)

K. Huang, *Statistical Mechanics*, John Wiley & Sons, New York, (1967)

G. Mie, "A Contribution to Optical Extinction by Metallic Colloidal Suspensions," *Ann. Physik.* vol. 25, p. 377 (1908)

W.K.H. Panofsky, M. Phillips, *Classical Electricity and Magnetism*, 2nd ed., Addison-Wesley (1962)

M. Planck, "The Origin and Development of the Quantum Theory," Nobel Prize in Physics Address, 1919, in *The World of the Atom*, H. A. Boorse and L. Motz, eds., Basic Books (1966) pp. 491–501

M. Planck, *The Theory of Heat Radiation*, trans. By M. Masius, Dover Publications (1959)

R.C. Tolman, *The Principles of Statistical Mechanics* (1938), reprinted by Dover Publications (1979)

Chapter 7

Matched Multipolar Sources

$\mathcal{R}$adiation properties of matched pairs of multipolar sources are quite different from those of individual modes. This rather obscure fact played a significant role in the early history of the quantum theory of radiation. In his historic 1917 paper Einstein stated that Maxwell's equations show the radiation has spherical (sic) symmetry and hence classical results are unacceptable. Harrington showed, 43 years later and in quite a different context, that properly matched multipoles produce partially directed radiation. Although Einstein's logic was impeccable, the then accepted wisdom upon which he based his conclusion was not. In this chapter we detail a continuum radiation field with the kinematic photon properties Einstein thought unobtainable.

In Sec. 7.1 we detail the radiation Q of an electric dipole and conclude the Q of optical frequency radiation emitted from an atom-sized region is about 10^9. In Sec. 7.2 we include reactive power in the calculation of radiation reaction force and find it is surely large enough to squelch all dipolar radiation. In Sec. 7.3 we discuss the stresses inherent to modal radiation fields. In Sec. 7.4 we replace the spherical Bessel functions in the multipolar expansion for the z-directed plane wave used in Chap. 2,3, and 4 by spherical Hankel functions; the new expansion describes a field emitted from an enclosed source that consists of matched modal pairs. Each pair consists of superimposed electric and magnetic moments of the same order; the moments are spatially orthogonal, lie in the basal plane, and the plane rotates about a normal axis at radian frequency ω. In Sec. 7.5 we determine exact field values at infinite range and in Sec. 7.6 we apply to that field the method of self-consistent fields, which provides an acceptable field form at lesser ranges, and from it we determine the field set is fully z-directed. In Sec. 7.7 we calculate the exchange rates of energy, linear momentum, and angular momentum and note their ratios are the same as for photons. In Sec. 7.8 we present computer-based far-field summations that confirm the far field is fully directed. In Sec. 7.9 we examine stresses in the field and show the energy exchanges satisfy Newton's second law. In Sec. 7.10 we examine near-field energy exchanges and in Sec. 7.11 we calculate the power output when summed over all orders. We find the field set creates a zero-Q porthole on the positive z-axis through which the full-field radiation can pass with no accompanying reactive power!

7.1 Radiating Electric Dipole

A dipole is the single moment with the smallest Q and therefore we detail its properties. To avoid situation-specific details we circumscribe the smallest possible encompassing virtual sphere of radius a about the dipole. Although the generated field occupies all available space both inside and outside the virtual sphere we replace the actual source with the equivalent sources it creates on that surface. The process leaves exterior fields unchanged and ignores interior fields. Since interior energy is not included calculated Qs are least possible values.

The exterior fields created by a z-directed electric dipole have rotational symmetry about the z-axis and, in phasor form, are:

$$\tilde{\mathbf{E}}_\ell = \left\{ \frac{2}{\sigma} h_1(\sigma)\cos\theta\hat{r} - h_\ell^\bullet(\sigma)\sin\theta\hat{\theta} \right\} e^{i\omega t} \tag{7.1.1}$$

$$\eta\tilde{\mathbf{H}}_\ell = ih_\ell(\sigma)\sin\theta\hat{\phi}e^{i\omega t}$$

With $t_R = t - R/c$ where R is the source-to-field point distance and R/c is the transit time for the radiation, the corresponding actual fields are:

$$\mathbf{E} = -\left[\frac{2}{\sigma^3}\sin(\omega t_R) + \frac{2}{\sigma^2}\cos(\omega t_R) \right]\cos\theta\hat{r}$$

$$+ \left[\frac{1}{\sigma}\left(1 - \frac{1}{\sigma^2}\right)\sin(\omega t_R) - \frac{1}{\sigma^2}\cos(\omega t_R) \right]\sin\theta\hat{\theta} \tag{7.1.2}$$

$$\eta\mathbf{H} = \left[\frac{1}{\sigma}\sin(\omega t_R) - \frac{1}{\sigma^2}\cos(\omega t_R) \right]\sin\theta\hat{\phi}$$

The Poynting vector is:

$$N = \frac{\hat{r}}{2\eta} \left\{ \begin{array}{l} \dfrac{1}{\sigma^2}\left[1 - \cos(2\omega t_R)\right] \\[2mm] + \dfrac{2}{\sigma^4}\sin(2\omega t_R) - \left(\dfrac{2}{\sigma^3} - \dfrac{1}{\sigma^5}\right)\cos(2\omega t_R) \end{array} \right\} \sin^2\theta$$

$$- \frac{\hat{\theta}}{2\eta}\left\{ \frac{4}{\sigma^4}\cos(2\omega t_R) - \left(\frac{3}{\sigma^3} - \frac{2}{\sigma^5}\right)\sin(2\omega t_R) \right\}\sin\theta\cos\theta \qquad (7.1.3)$$

The inverse square term describes energy that flows permanently away from the region. All others describe energy that oscillates about fixed positions; at each point the time average of the oscillating energies forms a standing energy field. At small radii the oscillating energies are much larger than the radiated energy and the standing energy times the radian frequency forms the numerator of the Q expression.

The integral of Eq. 7.1.3 over the surface of any sphere with radius σ/k shows the time dependent surface power is:

$$P = \frac{4\pi}{3\eta k^2}\left\{ \left[1 - \cos(2\omega t_R)\right] + \frac{2}{\sigma^2}\sin(2\omega t_R) - \left(\frac{2}{\sigma} - \frac{1}{\sigma^3}\right)\cos(2\omega t_R) \right\} \qquad (7.1.4)$$

The center and last term are reactive power.

The first term is the output power. Typical of electromagnetic power it pulses between zero and twice the average value twice each field cycle. The time-average output power is the denominator of the Q expression. It is in phase with the last term and hence, quite differently from electric circuits, the real and reactive powers within radiation fields are not in time quadrature.

Directivity is defined as the maximum-to-average surface power density and, for the dipole, is 3/2. The radiation pattern forms a torus around the z-axis:

$$D = \lim_{\sigma\to\infty} \frac{4\pi\sigma^2}{k^2}\frac{\left[N_r\right]_{max}}{P_{av}} = \frac{3}{2}; \quad \text{Pattern} = \frac{3}{2}\sin^2\theta \qquad (7.1.5)$$

With subscript 'T' representing 'total', the total time-dependent energy density w_T at each point for field set Eq. 7.1.2 is:

$$w_T = \frac{\varepsilon}{4}\left\{ \begin{cases} \left[\frac{4}{\sigma^4}\left[1+\cos\left(2\omega t_R\right)\right] + \frac{8}{\sigma^5}\sin\left(2\omega t_R\right) \\ + \frac{4}{\sigma^6}\left[1-\cos\left(2\omega t_R\right)\right] \right] \cos^2\theta \\ + \begin{cases} \left(\frac{2}{\sigma^2}+\frac{1}{\sigma^6}\right)\left[1-\cos\left(2\omega t_R\right)\right] + \frac{4}{\sigma^4}\cos\left(2\omega t_R\right) \\ -\left(\frac{4}{\sigma^3}-\frac{2}{\sigma^5}\right)\sin\left(2\omega t_R\right) \end{cases} \sin^2\theta \end{cases} \right\} \tag{7.1.6}$$

The inverse square terms describe energy present even at limitlessly large distances from the source and, therefore, the energy is infinite for a steady state field that as been operational since time $t = -\infty$. That energy, however, is permanently separated from and independent of the source. The significant value of field energy is that which remains attached to the source; to evaluate it we apply the continuity equation to the energy field:

$$\nabla \cdot N + \frac{\partial W_T}{\partial t} = 0 \tag{7.1.7}$$

At each differential volume within the field the first term on the left side of Eq. 7.1.7 measures the change in outbound power density and the second term measures energy change within it: energy conservation requires the equality.

Next consider a differential volume that remains at a fixed position within the wave; that is, it travels outward with the wave at speed c. The change in energy density at retarded time is the rate at which energy changes within the traveling differential volume; the separated energy is the standing energy. Solving for it, the partial of the energy density with respect to retarded time gives the rate of change of standing energy density at each point, subscript 'S' indicates 'standing'.

$$\nabla \cdot N + \frac{\partial W_S}{\partial t_R} = 0 \tag{7.1.8}$$

Taking the retarded time integral shows the standing energy density:

$$w_S = \frac{\varepsilon}{4}\left\{ \begin{aligned} &\left\{ \frac{4}{\sigma^4}\left[1+\cos\left(2\omega t_R\right)\right]+\frac{8}{\sigma^5}\sin\left(2\omega t_R\right) \right\}\cos^2\theta \\ &\quad +\frac{4}{\sigma^6}\left[1-\cos\left(2\omega t_R\right)\right] \\ &+\frac{1}{\sigma^6}\left[1-\cos\left(2\omega t_R\right)\right]\sin^2\theta \end{aligned} \right\} \qquad (7.1.9)$$

Comparison shows the total and standing energy densities are the same on the dipolar axis but quite different in the basal plane.

Defining the traveling energy w_0 as the difference between total and standing energy densities of Eq. 7.1.9 and 7.1.6:

$$w_0 = \frac{\varepsilon}{2}\left\{ \begin{aligned} &\frac{1}{2\sigma^2}\left[1-\cos\left(2\omega t_R\right)\right]+\frac{2}{\sigma^4}\cos\left(2\omega t_R\right) \\ &-\left(\frac{2}{\sigma^3}-\frac{1}{\sigma^5}\right)\sin\left(2\omega t_R\right) \end{aligned} \right\}\sin^2\theta \qquad (7.1.10)$$

This expression is c times the radial Poynting vector.

Radiation Q

The last two terms of Eq. 7.1.10 describe oscillating energy streams that dynamically support the standing energy densities of Eq. 7.1.9, and they form the reactive input impedances at the dipolar driving terminals.

The total standing energy, W_S, equals the volume integral of Eq. 7.1.10 over all exterior space:

$$W_S = \frac{1}{k^3}\int_{ka}^{\infty}\sigma^2 d\sigma\int_0^{2\pi}d\phi\int_0^{\pi}w_S\sin\theta d\theta \qquad (7.1.11)$$

The integrated value is:

$$W_S = \frac{2\pi\varepsilon}{3k^2}\left\{ \left(\frac{1}{(ka)^3}+\frac{2}{(ka)}\right)-\left(\frac{1}{(ka)^3}-\frac{2}{(ka)}\right)\cos\left(2\omega t_R\right)+\frac{2}{(ka)^2}\sin\left(2\omega t_R\right) \right\}$$

$$(7.1.12)$$

The peak value is:

$$W_{Spk} = \frac{2\pi\varepsilon}{3k^3}\left\{\left(\frac{1}{(ka)^3}+\frac{2}{(ka)}\right)+\frac{1}{(ka)^3}\sqrt{1+4(ka)^4}\right\} \qquad (7.1.13)$$

Combining the above gives the radiation Q:

$$Q = \frac{1}{2(ka)^3}\left(1+\sqrt{1+4(ka)^2}\right)+\frac{1}{(ka)} \approx \frac{1}{(ka)^3}+\frac{1}{(ka)} \qquad (7.1.14)$$

Dynamic parameters are summarized in Table 7.1.1; interior fields are ignored.

Table 7.1.1. Dynamic properties of an electric dipole radiating at frequency ω.

$$w_T = \frac{\varepsilon}{4}\left\{\begin{array}{l}\left\{\frac{4}{\sigma^4}\left[1+\cos(2\omega t_R)\right]+\frac{8}{\sigma^5}\sin(2\omega t_R)+\frac{4}{\sigma^6}\left[1-\cos(2\omega t_R)\right]\right\}\cos^2\theta \\[2mm] +\left\{\left(\frac{2}{\sigma^2}+\frac{1}{\sigma^6}\right)\left[1-\cos(2\omega t_R)\right]+\frac{4}{\sigma^4}\cos(2\omega t_R)-\left(\frac{4}{\sigma^3}-\frac{2}{\sigma^5}\right)\sin(2\omega t_R)\right\}\sin^2\theta\end{array}\right\}$$

$$w_S = \frac{\varepsilon}{4}\left\{\begin{array}{l}\left\{\frac{4}{\sigma^4}\left[1+\cos(2\omega t_R)\right]+\frac{8}{\sigma^5}\sin(2\omega t_R)+\frac{4}{\sigma^6}\left[1-\cos(2\omega t_R)\right]\right\}\cos^2\theta \\[2mm] +\frac{1}{\sigma^6}\left[1-\cos(2\omega t_R)\right]\sin^2\theta\end{array}\right\}$$

$$w_0 = \frac{\varepsilon}{2}\left\{\frac{1}{\sigma^2}\left[1-\cos(2\omega t_R)\right]+\frac{2}{\sigma^4}\cos(2\omega t_R)-\left(\frac{2}{\sigma^3}-\frac{1}{\sigma^5}\right)\sin(2\omega t_R)\right\}\sin^2\theta$$

$$W_S = \frac{2\pi\varepsilon}{3k^3}\left\{\frac{1}{(ka)^3}\left[1-\cos(2\omega t_R)\right]+\frac{2}{(ka)^2}\sin(2\omega t_R)+\frac{2}{(ka)}\left[1+\cos(2\omega t_R)\right]\right\}$$

$$P = \frac{4\pi\varepsilon}{3\eta k^2}\left\{\left[1-\cos(2\omega t_R)\right]+\left(\frac{1}{(ka)^3}-\frac{2}{(ka)}\right)\sin(2\omega t_R)+\frac{2}{(ka)^2}\cos(2\omega t_R)\right\}$$

$$Q = \frac{1}{2(ka)^3}\left(1+\sqrt{1+4(ka)^4}\right)+\frac{1}{(ka)} \qquad\qquad Gain = \frac{3}{2}$$

7.2 Radiation Reaction Force

In this section we examine radiation reaction forces: the forces created on a radiating charge by the radiation it generates. For that purpose we first consider the fields produced by an accelerating point charge q as it spirals towards an attractive charge that remains fixed at the origin. The charge is located at distance r and its velocity v is much less than the speed of light. By Eq. 1.7.3 the acceleration fields for this case are:

$$E = \frac{\mu q}{4\pi r}\left[\hat{r}\times\left(\hat{r}\times\frac{\partial v}{\partial t}\right)\right]; \quad H = \frac{q}{4\pi c r}\left(\frac{\partial v}{\partial t}\times\hat{r}\right) \tag{7.2.1}$$

Although the acceleration is radially directed, to simplify the algebra we consider it to be $\pm z$-directed. A more realistic radial acceleration model complicates matters and adds nothing essential to the discussion.

With $\pm z$-directed acceleration the generated far-fields are:

$$E = \frac{mq\,\hat{\theta}}{4\pi r}\frac{\partial v}{\partial t}\sin\theta; \quad H = \frac{q\,\hat{\phi}}{4\pi c r}\frac{\partial v}{\partial t}\sin\theta \tag{7.2.2}$$

The radial Poynting vector is:

$$N_r = \frac{\mu}{c}\left[\frac{q}{4\pi R}\frac{\partial v}{\partial t}\right]^2\sin^2\theta \tag{7.2.3}$$

The radiated power is:

$$P = \frac{\mu q^2}{6c\pi}\left[\frac{\partial v}{\partial t}\right]^2 \tag{7.2.4}$$

Energy conservation requires equal time-average radiated and generated energies. This in turn requires a radiation reaction braking force, F_{RR} that satisfies the condition:

$$\int_0^\tau F_{RR}\bullet v\,dt + \frac{\mu q^2}{6\pi c}\int_0^\tau\left[\frac{\partial v}{\partial t}\right]^2 dt = 0 \tag{7.2.5}$$

Time τ is the period required for an integer number of rotations of the two charges. Doing the integral by parts leads to:

$$\int_0^\tau \left(\boldsymbol{F}_{RR} - \frac{\mu q^2}{6\pi c} \frac{\partial^2 \boldsymbol{v}}{\partial t^2} \right) \bullet \boldsymbol{v} dt = \frac{\mu q^2}{6\pi c} \frac{\partial \boldsymbol{v}}{\partial t} \bullet \boldsymbol{v} \qquad (7.2.6)$$

With rotational motion the right side vanishes and with the usual definition of dipole moment, $\boldsymbol{p}$, the result is:

$$\boldsymbol{F}_{RR}\Big|_{real} = \frac{\mu q^2}{6\pi c} \frac{\partial^2 \boldsymbol{v}}{\partial t^2} = \frac{\mu q}{6\pi c} \frac{\partial^3 \boldsymbol{p}}{\partial t^3} = \frac{\mu q \omega^3}{6\pi c} \boldsymbol{p} \qquad (7.2.7)$$

Equation 7.2.7 expresses the time-average radiation reaction force on the charge due to dipole-generated energy leaving the system. This radiation reaction is a braking force that acts to reduce the energy; it acts uniformly on the entire charge and hence does not affect its shape or size.

Results from Sec. 7.1 show that an electrically small radiating dipole of radius a supports a reactive power that is larger than the real power by a factor of about $1/(ka)^3$. An atomic radius is typically sized on the order of 0.1 nm. If we apply Eq. 7.2.7 to such an atom at a frequency in the mid-optical range the factor is on the order of $1/(ka)^3 \cong 10^9$. That is, the reactive radiation reaction force is 10^9 times larger than the braking force of Eq. 7.2.7. Multiplying Eq. 7.2.7 by $1/(ka)^3$ shows that the reactive radiation reaction force is:

$$\boldsymbol{F}_{RR}\Big|_{reactive} = \frac{q}{6\pi \varepsilon a^3} \boldsymbol{p} \qquad (7.2.8)$$

This is the time-average reactive radiation reaction force on the electron. Although the magnitude is on the order of the Coulomb attractive force, it is not a braking force and hence has no affect on the electron's energy. Instead it is normal to the direction of motion and acts to distort the charge.

Real and Reactive Radiation Reaction Forces

A more in-depth analysis of radiation reaction forces considers a sinusoidally time-varying voltage as a generalized force and current as a generalized flow; the force-flow product is power. All responses to the force vary at the same frequency but, generally speaking, differ in phase. Some of the flow is unidirectional and the remainder oscillates twice each field cycle. It is typical of oscillating sources that the return of an emitted flow has a major impact upon it. In spite of this generality the textbook account of radiation reaction force is based solely on the outgoing power. As we have seen, a radiating multipole is enmeshed within a standing energy field of its own making and the standing field produces a reactive radiation reaction force on the source that is $(1/ka)^3$ times larger than that due to outbound power. We show in what follows, for a multipolar radiator of order 'ℓ', the reactive-to-real radiation reaction force ratio is on the order of $(1/ka)^{2\ell+1}$.

Expanding the argument to consider the effect of an arbitrary multipolar order we re-state the TM fields and, without loss of generality, use rotational symmetry so $m = 0$. For this case the phasor field components of mode ℓ are:

$$\tilde{E}_r = \frac{F_\ell}{\sigma^2}\ell(\ell+1)(B_\ell + i\,A_\ell)P_\ell(\cos\theta)e^{i\omega t_R}$$

$$\tilde{E}_\theta = \frac{1}{\sigma}F_\ell(D_\ell + i\,C_\ell)\frac{d}{d\theta}P_\ell(\cos\theta)e^{i\omega t_R} \tag{7.2.9}$$

$$\tilde{H}_\phi = \frac{1}{\eta\sigma}F_\ell(A_\ell - i\,B_\ell)\frac{d}{d\theta}P_\ell(\cos\theta)e^{i\omega t_R}$$

As in Sec. 7.1 we construct the smallest virtual sphere of radius a that circumscribes the emitter and then replace the actual emitter with virtual surface sources. The complex power on that surface follows from Eq. 7.2.9 and the complex Poynting theorem:

$$P_c = \frac{2\pi}{\eta k^2}\sum_{\ell=1}^{\infty}F_\ell^2\frac{\ell(\ell+1)}{(2\ell+1)}(A_\ell + iB_\ell)(D_\ell + iC_\ell) \tag{7.2.10}$$

Next introduce generalized voltage V_ℓ and current I_ℓ as variables proportional, respectively, to E_θ and H_ϕ. In these terms with the aid of Eq. 7.2.10 the complex surface power per mode is:

$$P_{c\ell} = V_\ell I_\ell^* / 2 \tag{7.2.11}$$

The same power expression as a function of impedance, Z, and field values is:

$$P_{c\ell} = \frac{2\pi Z}{\eta^2 k^2} F_\ell^2 \frac{\ell(\ell+1)}{(2\ell+1)}(A_\ell + iB_\ell)(A_\ell - iB_\ell) \tag{7.2.12}$$

Combining Eq. 7.2.11 and 7.2.12 yields an expression for the generalized impedance:

$$Z_\ell = \frac{V_\ell}{I_\ell} = \eta \frac{(D_\ell + iC_\ell)}{(A_\ell - iB_\ell)} \tag{7.2.13}$$

Separating multipolar impedance Eq. 7.2.13 into real and imaginary parts gives $Z(\sigma) = R(\sigma) + iX(\sigma)$ where:

$$R_\ell(\sigma) = \eta \left[\frac{A_\ell(\sigma)D_\ell(\sigma) - B_\ell(\sigma)C_\ell(\sigma)}{A_\ell(\sigma)^2 + B_\ell(\sigma)^2} \right]$$
$$X_\ell(\sigma) = \eta \left[\frac{A_\ell(\sigma)C_\ell(\sigma) + B_\ell(\sigma)D_\ell(\sigma)}{A_\ell(\sigma)^2 + B_\ell(\sigma)^2} \right] \tag{7.2.14}$$

The numerator of the resistance expression is one for all modes. Numerical values of the other functions are listed in Table 7.2.1 for the four lowest modes. Note the reactances are negative at all ranges and hence have no inherent resonances; a similar analysis of TE modes shows the reactances are positive at all ranges and hence have no inherent resonances.

The modal impedance may be used to synthesize modal equivalent circuits that simulate affects of ideal, single-mode antennas upon their sources.

Table 7.2.1. Functions that determine surface reactance.

ℓ	$A_\ell C_\ell + B_\ell D_\ell$	$A_\ell^2 + B_\ell^2$
1	$-\dfrac{1}{\sigma^3}$	$1+\dfrac{1}{\sigma^2}$
2	$-\dfrac{3}{\sigma^3}-\dfrac{18}{\sigma^5}$	$1+\dfrac{3}{\sigma^2}+\dfrac{9}{\sigma^4}$
3	$-\dfrac{6}{\sigma^3}-\dfrac{75}{\sigma^5}-\dfrac{675}{\sigma^7}$	$1+\dfrac{6}{\sigma^2}+\dfrac{45}{\sigma^4}+\dfrac{225}{\sigma^6}$
4	$-\dfrac{10}{\sigma^3}-\dfrac{220}{\sigma^5}-\dfrac{4725}{\sigma^7}-\dfrac{44{,}100}{\sigma^9}$	$1+\dfrac{10}{\sigma^2}+\dfrac{135}{\sigma^4}+\dfrac{1575}{\sigma^6}+\dfrac{11{,}025}{\sigma^8}$

The Dipole Case

From Eq. 7.2.13 the impedance of a spherical shell of radius a centered on and circumscribing an electric dipolar source is:

$$Z_1(\sigma)=\frac{1}{i\omega(\varepsilon a)}+\frac{1}{1/i\omega(\mu a)+1/\eta} \qquad (7.2.15)$$

This impedance is that of an εa-farad capacitor in series with a parallel connection of a μa-henry inductor and an η-ohm resistor. For small values of ka the input impedance is large and dominated by the capacitance. Far-field power loss from the system is represented by power loss in the resistor.

A related expression is valid for the ℓth mode and found similarly; it results in a reactive ladder network with a single terminating resistor and $(\ell+1)$ reactive elements, as illustrated in Fig. 7.2.1. The circuit emulates changes in an electromagnetic field due to the changing geometry with changing radius. The dual of this circuit emulates magnetic multipoles. It has the same input impedance as a spherical shell of radius 'a' radiating an electric multipolar mode of order ℓ.

For what lies ahead it is convenient to use partial division to express Eq. 7.2.15 in series form:

$$Z_1=\left(\frac{1}{i\omega\varepsilon a}+\mu\left[i\omega a+\omega^2\frac{a^2}{c}-i\omega^3\frac{a^3}{c^2}-\omega^4\frac{a^4}{c^3}+i\omega^5\frac{a^5}{c^4}+...\right]\right) \qquad (7.2.16)$$

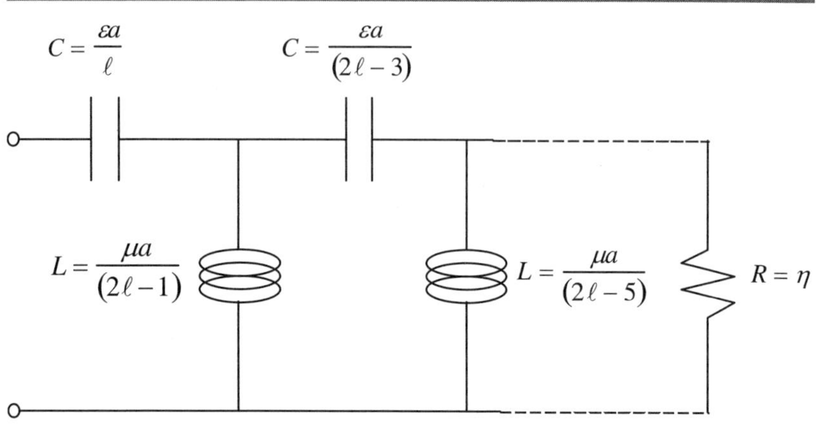

Figure 7.2.1. TM multipolar equivalent circuit.

Next define generalized force, F_1, and generalized flow, I_1, proportional respectively to the driving field and dipole moment, p_1. The actual current is proportional to the rate of change of dipole moment and, since we need only evaluate a proportionality constant, we put the generalized flow equal the actual current and then introduce unknown constant 'g' by the equation:

$$V_1 = 6\pi i a^2 F_1 / eg \quad \text{and} \quad I_1 = i\omega p_1 \tag{7.2.17}$$

Combining Eq. 7.2.16 and 7.2.17 gives:

$$F_1 = \frac{ieg}{6\pi c^2}\left(-i\frac{\eta c^3}{a^3} + i\frac{\eta \omega^2 c}{a} + \eta \omega^3 - i\frac{\eta \omega^4 a}{c} - \frac{\eta \omega^5 a^2}{c^2} + \ldots\right)p \tag{7.2.18}$$

This expresses the complete reaction force on a radiating electric dipole. With each succeeding power of ω the term magnitude drops by (ka/i) and hence, in the mid-optical frequency range, is smaller by a factor of about 1000. Even and odd powers of (ka) are respectively real and imaginary and hence associated respectively with traveling and oscillating energy. Comparison of the $\eta \omega^3$ term with Eq. 7.2.8 shows $g = 1$; it describes a braking force on the oscillator and it is the only term independent of source size. The first term is a factor of $(1/ka)^3$ larger and it acts to distort the source in a way that decreases the configurational, but not kinetic, energy. That is, it acts to extend and distribute an initially small but malleable charge around a closed loop thereby, if completed, dropping the radiation field energy to zero.

Making the substitution $g = 1$ and switching to time notation by replacing $i\omega p$ with a time derivative of the dipole moment changes Eq. 7.2.18 to:

$$F_1(t) = \frac{\mu e}{6\pi c}\left(\frac{c^3}{a^3} + \frac{c}{a}\frac{\partial^2}{\partial t^2} - \frac{\partial^3}{\partial t^3} + \frac{a}{c}\frac{\partial^4}{\partial t^4} - \ldots\right)p(t) \tag{7.2.19}$$

The first term is frequency independent and expresses force on a stationary dipole.

Impedance at higher orders To sample how the situation changes with increasing modal order, return to Eq. 7.2.13 and note the quadrupole impedance on a circumscribing shell of radius a may be written:

$$Z_2(\sigma) = \frac{2}{i\omega\varepsilon a} + \cfrac{1}{\cfrac{3}{i\omega\mu a} + \cfrac{1}{\cfrac{1}{i\omega\varepsilon a} + \eta}} \tag{7.2.20}$$

On the surface of a spherical shell of radius a:

$$Z_2 = \frac{2\eta}{i\sigma} + \frac{i\eta\sigma}{3} + \frac{i\eta\sigma^3}{9} + \frac{\eta\sigma^4}{9} + \frac{2\eta\sigma^5}{27i} + \ldots \tag{7.2.21}$$

For this case the (first real term)-to-(first reactive term) ratio is $18(ka)^5$; solving for the impedance of higher order modes shows the same ratio is proportional to $(ka)^{2\ell+1}$.

7.3 Stress in a Dipole Radiation Field

Radiation fields create pressure and shear on surfaces. Stress values follow from the three spatial dimensions of four-dimensional electromagnetic stress tensor Eq. 1.8.6. In this section we use that three-dimensional tensor to obtain the pressure and shear on a virtual shell supporting the radiation field of a contained electric dipole. The technique extends to any mode and, by superposition of fields, to superpositions of all modal combinations.

For a resting sphere, the force equation is:

$$F_i^v = \partial T_{ij} / \partial x_j \tag{7.3.1}$$

F_i^v, the force per unit volume, and T_{ij}, the stress tensor, are given respectively by Eq. 1.6.14 and 1.8.6. Changing Eq. 1.8.6 from rectangular to spherical coordinates gives:

$$|T_{ij}| = \begin{vmatrix} \left(\dfrac{\varepsilon}{2}\left[E_r^2 - E_\theta^2 - E_\phi^2\right] + \dfrac{\mu}{2}\left[H_r^2 - H_\theta^2 - H_\phi^2\right] \right) & \left(\varepsilon E_r E_\theta + \mu H_r H_\theta\right) & \left(\varepsilon E_r E_\phi + \mu H_r H_\phi\right) \\[2em] \left(\varepsilon E_\theta E_r + \mu H_\theta H_r\right) & \left(\dfrac{\varepsilon}{2}\left[E_\theta^2 - E_\phi^2 - E_r^2\right] + \dfrac{\mu}{2}\left[H_\theta^2 - H_\phi^2 - H_r^2\right] \right) & \left(\varepsilon E_\theta E_\phi + \mu H_\theta H_\phi\right) \\[2em] \left(\varepsilon E_\phi E_r + \mu H_\phi H_r\right) & \left(\varepsilon E_\phi E_\theta + \mu H_\phi H_\theta\right) & \left(\dfrac{\varepsilon}{2}\left[E_\phi^2 - E_r^2 - E_\theta^2\right] + \dfrac{\mu}{2}\left[H_\phi^2 - H_r^2 - H_\theta^2\right] \right) \end{vmatrix} \tag{7.3.2}$$

Diagonal matrix element $T_{rr}(t_R)$ describes surface pressure and off-diagonal matrix elements $S_{r\theta}(t_R)$ and $S_{r\phi}(t_R)$ describe surface shear. The equations are applicable on both exterior and interior surfaces.

A virtual sphere of radius a generates the stress tensor components:

$$T_{rr}(t_R) = \frac{\varepsilon}{2}\left[E_r^2 - E_\theta^2 - E_\phi^2\right] + \frac{\mu}{2}\left[H_r^2 - H_\theta^2 - H_\phi^2\right]$$

$$S_{r\theta}(t_R) = \varepsilon E_r E_\theta + \mu H_r H_\theta; \quad S_{r\phi}(t_R) = \varepsilon E_r E_\phi + \mu H_r H_\phi \tag{7.3.3}$$

An important but elementary example is Coulomb's law. Let point charge q occupy the center of a virtual sphere; the field and the tensor components on the surface are:

$$E_r = \frac{q}{4\pi\varepsilon a^2}; \quad T_{rr} = \frac{q^2}{32\pi^2\varepsilon a^4} = p; \quad S_{r\theta} = 0 \tag{7.3.4}$$

T_{rr} is the pressure, p, on the spherical surface.

Consider the z-directed, electric dipole fields on a spherical shell supporting the charge density of Eq. A.8.1. The exterior and interior phasor fields are given by Eq. A.8.7 and A.8.8; with subscripts 'e' and 'i' representing respectively exterior and interior regions the fields are:

$$\tilde{\boldsymbol{E}}_e = \frac{3}{2}\left(\frac{q}{4\pi\varepsilon a^2}\right)(ka)^3\left\{2\hat{r}\left(\frac{1}{\sigma^3}+\frac{i}{\sigma^2}\right)\cos\theta + \hat{\theta}\left(\frac{1}{\sigma^3}+\frac{i}{\sigma^2}-\frac{1}{\sigma}\right)\sin\theta\right\}e^{i\omega t_R}$$

$$\eta\tilde{\boldsymbol{H}}_e = \frac{3}{2}\left(\frac{3q}{4\pi\varepsilon a^2}\right)(ka)^3\,\hat{\phi}\left(\frac{i}{\sigma^2}-\frac{1}{\sigma}\right)\sin\theta\, e^{i\omega t_R}$$

$$\tag{7.3.5}$$

$$\tilde{\boldsymbol{E}}_i = \left(\frac{q}{4\pi\varepsilon a^2}\right)\left\{\hat{r}\cos\theta - \hat{\theta}\sin\theta\right\}e^{i\omega t}$$

$$\tag{7.3.6}$$

$$\eta\tilde{\boldsymbol{H}}_i = i\left(\frac{q}{8\pi\varepsilon a^2}\right)\hat{\phi}\sigma\sin\theta\, e^{i\omega t}$$

Shear arises from only the electric field since the magnetic field has no radial component. Also, since the magnetic field strength is less than the electric field by a factor of (ka), if the source is electrically small the pressure due to the magnetic field is small. Under these conditions the unbalanced, exterior minus interior, surface pressure normalized by the Coulomb surface pressure, indicated by subscript 'N', produced by Eq. 7.3.4, is:

$$p_N = \left[2\cos^2\theta - \frac{5}{16}\sin^2\theta\right]\left[1+\cos(2\omega t_R)\right] \tag{7.3.7}$$

This pressure is compressive at angles between about 68.5° to greater than 111.5° and expansive at both ends. Since the time-average to time-dependent pressure ratio is about one it follows that dipole and monopole effects are equal. Since the pressure is expansive on the z-axis and compressive in the basal plane, the pressure tends to distort the original sphere into something resembling a radiating bicone with extended caps.

Integrating over the surface gives the ratio of total radiation expansion force to Coulomb attraction force:

$$\frac{\text{Expansive dipole force}}{\text{Attractive Coulomb foce}} = \frac{11}{6}\left[1 + \cos\left(2\omega t_R\right)\right] \tag{7.3.8}$$

A plot of Eq. 7.3.8 is shown in Fig. 7.3.1; inspection shows the expansive pressure at the poles is four times the Coulomb compression.

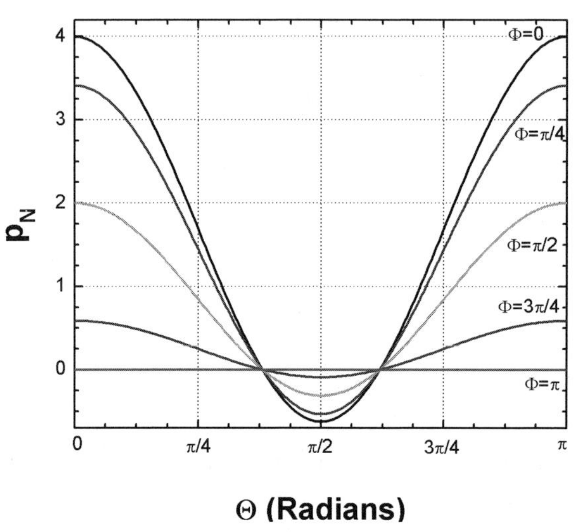

Figure 7.3.1. Normalized surface pressure p_N on a spherical shell radiating as a z-directed electric dipole versus zenith angle θ of radius $ka = 1$ at $\Phi = 2\omega t_R$ equal: 0, $\pi/4$, $\pi/2$, $3\pi/4$, and π.

The normalized exterior and interior surface shears calculated using Eq. 7.3.3, 7.3.5, and 7.3.6 are:

$$S_{eN} = \frac{9}{2}\sin\theta\cos\theta\left[1 + \cos\left(2\omega t_R\right)\right]$$
$$S_{iN} = -\sin\theta\cos\theta\left[1 + \cos\left(2\omega t_R\right)\right] \tag{7.3.9}$$

The two shears have identical time and angular dependence but are oppositely directed; the directional opposition acts to form θ-directed eddy currents on both interior and exterior sides of the shell. The magnitudes of the eddy-producing surface shear and the binding Coulomb force are approximately the same.

7.4 Pairs of Radiating Multipoles

Any and all electromagnetic fields in free space, described using spherical coordinates, may be expressed in the form of Eq. 1.13.9 with integer values of separation constants ℓ and m. The radial expressions are solutions of spherical Bessel differential equation, Eq. 1.12.7, with ν replaced by integer ℓ. Solutions are developed in Sec. A.18 and A.19.

Since this chapter is based upon fields with sources we impose arbitrary upper limit L on the sums. A multipolar source of mode ℓ requires at least $2^{\ell-1}$ separate charges, see Sec. A.22. Source granularity, in some cases, imposes an upper limit on L since at least 2^{L-1} separate charges are required.

Spherical surfaces that support either TE or TM multipolar radiation fields, but not both, have no intrinsic resonances. A surface supporting a TM mode is capacitive and a surface supporting a TE mode is inductive. With half-wave antennas producing TM modes a circumscribing sphere is always capacitive, as illustrated by the biconical antenna of Chap. 3. The antenna arms, however, act as a transmission line between the source and the circumscribing sphere and the transmission line effect enables resonance at the source.

We begin a characterization of mixed TE and TM fields by analyzing collocated, equal-order electric and magnetic multipoles with spatially orthogonal multipolar axes; the system rotates at radian frequency ω about an axis normal to the plane of the moments, *i.e.,* the basal plane. The rotating multipoles produce an output power maximum in the direction of a right hand screw rotating with the source; a power null exists in the opposite direction. We assign the $+z$-axis to the maximum power direction and, with that assignment and using $\exp(i\omega t)$ time dependence, the generated fields are those of Eq. 1.13.9 after putting $j = i$, using spherical Hankel functions of the second kind for the radial functions, putting $G(\ell,m) = -F(\ell,m)$, and $F(\ell,m) = 0$ for $m \neq 1$. The remaining fields are:

$$\eta \tilde{\mathbf{H}}_\ell = i\tilde{\mathbf{E}}_\ell; \quad F_\ell = F_0 \frac{(2\ell+1)}{\ell(\ell+1)}; \quad e^{i(\omega t - \phi)} \text{ is suppressed}$$

$$\tilde{\mathbf{E}} = \sum_{\ell=1}^{L} i^{-\ell} F_\ell \left\{ \begin{array}{l} i\ell(\ell+1)\dfrac{h_\ell(\sigma)}{\sigma} P_\ell^1(\cos\theta)\hat{r} \\[2ex] +\left[ih_\ell^\bullet(\sigma)\dfrac{d}{d\theta}P_\ell^1(\cos\theta) + h_\ell(\sigma)\dfrac{P_\ell^1(\cos\theta)}{\sin\theta}\right]\hat{\theta} \\[2ex] -i\left[ih_\ell^\bullet(\sigma)\dfrac{P_\ell^1(\cos\theta)}{\sin\theta} + h_\ell(\sigma)\dfrac{d}{d\theta}P_\ell^1(\cos\theta)\right]\hat{\phi} \end{array} \right. \tag{7.4.1}$$

Recursion relationship F_ℓ is the same as that of Eq. 2.8.11.

Inserting large-range values of Hankel functions from Eq. A.19.17 into Eq. 7.4.1 shows that at limitlessly large ranges the electric field of mode ℓ is:

$$\lim_{\sigma\to\infty} \tilde{E}_\ell(\sigma,\theta,\phi) = F_0 \frac{(2\ell+1)}{\ell(\ell+1)} \left\{ \begin{array}{l} \dfrac{i}{\sigma^2}\ell(\ell+1)P_\ell^1 \hat{r} \\[2ex] +\dfrac{1}{\sigma}\left[\dfrac{dP_\ell^1}{d\theta} + \dfrac{P_\ell^1}{\sin\theta}\right](i\hat{\theta}+\hat{\phi}) \end{array} \right\} e^{i(\omega t_R - \phi)} \tag{7.4.2}$$

After substituting for the angular functions using Table A.12.1 and then converting to rectangular coordinates the radiation on the positive z-axis shows the radiation is circularly polarized:

$$\lim_{\sigma\to\infty} \tilde{E}_\ell(\sigma,0,\phi) = \frac{i}{\sigma}(2\ell+1)(\hat{x}-i\hat{y})e^{i\omega t_R} \tag{7.4.3}$$

The phasor fields of Eq. 7.4.1 generate the time-average Poynting vector which, using letter symbols of Sec. A.20, is:

$$N = \frac{F_\ell^2}{\eta\sigma^2}\left\{\left[\left(\frac{dP_\ell^1}{d\theta}\right)^2 + \left(\frac{P_\ell^1}{\sin\theta}\right)^2\right] + \left(A_\ell^2 + B_\ell^2 + C_\ell^2 + D_\ell^2\right)\frac{dP_\ell^1}{d\theta}\frac{P_\ell^1}{\sin\theta}\right\}\hat{r}$$

$$-\frac{F_\ell^2}{\eta\sigma^3}\ell(\ell+1)\left\{\left(A_\ell C_\ell + B_\ell D_\ell\right)\frac{\left[P_\ell^1\right]^2}{\sin\theta}\hat{\theta} - \left[P_\ell^1\frac{dP_\ell^1}{d\theta} + \left(A_\ell^2 + B_\ell^2\right)\frac{\left[P_\ell^1\right]^2}{\sin\theta}\right]\hat{\phi}\right\}$$

$$(7.4.4)$$

Only the terms in the square bracket of the top line contribute to total surface power on a virtual circumscribing sphere, and from it comes the time-average output power:

$$P = \frac{4\pi F_0^2}{\eta k^2}(2\ell+1) \tag{7.4.5}$$

Directivity

Directivity is defined as the maximum-to-average power density ratio on the surface of a limitlessly large, virtual circumscribing sphere. The top line of Eq. 7.4.4 gives the outbound power density at each point; the maximum value occurs at $\theta = 0$. Directionality per mode ℓ equals the ratio of $4\pi r^2 N_r$ times that maximum value to the total output power:

$$D_\ell = (2\ell+1) \tag{7.4.6}$$

In 1960, 43 years after Einstein's seminal paper on the quantum theory of radiation, Harrington published this gain-per-mode expression showing the proper combination of paired, spherical modes produce directed radiation that, in turn, produces a push on the emitter.

Field Energy

The total time-average modal energy density, w_T, at each field point follows from phasor field Eq. 7.4.1:

$$w_T = \frac{\varepsilon F_\ell^2}{2\sigma^2}\left\{ \begin{array}{l} \dfrac{1}{\sigma^2}\ell^2(\ell+1)^2\left(A_\ell^2+B_\ell^2\right)\left[P_\ell^1\right]^2 \\[2ex] +\left(A_\ell^2+B_\ell^2+C_\ell^2+D_\ell^2\right)\left[\left(\dfrac{dP_\ell^1}{d\theta}\right)^2+\left(\dfrac{P_\ell^1}{\sin\theta}\right)^2\right] \\[2ex] +4\left(A_\ell D_\ell-B_\ell C_\ell\right)\dfrac{P_\ell^1}{\sin\theta}\dfrac{dP_\ell^1}{d\theta} \end{array} \right\} \tag{7.4.7}$$

Only stored energy reactively affects the source and Eq. 7.4.7 includes energy permanently separated from the source. Therefore to calculate Q we subtract the energy density of radiation permanently leaving the local region. The source of the input resistance is the energy:

$$w_0 = \frac{\varepsilon F_\ell^2}{\sigma^2}\left\{ \left(\frac{dP_\ell^1}{d\theta}\right)^2+\left(\frac{P_\ell^1}{\sin\theta}\right)^2+\left(A_\ell^2+B_\ell^2+C_\ell^2+D_\ell^2\right)\frac{dP_\ell^1}{d\theta}\frac{P_\ell^1}{\sin\theta}\right\} \tag{7.4.8}$$

The difference is the local standing energy density, some of which returns to the source twice each field cycle and thereby creates the input reactance:

$$w_S = \frac{\varepsilon F_\ell^2}{2}\left\{ \begin{array}{l} \dfrac{\ell^2(\ell+1)^2}{\sigma^4}\left(A_\ell^2+B_\ell^2\right)\left(P_\ell^1\right)^2 \\[2ex] +\dfrac{1}{\sigma^2}\left(A_\ell^2+B_\ell^2+C_\ell^2+D_\ell^2-2\right)\left(\dfrac{dP_\ell^1}{d\theta}-\dfrac{P_\ell^1}{\sin\theta}\right)^2 \end{array} \right\} \tag{7.4.9}$$

The volume integral of this standing energy density over exterior space gives the total standing energy:

$$W_S = \frac{2\pi\varepsilon F_0^2}{k^3}(2\ell+1)\left\{\left(A_\ell C_\ell+B_\ell D_\ell\right)+2\int\left(A_\ell^2+B_\ell^2-1\right)d\sigma\right\}_{ka}^{\infty} \tag{7.4.10}$$

The limits apply to all terms within the curly bracket.

Radiation Q

The radiation Q follows from the above by entering Eq. 7.4.10 into defining equation Eq. 2.4.1:

$$Q = \left\{ \frac{1}{2}\left(A_\ell C_\ell + B_\ell D_\ell\right) + \int\left(A_\ell^2 + B_\ell^2 - 1\right)d\sigma \right\}_{ka}^{\infty} \tag{7.4.11}$$

Comparison of Eq. 7.4.11 and Eq. 7.1.14 shows the radiation Q of a dipole pair is approximately half that of a single dipole; the pair emits twice as much power as a single mode yet the maximum standing energy is nearly unchanged.

In the limit of vanishingly small radius Eq. 7.4.11 shows the Q of a multipolar pair is:

$$\lim_{ka \to 0} Q_\ell(ka) = \frac{\ell\left[(2\ell+1)!!\right]^2}{2(ka)^{2\ell+1}} \tag{7.4.12}$$

This significant equation confirms that Qs of electrically small, matched sources and large orders are so large that radiation is squelched and the lowest order mode provides a field description sufficiently accurate for many purposes. The complete expansions for Q versus modal order for the matched TE and TM modes of Eq. 7.4.11 are shown in Table 7.4.1 for orders 1, 3, and 5.

Table 7.4.1. Dependence of modal Q on electrical size ka.

$$Q_1(ka) = \frac{1}{2(ka)^3} + \frac{1}{(ka)}$$

$$Q_3(ka) = \frac{675}{2(ka)^7} + \frac{90}{(ka)^5} + \frac{18}{(ka)^3} + \frac{6}{(ka)}$$

$$Q_5(ka) = \frac{4,465,125}{2(ka)^{11}} + \frac{297,675}{(ka)^9} + \frac{23,635}{(ka)^7} + \frac{1575}{(ka)^5} + \frac{225}{2(ka)^3} + \frac{15}{(ka)}$$

For an atom emitting optical wavelengths the initial value of ka is on the order of 10^{-3} and hence by Table 7.4.1 the Q of modes one, three and five are respectively on the order of 5×10^{10}, 3×10^{19} and 2×10^{39}. As the modal number increases further, Qs soon reach googol sizes: the atom is a Q-trap for exchanged radiation.

Linear Momentum

As shown by Eq. 1.9.8, streams of radiated energy are also streams of radiated momentum. Emission from a single dipole produces a circularly symmetric radiation pattern that creates a compressive force but no net transfer of momentum. In contrast, the collocated multipoles of Eq. 7.4.1 create a net flow of energy and momentum and, by Newton's law, transfer an oppositely directed momentum to the source. By Eq. 1.9.8 the rate at which linear momentum passes through a sphere circumscribing an emitter of z-directed power is:

$$\frac{dp_z}{dt} = \frac{2\pi}{ck^2} \int_0^\pi \sigma^2 \left(N_r \cos\theta - N_\theta \sin\theta \right) \sin\theta \, d\theta \tag{7.4.13}$$

Combining Eq. 7.4.4 and 7.4.13 shows the rate of z-directed linear momentum transfer per modal pair is:

$$\frac{dp_z}{dt}\bigg|_\ell = \frac{2\pi\varepsilon F_\ell^2}{k^2} \int_0^\pi \sin\theta \cos\theta \, d\theta \left\{ \begin{array}{l} \left(\dfrac{dP_\ell^1}{d\theta} \right)^2 + \left(\dfrac{P_\ell^1}{\sin\theta} \right)^2 \\[2mm] + \left(A_\ell^2 + B_\ell^2 + C_\ell^2 + D_\ell^2 \right) \dfrac{dP_\ell^1}{d\theta} \dfrac{P_\ell^1}{\sin\theta} \\[2mm] - \ell(\ell+1)\left(A_\ell C_\ell + B_\ell D_\ell \right) \left[P_\ell^1 \right]^2 \end{array} \right\} \tag{7.4.14}$$

Evaluating the integral with the aid of Table A.16.1 gives:

$$\frac{dp_z}{dt} = \frac{2\pi\varepsilon}{k^2} F_\ell^2 \frac{\ell(\ell+1)}{(2\ell+1)} \left(A_\ell^2 + B_\ell^2 + C_\ell^2 + D_\ell^2 \right) \tag{7.4.15}$$

At limitlessly large distances this becomes:

$$\lim_{\sigma \to \infty} \frac{dp_z}{dt} = \frac{4\pi\varepsilon}{k^2} F_0^2 \frac{(2\ell+1)}{\ell(\ell+1)} \qquad (7.4.16)$$

This equation shows that for large modal numbers the exchanged linear momentum decreases inversely with modal number.

An independent measure of field alignment is the energy-to-linear momentum ratio. Combining Eq. 7.4.5 and 7.4.16 gives:

$$\frac{dW_\ell}{dt} \bigg/ \frac{dp_{z\ell}}{dt} = \ell(\ell+1)c \qquad (7.4.17)$$

The matched electric and magnetic multipoles of Eq. 7.4.1 radiate an energy-to-linear momentum ratio that increases as the square of increasing modal number.

Radiated Angular Momentum

The fields of Eq. 7.4.1 also carry angular momentum away from the source and, by Newton's law, impart an oppositely directed angular momentum on the source. The rate at which angular momentum exits a closed volume is:

$$\frac{dl_z}{dt} = \frac{2\pi\sigma^3 F_\ell^2}{ck^3} \int_0^\pi N_\phi \sin^2 \theta d\theta \qquad (7.4.18)$$

Using the phi component of Poynting vector Eq. 7.4.4 and an integral from Table A.16.1:

$$\frac{dl_z}{dt} = \frac{4\pi F_\ell^2}{\eta\omega k^2} \frac{\ell^2(\ell+1)^2}{(2\ell+1)}\left(A_\ell^2 + B_\ell^2\right) \qquad (7.4.19)$$

It follows from Table A.20.1 that the radiated angular momentum at a limitlessly large range is:

$$\frac{dl_z}{dt} = \frac{4\pi F_0^2}{\eta\omega k^2}(2\ell+1) \qquad (7.4.20)$$

Independently of the emitted waveform, taking time integrals of Eq. 7.4.5 and 7.4.20 and then taking the ratio gives:

$$dW_\ell / dl_{\ell z} = \omega \tag{7.4.21}$$

Each of the matched multipoles of Eq. 7.4.1 supports a radiated energy-to-angular momentum ratio pair of ω.

7.5 Characterization of Sums over Matched Modes

The purpose of this section is to evaluate and apply sums over the matched modal pairs discussed in Sec. 7.4. There are closed form expressions for sums over spherical Bessel functions but not for spherical Neumann functions. Therefore we begin with the source-free field set obtained by replacing the Hankel functions of Eq. 7.4.1 with Bessel functions:

$$\eta\tilde{\mathbf{H}} = i\tilde{\mathbf{E}}; \quad F_0 e^{i(\omega t - \phi)} \text{ is suppressed}$$

$$\tilde{\mathbf{E}} = F_0 \sum_{\ell=1}^{\infty} i^{-\ell} \frac{(2\ell+1)}{\ell(\ell+1)} \left\{ \begin{array}{l} i\ell(\ell+1)\dfrac{j_\ell(\sigma)}{\sigma} P_\ell^1(\cos\theta)\hat{r} \\[2ex] + \left[ij_\ell^\bullet(\sigma)\dfrac{d}{d\theta}P_\ell^1(\cos\theta) + j_\ell(\sigma)\dfrac{P_\ell^1(\cos\theta)}{\sin\theta} \right]\hat{\theta} \\[2ex] -i\left[ij_\ell^\bullet(\sigma)\dfrac{P_\ell^1(\cos\theta)}{\sin\theta} + j_\ell(\sigma)\dfrac{d}{d\theta}P_\ell^1(\cos\theta) \right]\hat{\phi} \end{array} \right\} \tag{7.5.1}$$

This equation describes a circularly polarized, z-directed plane wave. Another expression for the same field is:

$$\tilde{\mathbf{E}} = \left[\sin\theta\hat{r} + \cos\theta\hat{\theta} - i\hat{\phi} \right] e^{i(\omega t - \phi - \sigma\cos\theta)}; \quad \eta\tilde{\mathbf{H}} = i\tilde{\mathbf{E}} \tag{7.5.2}$$

The two expressions describe the same field set and, by the uniqueness theorem, are equal. Since the full directivity of Eq. 7.5.2 is obvious but that of Eq. 7.5.1 is not we use the equality to characterize Eq. 7.5.1. Results are obtained in Sec. A.21 and selected ones listed in Table 7.5.1.

Table 7.5.1. Sums over spherical Bessel and related functions; sums are exact.

1.
$$\sigma \sin\theta\, e^{-i\sigma\cos\theta} = \sum_{\ell=1}^{\infty} i^{1-\ell}(2\ell+1)\, j_\ell(\sigma)\, P_\ell^1(\cos\theta)$$

2.
$$-iU = \sum_{\ell=1}^{\infty} i^{-\ell}\frac{(2\ell+1)}{\ell(\ell+1)}\, j_\ell(\sigma)\frac{P_\ell^1(\cos\theta)}{\sin\theta}$$

3.
$$-iV = \sum_{\ell=1}^{\infty} i^{1-\ell}\frac{(2\ell+1)}{\ell(\ell+1)}\, j_\ell^\bullet(\sigma)\frac{P_\ell^1(\cos\theta)}{\sin\theta}$$

4.
$$e^{-i\sigma\cos\theta}+iV = \sum_{\ell=1}^{\infty} i^{-\ell}\frac{(2\ell+1)}{\ell(\ell+1)}\, j_\ell(\sigma)\frac{dP_\ell^1}{d\theta}$$

5.
$$e^{-i\sigma\cos\theta}\cos\theta+iU = \sum_{\ell=1}^{\infty} i^{1-\ell}\frac{(2\ell+1)}{\ell(\ell+1)}\, j_\ell^\bullet(\sigma)\frac{dP_\ell^1}{d\theta}$$

Functions $U(\sigma,\theta)$ and $V(\sigma,\theta)$ are defined in A.21.16 and equal to:

$$U(\sigma,\theta) = \frac{1}{\sigma\sin^2\theta}\left[e^{-i\sigma\cos\theta}-\left(\cos\sigma-i\sin\sigma\cos\theta\right)\right]$$
$$V(\sigma,\theta) = \frac{1}{\sigma\sin^2\theta}\left[e^{-i\sigma\cos\theta}\cos\theta-\left(\cos\sigma\cos\theta-i\sin\sigma\right)\right]$$
(7.5.3)

Sum $(U+V)$ consists of equal parts z- and r-directed waves:

$$U(\sigma,\theta)+V(\sigma,\theta) = \frac{(1+\cos\theta)}{\sigma\sin^2\theta}\left[e^{-i\sigma\cos\theta}-e^{-i\sigma}\right]$$
(7.5.4)

We wish to establish a relationship similar to that of Table 7.5.1 for the fields of Eq. 7.4.1. For that purpose we note from Eq. A.18.14 and A.19.15 that if $\sigma \gg 1$ the relationship between spherical Bessel and Hankel functions is:

$$\lim_{\sigma\to\infty} y_\ell(\sigma) = -j_\ell^\bullet(\sigma); \quad \lim_{\sigma\to\infty} h_\ell(\sigma) = j_\ell(\sigma)+\frac{i}{\sigma}\frac{d}{d\sigma}\left[\sigma j_\ell(\sigma)\right]$$
(7.5.5)

Applying Eq. 7.5.5 to Table 7.5.1 yields the asymptotic, large range sums over spherical Hankel functions of Table 7.5.2. Define the maximum range of interest to be σ_0 and an additional requirement is the maximum modal number must exceed the maximum normalized range: $L \gg \sigma_0$.

Table 7.5.2. Sums over spherical Hankel and related functions, valid for
inequality set $L \gg \sigma_0 \gg 1$.

1.
$$\sigma \sin\theta \left[(1+\cos\theta) + \frac{2i}{\sigma} \right] e^{-i\sigma\cos\theta}$$
$$= \sum_{\ell=1}^{L} i^{1-\ell} (2\ell+1) h_\ell(\sigma) P_\sigma^1(\cos\theta)$$

2.
$$-i(U+V) = \sum_{\ell=1}^{L} i^{1-\ell} \frac{(2\ell+1)}{\ell(\ell+1)} h_\ell(\sigma) \frac{P_\ell^1}{\sin\theta}$$

3.
$$\left[\frac{i}{\sigma} e^{-i\sigma\cos\theta} - i(U+V) \right] = \sum_{\ell=1}^{L} i^{1-\ell} \frac{(2\ell+1)}{\ell(\ell+1)} h_\ell^{\bullet}(\sigma) \frac{P_\ell^1}{\sin\theta}$$

4.
$$\left[(1+\cos\theta) e^{-i\sigma\cos\theta} + i(U+V) \right]$$
$$= \sum_{\ell=1}^{L} i^{1-\ell} \frac{(2\ell+1)}{\ell(\ell+1)} h_\ell(\sigma) \frac{dP_\ell^1}{d\theta}$$

5.
$$\left[\cos\theta \left((1+\cos\theta) + \frac{i}{\sigma} \right) e^{-i\sigma\cos\theta} + i(U+V) \right]$$
$$= \sum_{\ell=1}^{L} i^{1-\ell} \frac{(2\ell+1)}{\ell(\ell+1)} h_\ell^{\bullet}(\sigma) \frac{dP_\ell^1}{d\theta}$$

Four sums include $(U+V)$ and hence consist of both z- and r-directed waves. However, when combined as in Eq. 7.4.1 all such terms vanish and only z-directed fields remain, as seen below:

$$\eta \tilde{H}(\sigma,\theta,\phi) = i\tilde{E}(\sigma,\theta,\phi)$$

$$\lim_{\sigma\to\infty} \tilde{E}(\sigma,\theta,\phi) = F_0 \left\{ \begin{array}{l} \sin\theta \left((1+\cos\theta) + \dfrac{2i}{\sigma} \right) \hat{r} \\[2ex] (\cos\theta\,\hat{\theta} - i\hat{\phi}) \left((1+\cos\theta) + \dfrac{i}{\sigma} \right) \end{array} \right\} e^{i(\omega t - \phi - \sigma\cos\theta)} \qquad (7.5.6)$$

This expression confirms that the fields of Eq. 7.4.1 are fully z-directed wave at large ranges.

For the special cases of $\theta = 0$ and $\theta = \pi$ the fields are:

$$\lim_{\sigma \to \infty} \tilde{E}(\sigma, 0, \phi) = 2F_0(\hat{x} - i\hat{y})e^{i(\omega t - kz)}; \quad \lim_{\sigma \to \infty} \tilde{E}(\sigma, \pi, \phi) = 0 \qquad (7.5.7)$$

The magnitude of the $(1 + \cos\theta)$ portion of this circularly polarized wave, uniquely to the recursion relationship of Eq. 7.4.1, is independent of range: so long as inequality $L \gg \sigma_0$ is satisfied the field magnitude is unabated with increasing range.

Truncating the series at L shows the time-average output power is:

$$P_{av} = \frac{4\pi F_0^2}{\eta k^2} L(L+2) \qquad (7.5.8)$$

For large values of L, this expression shows a parabolic relationship between F_0 and L.

7.6 Self-Consistent Fields

We use the method of self-consistent field analysis to extend Eq. 7.5.6 to lesser ranges. Note that since self-consistent techniques introduce no new symmetries and since the limiting field is fully z-directed, so are all fields calculated by this means.

It is convenient to assign symbols to each of the sums of Eq. 7.4.1; we define three 'S' sums, S_n, and two '$\$$' sums, $\$_n$, where $n = 1,2,3$. The definitions are:

$$S_1(\sigma, \theta) = \sum_{\ell=1}^{\infty} i^{1-\ell}(2\ell+1)h_\ell(\sigma)P_\ell^1$$

$$S_2(\sigma, \theta) = \sum_{\ell=1}^{\infty} i^{-\ell}\frac{(2\ell+1)}{\ell(\ell+1)}h_\ell(\sigma)\frac{dP_\ell^1}{d\theta}; \quad \$_2(\sigma, \theta) = \frac{i}{\sigma}\frac{\partial}{\partial\sigma}\left[\sigma S_2(\sigma, \theta)\right]$$

$$S_3(\sigma, \theta) = \sum_{\ell=1}^{\infty} i^{-\ell}\frac{(2\ell+1)}{\ell(\ell+1)}h_\ell(\sigma)\frac{P_\ell^1}{\sin\theta}; \quad \$_3(\sigma, \theta) = \frac{i}{\sigma}\frac{\partial}{\partial\sigma}\left[\sigma S_3(\sigma, \theta)\right]$$

$$(7.6.1)$$

Inspection shows the sums are related as:

$$S_2 = \frac{\partial\left(S_3 \sin\theta\right)}{\partial\theta}; \quad \$_3 = \frac{i}{\sigma}\frac{\partial}{\sigma}\left(\sigma S_3\right); \quad \$_2 = \frac{\partial\left(\$_3 \sin\theta\right)}{\partial\theta} = \frac{i}{\sigma}\frac{\partial}{\sigma}\left(\sigma S_2\right) \qquad (7.6.2)$$

Separating Eq. 7.4.1 into TM and TE parts and suppressing $F_0 \exp[i(\omega t - \phi)]$ gives:

$$\tilde{\boldsymbol{E}}_{TM} = \left[S_1 \hat{r}/\sigma + \$_2\hat{\theta} - i\$_3\hat{\phi}\right]; \quad \eta\tilde{\boldsymbol{H}}_{TM} = \left[iS_3\hat{\theta} + S_2\hat{\phi}\right]$$

$$\tilde{\boldsymbol{E}}_{TE} = \left[S_3\hat{\theta} - iS_2\hat{\phi}\right]; \quad \eta\tilde{\boldsymbol{H}}_{TE} = \left[iS_1\hat{r}/\sigma + i\$_2\hat{\theta} + \$_3\hat{\phi}\right] \qquad (7.6.3)$$

Separating the Hankel functions into Bessel and Neumann function parts:

$$S_n = S_{n1}\left(\sigma,\theta\right) - iS_{n2}\left(\sigma,\theta\right); \quad \$_n\left(\sigma,\theta\right) = \$_{n1}\left(\sigma,\theta\right) - i\$_{n2}\left(\sigma,\theta\right) \qquad (7.6.4)$$

Second subscript '1' and '2' indicate respectively Bessel and Neumann functions.

Although in principle the sums of Eq. 7.4.1 contain all information about the fields, the expressions arose from an expansion about the origin. Here, as noted, we begin with the limitlessly large range values of Sec. 7.5. The method is a modified Taylor series expansion and assures that if the exact value is known at any point and if the function is continuous through all orders, the value and at any other point can be constructed. We begin with the asymptotic values at a limitlessly large range and descend towards lesser ones.

Suppressing $F_0 \exp[i(\omega t - \phi)]$ and applying the Maxwell curl equation to the magnetic field portion of Eq. 7.6.3 gives:

$$\eta\mathbf{H}_{TM} = \left[iS_3\hat{\theta} + S_2\hat{\phi}\right]; \quad \tilde{\mathbf{E}}_{TM} = -\frac{i}{k}\nabla\times\eta\tilde{\mathbf{H}}_{TM}$$

$$\tilde{\mathbf{E}}_{TM} = \left\{\frac{i\hat{r}}{\sigma\sin\theta}\left[S_3 - \frac{\partial\left(\sin\theta S_2\right)}{\partial\theta}\right] + i\frac{\hat{\theta}}{\sigma}\frac{\partial\left(\sigma S_2\right)}{\partial\sigma} - i\frac{\hat{\phi}}{\sigma}\frac{\partial\left(\sigma S_3\right)}{\partial\sigma}\right\}e^{i(\omega t - \phi)}$$

$$(7.6.5)$$

Equating terms of Eq. 7.6.3 and 7.6.5 gives:

$$\tilde{\boldsymbol{E}}_{TM} = \left[S_1 \hat{r}/\sigma + \$_2 \hat{\theta} - i\$_3 \hat{\phi} \right] \text{ where:}$$

$$S_1 = \frac{i}{\sin\theta}\left[S_3 - \frac{\partial(\sin\theta S_2)}{\partial\theta} \right]; \quad \$_2 = \frac{i}{\sigma}\frac{\partial(\sigma S_2)}{\partial\sigma}; \quad \$_3 = \frac{i}{\sigma}\frac{\partial(\sigma S_3)}{\partial\sigma} \qquad (7.6.6)$$

Applying Maxwell's curl equation to the electric field gives:

$$\tilde{\mathbf{E}}_{TM} = \left[S_1\hat{r}/\sigma + \$_2\hat{\theta} - i\$_3\hat{\phi} \right]; \quad \eta\tilde{\mathbf{H}}_{TM} = \frac{i}{k}\nabla\times\tilde{\mathbf{E}}_{TM}$$

$$\eta\tilde{\mathbf{H}}_{TM} = \left\{ \begin{array}{l} \dfrac{\hat{r}}{\sigma\sin\theta}\left[\dfrac{\partial(\sin\theta\$_3)}{\partial\theta} - \$_2 \right] + \hat{\theta}\left[\dfrac{S_1}{\sigma^2\sin\theta} - \dfrac{1}{\sigma}\dfrac{\partial(\sigma\$_3)}{\partial\sigma} \right] \\[1.5em] \quad + i\hat{\phi}\left[\dfrac{1}{\sigma}\dfrac{\partial(\sigma\$_2)}{\partial\sigma} - \dfrac{1}{\sigma^2}\dfrac{\partial S_1}{\partial\theta} \right] \end{array} \right\} \qquad (7.6.7)$$

Combining the magnetic parts of Eq. 7.6.3 and 7.6.7 gives:

$$\eta\tilde{\mathbf{H}}_{TM} = \left[iS_3\hat{\theta} + S_2\hat{\phi} \right] \quad \text{where:} \quad 0 = \left[\frac{\partial(\sin\theta\$_3)}{\partial\theta} - \$_2 \right]$$

$$S_3 = \left[\frac{i}{\sigma}\frac{\partial(\sigma\$_3)}{\partial\sigma} - i\frac{S_1}{\sigma^2\sin\theta} \right] \quad S_2 = \left[\frac{i}{\sigma}\frac{\partial(\sigma\$_2)}{\partial\sigma} - \frac{i}{\sigma^2}\frac{\partial S_1}{\partial\theta} \right] \qquad (7.6.8)$$

The technique is based upon the differentials of Eq. 7.6.6 and 7.6.8 and detailed in Table 7.6.1.

Table 7.6.1. Chart for evaluating field sums.

TM electric field sums :

$$\$_3^{\ell+1} = \frac{i}{\sigma}\partial\left[\sigma S_3^\ell\right]/\partial\sigma; \quad \$_2^{\ell+1} = \frac{i}{\sigma}\partial\left[\sigma S_2^\ell\right]/\partial\sigma$$

$$S_1^{\ell+1} = \frac{i}{\sin\theta}\left[S_3^\ell - \partial\left(\sin\theta S_2^\ell\right)/\partial\theta \right]$$

TM magnetic field sums :

$$S_3^{\ell+1} = \frac{i}{\sigma^2}\left[\sigma\partial\left(\sigma\$_3^\ell\right)/\partial\sigma - S_1^\ell/\sin\theta \right]$$

$$S_2^{\ell+1} = \frac{i}{\sigma^2}\left[\sigma\partial\left(\sigma\$_2^\ell\right)/\partial\sigma - \partial S_1^\ell/\partial\theta \right]$$

Superscripts show the number of iterations away from the asymptotic fields. For the first iteration substitute asymptotic sums $S_2^{\,0}$ and $S_3^{\,0}$ into Table 7.6.1 and obtain corrected ums $S_1^{\,1}$, $\$_2^{\,1}$ and $\$_3^{\,1}$; next substitute sums $S_1^{\,0}$, $\$_2^{\,0}$ and $\$_3^{\,0}$ into Table 7.6.1 and obtain $S_2^{\,1}$ and $S_3^{\,1}$. Table 7.6.2 contains a useful set of differentials.

The fields of Table 7.5.1 for spherical Bessel functions only are shown in Table 7.6.3.

Using flow chart Fig. 7.6.1 to operate on the values of Table 7.6.3 returns the same sums with unaltered values and hence the solution is correct for all iterations; this is expected since the Bessel portion of the solution is exact.

Table 7.6.2. A useful set of differentials.

$$\frac{\partial}{\partial\theta}\left[\sin\theta U(\sigma,\theta)\right] = -V(\sigma,\theta) + ie^{-i\sigma\cos\theta}$$

$$\frac{\partial}{\partial\theta}\left[\sin\theta V(\sigma,\theta)\right] = -U(\sigma,\theta) + i\cos\theta e^{-i\sigma\cos\theta}$$

$$\frac{i}{\sigma}\frac{\partial}{\partial\sigma}\left[\sigma U(\sigma,\theta)\right] = V(\sigma,\theta)$$

$$\frac{i}{\sigma}\frac{\partial}{\partial\sigma}\left[\sigma V(\sigma,\theta)\right] = U(\sigma,\theta) - \frac{1}{\sigma}e^{-i\sigma\cos\theta}$$

Table 7.6.3. Bessel sums.

$$S_1(\sigma,\theta) = \sigma\sin\theta e^{-i\sigma\cos\theta}$$

$$S_2(\sigma,\theta) = e^{-i\sigma\cos\theta} + iV; \qquad \$_3(\sigma,\theta) = -iV$$

$$S_3(\sigma,\theta) = -iU; \qquad \$_3(\sigma,\theta) = \cos\theta e^{-i\sigma\cos\theta} + iV$$

Table 7.6.4. Zero-Order Hankel Sums.

$$S_1^0(\sigma,\theta) = \sigma\sin\theta(1+\cos\theta)e^{-i\sigma\cos\theta}$$

$$S_2^0(\sigma,\theta) = (1+\cos\theta)e^{-i\sigma\cos\theta} + i(U+V)$$

$$S_3^0(\sigma,\theta) = -i(U+V)$$

$$\$_2^0(\sigma,\theta) = \cos\theta(1+\cos\theta)e^{-i\sigma\cos\theta} + i(U+V)$$

$$\$_3^0(\sigma,\theta) = -i(U+V)$$

Table 7.6.5. Electric fields after two iterations, symbol in $<>$ arise from Bessel functions and all others from Neumann functions.

2	$\eta\tilde{\mathbf{H}} = i\tilde{\mathbf{E}};\ F_0 e^{i(\omega t - \phi - \sigma\cos\theta)}$ suppressed, $<>$ indicates Bessel term
E_r^2	$\sin\theta\left\{(<1>+\cos\theta) + \dfrac{i}{\sigma}\left(1+2\sin^2\theta\right) + \dfrac{7}{\sigma^2}\cos\theta + \dfrac{2i}{\sigma^3}\right\}$
E_θ^2	$\left\{\cos\theta\left(<1>+\cos\theta\right) + \dfrac{i}{\sigma}2\sin^2\theta\cos\theta + \dfrac{1}{\sigma^2}\left(2-\sin^2\theta\right) - \dfrac{i}{\sigma^3}\cos\theta\right\}$
E_ϕ^2	$-i\left\{(<1>+\cos\theta) + \dfrac{i}{\sigma}2\sin^2\theta + \dfrac{1}{\sigma^2}\cos\theta - \dfrac{i}{\sigma^3}\right\}$

Table 7.6.6. Electric fields after six iterations, symbol in $<>$ arise from Bessel functions and all others from Neumann functions.

6	$\eta\tilde{\mathbf{H}} = i\tilde{\mathbf{E}};\ F_0 e^{i(\omega t - \phi - \sigma\cos\theta)}$ suppressed, $<>$ indicates Bessel term
E_r^6	$\sin\theta\left\{\begin{array}{l}(<1>+\cos\theta) + \dfrac{i}{\sigma}\left(1+6\sin^2\theta\right) + \dfrac{1}{\sigma^2}\left(19+36\sin^2\theta\right)\cos\theta \\[2ex] -\dfrac{i}{\sigma^3}\left(122-36\sin^2\theta-120\sin^4\theta\right) \\[2ex] -\dfrac{1}{\sigma^4}\left(204-900\sin^2\theta\right)\cos\theta + \dfrac{i}{\sigma^5}\left(360-600\sin^2\theta\right)\end{array}\right\}$
E_θ^6	$\left\{\begin{array}{l}\cos\theta\left(<1>+\cos\theta\right) + \dfrac{i}{\sigma}6\sin^2\theta\cos\theta + \dfrac{1}{\sigma^2}\left(6+35\sin^2\theta-36\sin^4\theta\right) \\[2ex] -\dfrac{i}{\sigma^3}\left(53+84\sin^2\theta-120\sin^4\theta\right)\cos\theta \\[2ex] -\dfrac{1}{\sigma^4}\left(204-360\sin^2\theta+60\sin^4\theta\right) + \dfrac{i}{\sigma^5}\left(240-900\sin^2\theta\right)\cos\theta\end{array}\right\}$
E_ϕ^6	$\left\{\begin{array}{l}(<1>+\cos\theta) + \dfrac{i}{\sigma}6\sin^2\theta + \dfrac{1}{\sigma^2}\left(6+36\sin^2\theta\right)\cos\theta \\[2ex] -\dfrac{i}{\sigma^3}\left(53+24\sin^2\theta-96\sin^4\theta\right) \\[2ex] -\dfrac{1}{\sigma^4}\left(204-288\sin^2\theta\right)\cos\theta + \dfrac{i}{\sigma^5}\left(240-300\sin^2\theta\right)\end{array}\right\}$

The zero iteration fields of Table 7.5.2 for spherical Hankel functions are shown in Table 7.6.4. Using flow chart Fig. 7.6.1 to operate on the values of Table 7.6.4 returns modified values that include the desired higher inverse orders of σ.

Table 7.6.5 and 7.6.6 show the electric field portion of the total fields after, respectively, two and six iterations. As noted in Sec. A.21 the validity of the form is limited by the inequality $L(L+1) >> \sigma_0^2$, where σ_0 is the largest range of interest.

The suppressed multiplier $\exp[i(\omega t - \phi - kz)]$ confirms the fields are fully $+z$-directed.

7.7 Far-Field Kinematics

The products of some field terms are orthogonal and some are not. Integrals over orthogonal products leave no intermodal terms but nonorthogonal products do. Nonorthogonal products are, therefore, directly dependent upon the modal recursion relationship. Power and energy expressions are orthogonal but momenta, directivity, and radiation Q are not. In this section we characterize kinematic and radiation parameters associated with both orthogonal and nonorthogonal products of Eq. 7.4.1.

By Poynting's theorem field energy is emitted at the rate:

$$\frac{dW}{dt} = \frac{\sigma^2}{k^2} \int_0^{2\pi} d\phi \int_0^{\pi} \mathrm{Re}[N_r] \sin\theta d\theta \qquad (7.7.1)$$

By Eq. 1.9.7 the momentum contained within a volume is $1/c^2$ times the volume integral of the Poynting vector and hence the momentum contained within a spherical shell is:

$$\mathbf{p} = \frac{1}{c^2 k^3} \int \sigma^2 d\sigma \int_0^{2\pi} d\phi \int_0^{\pi} \mathrm{Re}[\mathbf{N}] \sin\theta d\theta \qquad (7.7.2)$$

The rate at which linear momentum p_z exits a closed volume equals the surface integral of the z-component of the Poynting vector:

$$\frac{dp_z}{dt} = \frac{\sigma^2}{ck^2} \int_0^{2\pi} d\phi \int_0^{\pi} \text{Re}\left[N_r \cos\theta - N_\theta \sin\theta \right] \sin\theta \, d\theta \qquad (7.7.3)$$

Similarly, the rate at which angular momentum l_z exits a closed volume is:

$$\frac{dl_z}{dt} = \frac{\sigma^3}{ck^3} \int_0^{2\pi} d\phi \int_0^{\pi} \text{Re}\left[N_\phi \right] \sin^2\theta \, d\theta \qquad (7.7.4)$$

Using Eq. 7.4.1 the three time-average vector components of the Poynting vector are:

$$N_r = \frac{\text{Re}}{2\eta} \sum_{\ell=1}^{\infty} \sum_{n=1}^{\infty} F_\ell F_n^* i^{n-\ell} \left\{ \begin{aligned} &\left(h_\ell h_n^* + \dot{h}_\ell \dot{h}_n^* \right) \left(\frac{P_\ell^1}{\sin\theta} \frac{dP_n^1}{d\theta} + \frac{P_n^1}{\sin\theta} \frac{dP_\ell^1}{d\theta} \right) \\ &-i\left(h_\ell \dot{h}_n^* - h_n^* \dot{h}_\ell \right) \left(\frac{P_\ell^1}{\sin\theta} \frac{P_n^1}{\sin\theta} + \frac{dP_\ell^1}{d\theta} \frac{dP_n^1}{d\theta} \right) \end{aligned} \right\}$$

$$(7.7.5)$$

$$N_\theta = -\frac{\text{Re}}{2\sigma\eta} \sum_{\ell=1}^{\infty} \sum_{n=1}^{\infty} F_\ell F_n^* i^{n-\ell} \left[n(n+1) h_n^* \dot{h}_\ell + \ell(\ell+1) h_\ell \dot{h}_n^* \right] \frac{P_\ell^1 P_n^1}{\sin\theta}$$

$$(7.7.6)$$

$$N_\phi = \frac{\text{Re}}{2\sigma\eta} \sum_{\ell=1}^{\infty} \sum_{n=1}^{\infty} F_\ell F_n i^{n-\ell} \left\{ \begin{aligned} &\left[n(n+1) h_n^* \dot{h}_\ell P_n^1 \frac{dP_\ell^1}{d\theta} - \ell(\ell+1) h_\ell \dot{h}_n^* P_\ell^1 \frac{dP_n^1}{d\theta} \right] \\ &-i\left(\left[n(n+1) + \ell(\ell+1) \right] h_\ell h_n^* \frac{P_\ell^1 P_n^1}{\sin\theta} \right) \end{aligned} \right\}$$

$$(7.7.7)$$

Although the above expressions would be correct with limitlessly large modal numbers, actual series that include sources terminate. Whatever that value may be, we designate it as L. Substituting the Poynting vector components as needed into the integrals of Eq. 7.7.1, 7.7.3, and 7.7.4, using integral Table A.16.1, and limiting results to far-fields gives:

$$\frac{dW}{dt} = \frac{4\pi}{\eta k^2} \sum_{\ell=1}^{L} F_\ell^2 \frac{\ell^2(\ell+1)^2}{(2\ell+1)}$$

$$\frac{dp_z}{dt} = \frac{4\pi}{\eta c k^2} \sum_{\ell=1}^{L} \frac{\ell(\ell+1)}{(2\ell+1)} \left\{ F_\ell^2 + F_\ell F_{\ell+1} \frac{\ell(\ell+2)^2}{(2\ell+3)} + F_\ell F_{\ell-1} \frac{(\ell-1)^2(\ell+1)}{(2\ell-1)} \right\}$$

$$\frac{dl_z}{dt} = \frac{4\pi}{\eta \omega k^2} \sum_{\ell=1}^{L} F_\ell^2 \frac{\ell^2(\ell+1)^2}{(2\ell+1)}$$

$$(7.7.8)$$

The center equation shows that much of the linear momentum is carried jointly by nearest-neighbor modes $\ell \pm 1$.

Entering the recursion relationship of Eq. 7.4.1 into Eq. 7.7.8 gives:

$$\lim_{\sigma \to \infty} \frac{dW}{dt} = \frac{4\pi F_0^2}{\eta k^2} \sum_{\ell=1}^{L} (2\ell+1)$$

$$\lim_{\sigma \to \infty} \frac{dp_z}{dt} = \frac{4\pi F_0^2}{\eta c k^2} \sum_{\ell=1}^{L} (2\ell+1) \qquad (7.7.9)$$

$$\lim_{\sigma \to \infty} \frac{dl_z}{dt} = \frac{4\pi F_0^2}{\eta \omega k^2} \sum_{\ell=1}^{L} (2\ell+1)$$

It follows that:

$$\frac{dW}{dt} \bigg/ \frac{dp_z}{dt} = c; \quad \frac{dW}{dt} \bigg/ \frac{dl_z}{dt} = \omega; \quad \frac{dp_z}{dt} \bigg/ \frac{dl_z}{dt} = k \qquad (7.7.10)$$

The time integral of Eq. 7.7.10 yields the kinematic properties of this radiation; the energy-to-linear momentum ratio shows the energy is fully z-directed and the energy-to-angular momentum ratio shows the field is fully rotational. The energy-to-linear momentum ratio is the value Einstein showed is necessary for photons.

We also note from Eq. 7.7.9 and 7.7.10 that the outbound power carried by the far field terms is:

$$\frac{dW}{dt} = c\frac{dp_z}{dt} = \omega\frac{dl_z}{dt} = \frac{4\pi F_0^2}{\eta k^2}L(L+2) \qquad (7.7.11)$$

The time integral of Eq. 7.7.11 over pulse period τ gives the output energy carried by the far field terms:

$$W = \frac{4\pi F_0^2}{\eta k^2}L(L+2)\tau \qquad (7.7.12)$$

The directivity of individual modes is given in Eq. 7.4.6. Since modal power is described by orthogonal functions, it also applies to mixed, multimodal TE and TM fields. The total directivity, therefore, is the simple sum of Eq. 7.4.6 over all modes:

$$D = \sum_{\ell=1}^{L}(2\ell+1) = L(L+2) \qquad (7.7.13)$$

Directivity increases as the square of maximum modal number L.

7.8 Computer Evaluated Far Field Sums

As a first estimate of L, note a biconical antenna requires a maximum modal number not less than 100 to match fields at the cone-cap junction. However, the magnitudes are so small the high orders have negligible effect away from the antenna. Quite differently, high order modes are large in field set Eq. 7.4.1 and the directivity expression Eq. 7.7.13 shows even the highest orders strongly influence the output power. Of course, the full directivity conclusion reached in Sec. 7.6 and the energy-to-linear momentum ratio of Eq. 7.7.10 is valid only with a limitlessly large value of L.

A plot of angular field magnitudes as a function of zenith angle illustrates the field dependence upon maximum modal number. The angular dependence of the Eq. 7.4.2 angular electric and magnetic field intensities is shown in Fig. 7.8.1. The plot is over the full range of zenith angle, uses upper limit $L = 1000$, and is normalized by $L(L+2)$. Computer sums were obtained using a MathCad software package.

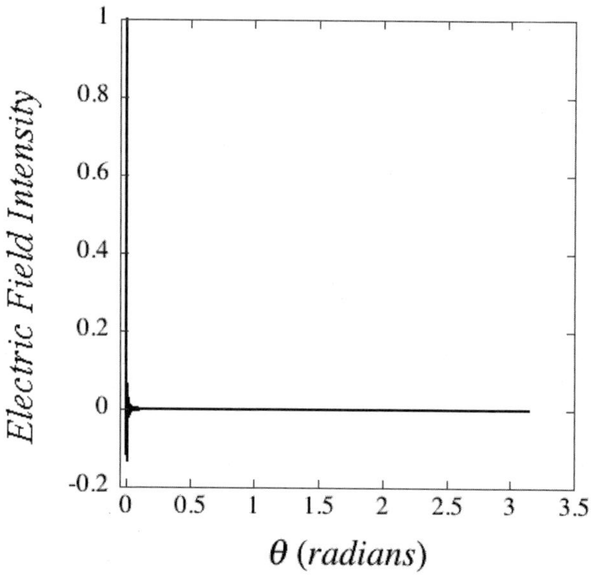

Figure 7.8.1. Zenith and azimuth far-fields of Eq. 7.4.2 versus zenith angle from 0 to π radians; $L = 1000$, normalization is $L(L+2)$.

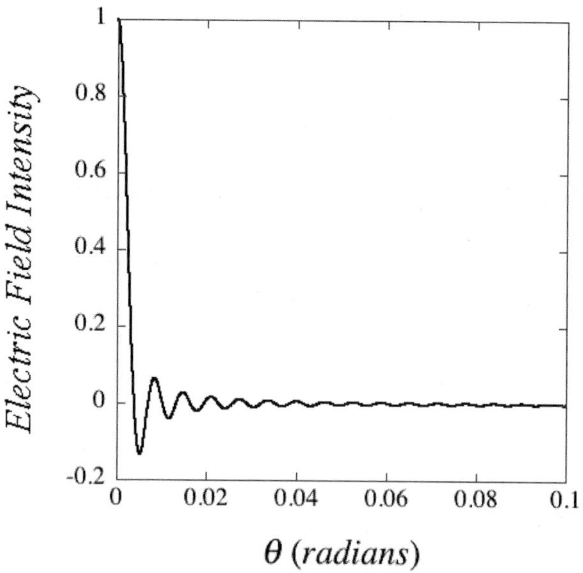

Figure 7.8.2. Zenith and azimuth far-fields of Eq. 7.4.2 versus zenith angle from 0 to 0.1 radians; $L = 1000$, normalization is $L(L+2)$.

Details for the portion within $\pi/10$ radians of the z-axis are shown in Fig. 7.8.2. Since the zenith and azimuth fields are vanishingly small everywhere except on the z-axis the output power approaches being fully directional. The plot shows a z-axis Dirac delta function $\delta(\theta,0)$.

We note that upper limit L requires a minimum of 2^{L-1} separate charges; that is, with $L=1000$ at least 10^{301} charged units are needed to produce the appropriate multipole. If the diameter of an isolated electron is about equal a Compton wavelength, 2.4×10^{-12} m, and if charge units are hard-packed inside it at the most each is allotted about 10^{-100} m linear dimension. Yet on some dimensional scale an electromagnetic origin of the exclusion principle requires discrete charges.

Curves showing the output power emitted by Eq. 7.4.2 versus zenith angle are shown in Fig. 7.8.3 and 7.8.4. These, and other related plots, show the relationship between the zenith angle of the first power null and maximum modal number:

$$L\Delta\theta \cong 5.13 \tag{7.8.1}$$

As L increases without limit the angle of the first null approaches zero parabolically. With Ψ to be determined, the figure shows that the z-directed Poynting vector has the form:

$$N_r = \frac{F_0^2}{\eta\sigma^2}\Psi\,\delta(\theta,0) \tag{7.8.2}$$

Integrating Eq. 7.7.15 over the full range of solid angle gives the power output:

$$P = \frac{2\pi F_0^2}{\eta k^2}\int_0^\pi \sin\theta d\theta\,\Psi\delta(\theta,0) = \frac{2\pi}{\eta k^2}F_0^2\Psi \tag{7.8.3}$$

Combining Eq. 7.7.11 and 7.8.3 gives:

$$\Psi = 2L(L+2) \tag{7.8.4}$$

Substituting back into Eq. 7.8.2 shows the radial component of the Poynting vector is:

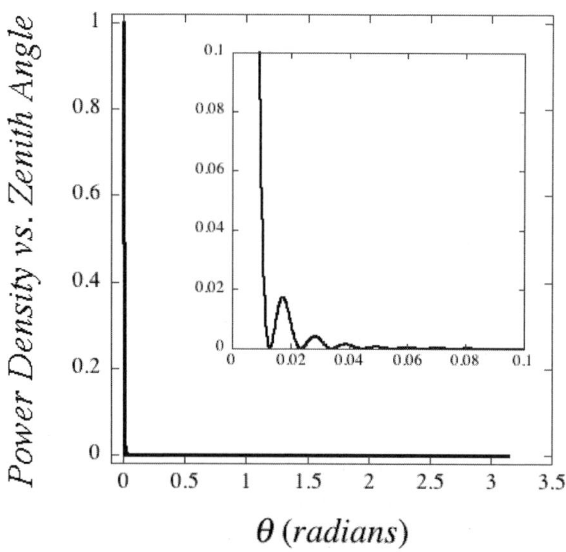

Figure 7.8.3. Variation of far-field power with zenith angle, 0 to π radians; inset is 0 to 0.1 radians; $L = 300$, normalization is $L(L+2)$.

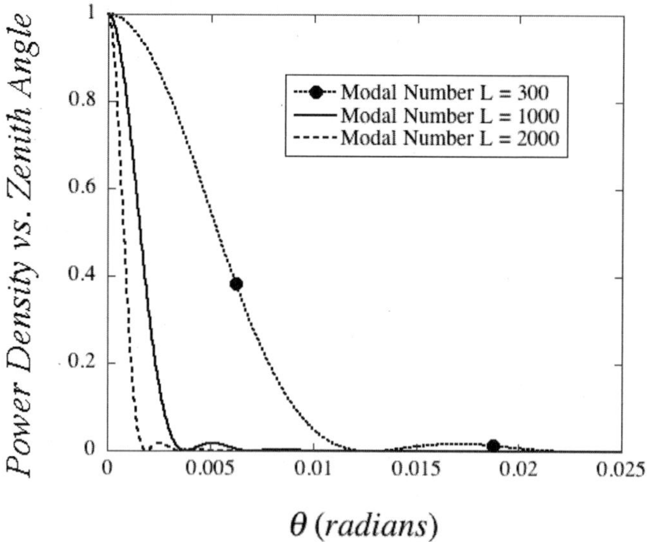

Figure 7.8.4. Variation of far-field power with zenith angle from 0 to 0.025 radians; maximum modal number: dotted line 300, solid line 1000, and dashed line 2000, each curve normalized to $L(L+2)$.

$$N_r = \frac{2F_0^2}{\eta\sigma^2}L(L+2)\delta(\theta,0) \qquad (7.8.5)$$

The radiated energy over period τ is:

$$Energy\,out = P\tau = \frac{4\pi F_0^2}{\eta k^2}L(L+2)\tau \qquad (7.8.6)$$

A parabolic relationship exists between the maximum modal number and the energy output.

7.9 Surface Tensile and Shear Pressures

The surface pressure on a sphere circumscribing the source of Eq. 7.4.1 is time independent. Although the electric and magnetic fields both generate time-dependent pressures the time dependencies have equal magnitude and opposite phase and hence cancel. However the electric field portion acts on charge densities and the magnetic field portion on current densities and, therefore, results depend upon properties of the sphere.

The top line of Eq. 7.3.3 gives an expression for the electromagnetic pressure on any surface. Pressures are dependent upon both zenith angle and range and we seek to examine the pressure created by the field set of Eq. 7.4.1 on the surface of a virtual sphere circumscribing the source.

To examine far field pressure consider the limit as the radius of the sphere becomes limitlessly large and, for that case, the radial field component vanishes. Combining Eq. 7.3.3 with the angular fields of Eq. 7.4.2 and summing over all modes gives the pressure on the sphere as a function of zenith angle:

$$\lim_{ka\to\infty} p_{r\ell} = -\frac{\varepsilon F_0^2}{2\sigma^2}\left\{\sum_{\ell=1}^{L}\frac{(2\ell+1)}{\ell(\ell+1)}\left(\frac{dP_\ell^1}{d\theta}+\frac{P_\ell^1}{\sin\theta}\right)\right\}^2 \qquad (7.9.1)$$

The angular dependence is the negative of the power expression and hence of Fig. 7.8.3 where, as shown there, the sum is a Dirac delta function at $\theta = 0$. On the positive z-axis the pressure is:

248   *Photon Creation-Annihilation*

$$p_{tot} = -\frac{4\pi F_0^2}{k^2} L(L+2)\delta(\theta,0) \tag{7.9.2}$$

Integrating the pressure over all solid angles gives the net radiation reaction force on the source:

$$force = -\frac{4\pi F_0^2}{k^2} L(L+2) \tag{7.9.3}$$

The negative sign indicates the source is pushed in the $-z$-direction. The change in its linear momentum is:

$$\Delta(momentum) = force \times \tau = \frac{4\pi F_0^2}{k^2} L(L+2)\tau \tag{7.9.4}$$

Comparing Eq. 7.8.6 and 7.9.4 shows the ratio of energy emitted by the source to the momentum it acquires is:

$$\frac{\Delta(energy)}{\Delta(momentum)} = c \tag{7.9.5}$$

Comparison with Eq. 7.7.10 shows it is also the energy-to-momentum ratio of the emitted radiation. This is the mechanism whereby Newton's second law is satisfied.

7.10 High Powers of Range, σ^{-s}

We characterized the portion of the fields of Eq. 7.4.1 with a magnitude that varies inversely with range in Sec. 7.7 and 7.8. In this section we extend that charcterization to fields with a magnitude that varies as an arbitrary power of range. For this purpose we begin with the fields of Eq. 7.4.1 expressed using the sums of Eq. 7.6.1:

$$\eta\tilde{\mathbf{H}} = i\tilde{\mathbf{E}}, \qquad e^{i(\omega t - \phi)} \text{ suppressed}$$

$$\tilde{\mathbf{E}} = F_0 \left\{ \begin{array}{l} \left[S_1(\sigma,\theta)/\sigma \right]\hat{r} \\ + \left[S_3(\sigma,\theta) + \$_2(\sigma,\theta) \right]\hat{\theta} - i\left[S_2(\sigma,\theta) + \$_3(\sigma,\theta) \right]\hat{\phi} \end{array} \right\} \tag{7.10.1}$$

The radial portion of the Poynting vector is:

$$N_2 = \frac{1}{2\eta} \left\{ \begin{array}{l} \left[S_3(\sigma,\theta) + \$_2(\sigma,\theta) \right] \left[S_2(\sigma,\theta) + \$_3(\sigma,\theta) \right]^* \\ + \left[S_2(\sigma,\theta) + \$_3(\sigma,\theta) \right] \left[S_3(\sigma,\theta) + \$_2(\sigma,\theta) \right]^* \end{array} \right\} \tag{7.10.2}$$

The single sum for a spherical Hankel function is expressed in Eq. A.19.14 and, substituting it into the sums of Eq. 7.10.1, gives the double sums:

$$S_1 = \frac{ie^{-i\sigma}}{\sigma} \sum_{\ell=1,s}^{L} (2\ell+1) P_\ell^1 \sum_{s=0}^{\ell} \left(\frac{1}{2i\sigma} \right)^s \frac{(\ell+s)!}{s!(\ell-s)!}$$

$$S_2 = \frac{ie^{-i\sigma}}{\sigma} \sum_{\ell=1,s}^{L} \frac{(2\ell+1)}{\ell(\ell+1)} \frac{dP_\ell^1}{d\theta} \sum_{s=0}^{\ell} \left(\frac{1}{2i\sigma} \right)^s \frac{(\ell+s)!}{s!(\ell-s)!} \tag{7.10.3}$$

$$S_3 = S_2 \left[\frac{P_\ell^1}{\sin\theta} \middle/ \frac{dP_\ell^1}{d\theta} \right]$$

A single sum for an associated spherical Hankel function is expressed in Eq. A.20.5 and, after doing some algebra, gives the double sums:

$$\$_2(\sigma,\theta) = \frac{ie^{-i\sigma}}{\sigma} \sum_{\ell=1}^{L} \frac{(2\ell+1)}{\ell(\ell+1)} \left(\frac{dP_\ell^1}{d\theta} \right)$$

$$\times \sum_{s=0}^{\ell} \left(\frac{1}{2i\sigma} \right)^s \frac{(\ell+s-1)!}{s!(\ell-s+1)!} \left[\ell(\ell+1) + s(s-1) \right] \tag{7.10.4}$$

$$\$_3 = \$_2 \left[\frac{P_\ell^1}{\sin\theta} \middle/ \frac{dP_\ell^1}{d\theta} \right]$$

With the arbitrary choice of $s = 5$ as an example, a plot of the Poynting vector versus zenith angle is shown in Fig. 7.10.1. As in Sec. 7.7, and as expected, it has the general form of a Dirac delta function. The normalization factor is 1.93×10^{63} and the ratio of this value to the equivalent for $s = 0$ is 1.83×10^{51}.

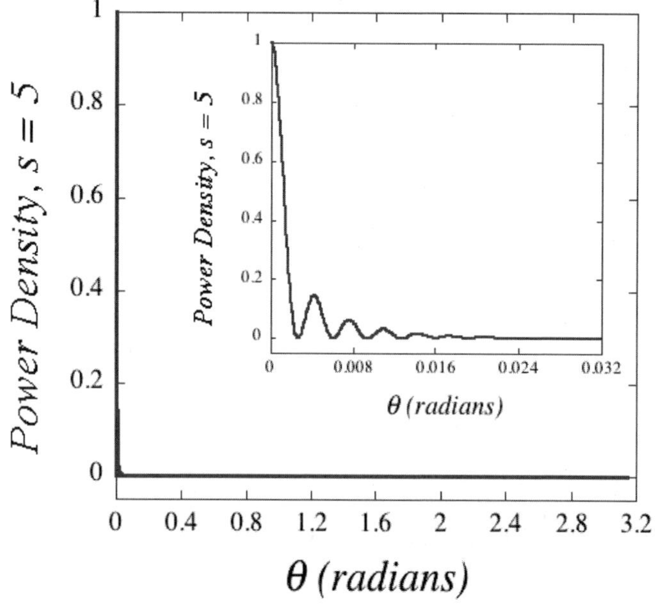

Figure 7.10.1. Power density for special case $L = 1000$ and $s = 5$, full range of θ from 0 to π, with inset showing range 0 to $\pi/100$ radians, normalization factor at $\theta = 0$ is 1.83×10^{63}, first zero at 2.61×10^{-3} radians.

The zero nearest the z-axis, as determined from the inset of Fig. 7.10.1, occurs at:

$$L\Delta\theta \cong 2.61 \tag{7.10.5}$$

Defining X to be the field values of Eq. 7.10.1 near the positive z-axis and combining with Eq. 7.10.3 and 7.10.4 gives:

$$S_1 = \frac{ie^{-i\sigma}}{\sigma^2} \sum_{\ell=1,s}^{L} (2\ell+1)\ell(\ell+1) \sum_{s=0}^{\ell} \left(\frac{1}{2i\sigma}\right)^s \frac{(\ell+s)!}{s!(\ell-s)!} \sin\theta$$

$$\left[S_3(\sigma,0) + S_2(\sigma,0) \right] = \left[S_2(\sigma,0) + S_3(\sigma,0) \right] \tag{7.10.6}$$

$$= \frac{ie^{-i\sigma}}{\sigma} \sum_{\ell=1}^{L} (2\ell+1)\ell(\ell+1) \sum_{s=0}^{\ell} \left(\frac{1}{2i\sigma}\right)^s \frac{(\ell+s-1)!}{s!(\ell-s+1)!}$$

The angular field terms have the form:

$$\tilde{\mathbf{E}} = X(\sigma)\left[\hat{\theta} - i\hat{\phi}\right]e^{-i\phi}e^{i(\omega t - kz)}; \quad \tilde{\mathbf{H}} = \frac{1}{\eta}X(\sigma)\left[i\hat{\theta} + \hat{\phi}\right]e^{-i\phi}e^{i(\omega t - kz)} \quad (7.10.7)$$

Since the field is entirely z-directed:

$$\frac{d\mathrm{W}}{dt} \bigg/ \frac{dp_z}{dt} = c \quad (7.10.8)$$

7.11 A Zero-Q Radiation Field

In this section we discover and explore characteristics of electromagnetic emission in the total absence of reactive energy. With no reactive energy there is no reactive power and the radiation Q is zero.

We begin by examining properties of the matched electric and magnetic modal pairs of Eq. 7.4.1. For convenience, and without loss of rigor, we examine modal pairs with the phasing of $\ell = 1, 5, 9, \ldots$ This is no loss of generality since we revert to the original phasing when summing over modal pairs. For convenience we use definitions $S_\ell = P_\ell^1/\sin\theta$ and $T_\ell = dP_\ell^1/d\theta$, separate the Hankel functions into rational and transcendental parts, see Sec. A.20, and find the resulting time-dependent fields per modal pair:

$$\sigma E_r = \frac{F_0}{\sigma}(2\ell+1)\left[B_\ell \cos(\omega t_R - \phi) - A_\ell \sin(\omega t_R - \phi)\right]P_\ell^1 \hat{r}$$

$$+ F_0 \frac{(2\ell+1)}{\ell(\ell+1)}\left\{\begin{array}{l}\left[D_\ell \cos(\omega t_R - \phi) - C_\ell \sin(\omega t_R - \phi)\right]T_\ell^1 \\ + \left[A_\ell \cos(\omega t_R - \phi) + B_\ell \sin(\omega t_R - \phi)\right]S_\ell^1\end{array}\right\}\hat{\theta}$$

$$+ F_0 \frac{(2\ell+1)}{\ell(\ell+1)}\left\{\begin{array}{l}\left[C_\ell \cos(\omega t_R - \phi) + D_\ell \sin(\omega t_R - \phi)\right]T_\ell^1 \\ + \left[B_\ell \cos(\omega t_R - \phi) - A_\ell \sin(\omega t_R - \phi)\right]S_\ell^1\end{array}\right\}\hat{\theta} \quad (7.11.1)$$

The radial component of the Poynting vector contains two polarizations: $E_\theta H_\phi$ and $E_\phi H_\theta$. They carry superimposed but independent energy flows. The power densities per polarization are shown in Eq. 7.11.2; the upper and lower signs give values for polarizations $E_\theta H_\phi$ and $E_\phi H_\theta$, respectively:

$$N_{r\ell} = \frac{F_0^2}{2\eta\sigma^2} \frac{(2\ell+1)}{\ell^2(\ell+1)^2} \left\{ \begin{array}{l} (A_\ell D_\ell - B_\ell C_\ell) \pm (A_\ell D_\ell + B_\ell C_\ell)\cos 2(\omega t_R - \phi) \\ \mp (A_\ell C_\ell - B_\ell D_\ell)\sin 2(\omega t_R - \phi) \end{array} \right\}$$

$$\times \left[\left(S_\ell^1\right)^2 + \left(T_\ell^1\right)^2 \right] + \frac{F_0^2}{2\eta\sigma^2} \frac{(2\ell+1)}{\ell^2(\ell+1)^2} \left(S_\ell^1\right)\left(T_\ell^1\right)$$

$$\times \left\{ \begin{array}{l} \left(A_\ell^2 + B_\ell^2 + C_\ell^2 + D_\ell^2\right) \pm \left(A_\ell^2 - B_\ell^2 - C_\ell^2 + D_\ell^2\right)\cos 2(\omega t_R - \phi) \\ \pm 2(A_\ell B_\ell - C_\ell D_\ell)\sin 2(\omega t_R - \phi) \end{array} \right\}$$

$$(7.11.2)$$

To assist in characterizing Eq. 7.11.2 note from Table A.21.1 the angular functions on the positive z-axis are:

$$S_\ell^1(\cos 0) = T_\ell^1(\cos 0) = \ell(\ell+1)/2 \qquad (7.11.3)$$

Since the radial field vanishes at $\theta = 0$ there is no angularly directed component of the Poynting vector and the complete Poynting vector is:

$$\mathbf{N}\big|_{\theta=0} = \frac{F_0^2 \hat{r}}{4\eta\sigma^2}(2\ell+1)^2 \left\{ \begin{array}{l} \left[\left(A_\ell^2 + B_\ell^2 + C_\ell^2 + D_\ell^2\right) + 2(A_\ell D_\ell - B_\ell C_\ell)\right] \\ \times\left[1 \pm \cos\left(2(\omega t_R - \phi + \zeta_\ell)\right)\right] \end{array} \right\}$$

$$d\mathbf{N}/d\theta\big|_{\theta=0} \sim \sin\theta$$

$$(7.11.4)$$

ζ_ℓ is the constant phase factor:

$$\tan \zeta_\ell = -\frac{2(A_\ell + D_\ell)(B_\ell - C_\ell)}{\left[\left((A_\ell + D_\ell)\right)^2 - \left((B_\ell - C_\ell)\right)^2\right]}$$

The unique property of Eq. 7.11.4 is that the energy flow consists solely of outgoing trigonometric pulses of outbound energy and none of it returns to the source. There is no reactive power! There is a window, or more accurately a porthole, on the $+z$-axis, and there is nothing to prohibit the full field energy of Eq. 7.4.1 from leaving the source. Since order number ℓ of Eq. 7.11.4 is arbitrary the equation applies to every modal order and since it applies to every order it applies to the sum over all modal combinations!

As an aside, and in an attempt to understand why this occurs, note the radial portion of a modal field plays a significant role since reactive energy oscillates between radial and angular directions. Under the conditions in which Eq. 7.11.4 applies there is no radial field. With TM modes the magnetic field has no radial component and the radial electric field vanishes in the plane perpendicular to the electric moment; of course the dual is true for TE modes. At the instant the electric moments are x-oriented there is no radial field in the yz-plane and at that instant the magnetic moments are y-oriented and there is no radial field in the xz-plane. The planes intersect on the z-axis and hence on it there is no z-directed field. As the ensemble rotates about the z-axis the conclusion remains unaltered: the z-axis supports no radial field, no reactive power, and no reactive energy. It does support outbound trigonometric pulses of power, with equal and out of phase pulses on the two polarizations the total output power is constant.

In summary, Eq. 7.4.12 shows the radiation Q of a modal pair is far larger than necessary to squelch all radiation. However, summing over all modal pairs of Eq. 7.11.4 shows the overall radiation Q vanishes. That radiation, without further ado, simply exits the system. There is, therefore, no electromagnetic barrier to the emission of a pulse of radiation that satisfies Eq. 7.4.1! The remaining questions are: How and why does nature produce the unique field set that creates or annihilates such radiation? We address the issue in the next two chapters.

References

L.J. Chu, "Physical Limitations of Omni-Directional Antennas," J. Appl. Phys., Vol. 19, 1163–1175 (1948)

R.E. Collin, S. Rothschild, "Evaluation of Antenna Q," IEEE Trans. Antennas and Propagation, Vol. 12, pp. 23–27 (1964)

C.A. Grimes, G. Liu, D.M. Grimes, K.G. Ong, "Characterization of a Wideband,

Low-Q, Electrically Small Antenna," Microwave and Optical Technology Letters, vol. 27, pp. 53–58 (2000)

R.F. Harrington, "Effect of Antenna Size on Gain, Bandwidth, and Efficiency," J. Research, Nat. Bureau of Standards, Vol. 64D, pp. 1–12 (1960)

J. D. Jackson, Classical Electrodynamics, John Wiley & Sons, 2nd ed., (1975)

W.R. Smythe, Static and Dynamic Electricity, 3rd ed., Summa Press, New York, NY, p. 39 (1989)

W.K.H. Panofsky, M. Phillips, *Classical Electricity and Magnetism*, 2nd ed., Addison-Wesley (1962)

Chapter 8

Spontaneous Emission

$\mathcal{T}$he legacy of previous chapters is a radiation pulse with the kinematic properties of a photon. The method of ejecting such a pulse is quite different from the operation of a transmitting antenna; most especially because the pulse supports no reactive energy from which there is, in turn, no distinction between near and far fields. At first glance it appears stupefyingly difficult to create such a pulse. In this chapter we address the issue of how it is readily accomplished on an atomic level.

A subject notable for its absence in discussions of atomic phenomena is nonlinear response to driving forces. Modern radio, on the other hand, arguably began in 1901 with Fessenden's invention of the heterodyne receiver. It was extended by Armstrong's 1918 invention of a superheterodyne receiver; it requires signal modulation, mixing, and detection, all of which require nonlinear circuits. In 1922 he wrote about regeneration and superregeneration in electrical circuits. Both an atom-photon pair and the first mixer of a superheterodyne receiver use nonlinearities to create the desired output frequency from two input frequencies. The similarity, however, was not recognized by Dirac when he wrote, eight years later, about an 'unexpected connexion' between atomic frequencies. Although frequency pulling was utilized by World War-II electronics, quantitative discussions were delayed until after the war. In 1954 Manley and Rowe published the power-frequency relationships that govern nonlinear systems and shortly afterward, 1957, Weiss pointed out that the MR equations correctly predict Dirac's 'unexpected connexion'. The MR equations provide a quantitative explanation of power-frequency relationships in nonlinear systems. Since the predicted properties match those of eigenstate electrons we develop, in the manner of their second paper, the MR equations in Sec. 8.1.

In Sec. 8.2 we analyze feedback in nonlinear systems. In Sec. 8.3 and 8.4 respectively we show that the modes of Eq. 7.4.1 support intermodal fields and energies that provide positive feedback for electromagnetic oscillations. These two feedback mechanisms drive an amenable system to ever-higher modal orders, in a way that satisfies the Manley Rowe equations. In Sec. 8.5 we discuss the circumstantial evidence that such forces create spontaneous photon emission.

A superregenerative modal buildup can only exist if the source maintains an ordered internal structure. Although Eq. 7.4.1 applies only to a charge-free region, external fields and internal sources have the same symmetry: only a highly ordered source could produce highly ordered field set Eq. 7.4.1. In Sec. 8.6 to 8.9 we obtain exact values of the sums on each of the coordinates axes, from which we conclude the source is a highly ordered structure.

8.1 Power-Frequency Relationship

The power-frequency relationships obeyed by quantized radiation were not understood in the early years of the 20th century. As late as 1930 Dirac referred to the Ritz power-frequency combination law as a 'new and unexpected connexion between the frequencies'. Some 24 years later, 1954, Manley and Rowe completed the connection as part of their attempt to understand and design nonlinear parametric amplifiers. The laws govern possible interactions of superimposed but otherwise independent frequencies within a nonlinear medium. Although the Manley Rowe equations also apply to linear media that result is moot since the rate of energy transfer in linear media is zero.

Let us examine possible power-frequency relationships in closed, lossless systems. For that purpose, consider a system that supports an eigenstate with its associated frequency ω_s. The system is then immersed in and perturbed by an applied electromagnetic wave of frequency ω; if the system is linear it contains oscillations at those two frequencies only. If the system response is nonlinear, however, we know from Manley and Rowe that it will also support receptive sum and difference frequencies.

Following Manley and Rowe, with A, B, and p representing constant, system-specific parameters we write the expression for all possible internal signals as:

$$S(t) = \left[A\cos(\omega_s t) + B\cos(\omega t) \right]^p \tag{8.1.1}$$

The system response may be expanded as a polynomial of trigonometric functions. An especially important example is the special case $p = 2$, for which:

$$S(t) = \frac{A^2 + B^2}{2} + AB\left\{ \cos\left[(\omega_s - \omega)t \right] + \cos\left[(\omega_s + \omega)t \right] \right\}$$

$$+ \frac{1}{2}\left\{ A^2 \cos(2\omega_s t) + B^2 \cos(2\omega t) \right\} \tag{8.1.2}$$

The constant term is not important for present purposes. The four generated frequencies, $[(\omega_S \pm \omega), 2\,\omega_S, 2\omega]$, remain within the system and, being part of it, also drive it and thereby produce additional frequencies. The ultimate series of generated frequencies continues and includes all frequencies of the form $(m\omega + n\omega_S)$, where m and n are integers. (This conclusion is not specific to the $p = 2$ response but follows for all powers p greater than one.)

Energy is conserved in lossless systems independently of the degree of nonlinearity and we restrict the discussion to lossless systems. Let $P_{m,n}$ represent the time average power out of an interacting system at frequency $(m\omega + n\omega_S)$. Energy conservation requires that:

$$\sum_{m=-\infty}^{\infty} \sum_{n=-\infty}^{\infty} P_{m,n} = 0 \tag{8.1.3}$$

It is helpful for what lies ahead to rewrite Eq. 8.1.3 as:

$$\omega \sum_{m=-\infty}^{\infty} \sum_{n=-\infty}^{\infty} \frac{mP_{m,n}}{m\omega + n\omega_s} + \omega_s \sum_{m=-\infty}^{\infty} \sum_{n=-\infty}^{\infty} \frac{nP_{m,n}}{m\omega + n\omega_s} = 0 \tag{8.1.4}$$

Equation 8.1.4 contains redundant information since with integer pair (m_0, n_0) the sum is identical to the sum obtained using $(-m_0, -n_0)$. The redundancy is removed and all information retained by writing the sums as:

$$\omega \sum_{m=0}^{\infty} \sum_{n=-\infty}^{\infty} \frac{mP_{m,n}}{m\omega+n\omega_s} + \omega_s \sum_{m=-\infty}^{\infty} \sum_{n=0}^{\infty} \frac{nP_{m,n}}{m\omega+n\omega_s} = 0 \tag{8.1.5}$$

Consider a nonlinear capacitor as an example of a reactive system; inductive systems may be analyzed in a similar way. The charge on the capacitor may be expressed as:

$$q(t) = \frac{1}{2} \sum_{m=-\infty}^{\infty} \sum_{n=-\infty}^{\infty} Q_{m,n} \exp\left[i\left(m\omega+n\omega_s\right)t\right] \tag{8.1.6}$$

The condition $Q^*_{-m,-n} = Q_{m,n}$ applies since $q(t)$ is real. Similarly the voltage across the capacitor, $v(t)$, is:

$$v(t) = \frac{1}{2} \sum_{m=-\infty}^{\infty} \sum_{n=-\infty}^{\infty} V_{m,n} \exp\left[i(m\omega+n\omega_s)t\right] \tag{8.1.7}$$

The values of $Q_{m,n}$ and $V_{m,n}$ depend upon capacitor size, shape, and permittivity but not upon the frequency of operation.

The capacitor current $i(t)$ is equal to the rate of change of charge, and may be written

$$i(t) = \frac{1}{2} \sum_{m=-\infty}^{\infty} \sum_{n=-\infty}^{\infty} I_{m,n} \exp\left[i\left(m\omega+n\omega_s\right)t\right]$$

$$= \frac{i}{2} \sum_{m=-\infty}^{\infty} \sum_{n=-\infty}^{\infty} \left(m\omega+n\omega_s\right)Q_{m,n} \exp\left[i\left(m\omega+n\omega_s\right)t\right] \tag{8.1.8}$$

The second equality follows by differentiation of Eq. 8.1.6 with respect to time and shows that, differently from either the charge or voltage, current $I_{m,n}$ does depend upon the frequency.

The time average power into the capacitor is:

$$-P_{m,n} = -\frac{1}{2}\mathrm{Re}\left(V_{m,n}I^*_{m,n}\right) = \frac{1}{2}\left(m\omega+n\omega_s\right)\mathrm{Re}\left(iV_{m,n}Q^*_{m,n}\right) \tag{8.1.9}$$

Combining gives:

$$\frac{P_{m,n}}{\left(m\omega + n\omega_s\right)} = -\frac{1}{2}\mathrm{Re}\left(i V_{m,n} Q^*_{m,n}\right) \tag{8.1.10}$$

Since the right side of Eq. 8.1.10 depends upon products $Q_{m,n}V_{m,n}$, which are frequency independent, the right side is frequency independent. Hence the left side is also frequency independent. Since a similar argument follows if the roles of m and n are reversed, with coefficients c_1 and c_2 as frequency independent parameters Eq. 8.1.5 has the form:

$$c_1\omega + c_2\omega_s = 0 \tag{8.1.11}$$

Since the frequencies are independent variables the equation can be satisfied only if both c_1 and c_2 equal zero.

Applying this result to Eq. 8.1.5 shows the sums separately vanish:

$$\sum_{m=0}^{\infty}\sum_{n=-\infty}^{\infty}\frac{mP_{m,n}}{m\omega + n\omega_m} = 0; \quad \sum_{m=-\infty}^{\infty}\sum_{n=0}^{\infty}\frac{nP_{m,n}}{m\omega + n\omega_s} = 0 \tag{8.1.12}$$

To illustrate the use of this equation, consider a system in which one of two possible atomic states is occupied by an electron. The state frequencies are ω_{initial} and ω_{final}, for initially occupied and empty states. The ensemble is then enmeshed in a plane wave of frequency ω where the frequency of the applied wave satisfies the relationship

$$\omega = \left|\omega_{\text{initial}} - \omega_{\text{final}}\right| \tag{8.1.13}$$

Only the driven frequency, ω, and the system frequency, ω_{initial}, are present in linear systems. With nonlinear systems, all frequencies ($m\omega_{\text{initial}} + n\omega$) are potentially present. If an oscillating linear system supports only the frequencies ω_{initial} and ω_{final}, only the three frequencies ω, ω_{initial}, and ω_{final} are present.

Consider the special case of a nonlinear system initially supporting frequencies ω_{initial} and ω_{final} that is driven at frequency ω. If $\omega_{\text{initial}} > \omega$

and $\omega_{\text{initial}} > \omega_{\text{final}}$, Eq. 8.1.12 is satisfied for integer pairs ($m = 1$, $n = 0$) and ($m = 1$, $n = -1$); if $\omega_{\text{initial}} > \omega$ and $\omega_{\text{initial}} < \omega_{\text{final}}$, Eq. 8.1.12 is satisfied for integer pairs ($m = 0$, $\pm n = 1$) and ($m = -1$, $\pm n = 1$). Changing the power subscripts to match the frequency subscripts, for these special cases Eq. 8.1.12 go to:

$$\frac{P_{\text{initial}}}{\omega_{\text{initial}}} + \frac{P_{\text{final}}}{\omega_{\text{final}}} = 0 \quad \text{and} \quad \frac{P_{\text{initial}}}{\omega_{\text{initial}}} \pm \frac{P}{\omega} = 0 \tag{8.1.14}$$

In the second equation, the sign is respectively positive or negative if ω_{initial} is greater or less than ω_{final}.

The energy that goes into the final state is:

$$W_{\text{final}} = \int P_{\text{final}} dt \tag{8.1.15}$$

Combining Eq. 8.1.14 and 8.1.15 gives:

$$\left| \frac{W_{\text{initial}}}{\omega_{\text{initial}}} \right| = \left| \frac{W_{\text{final}}}{\omega_{\text{final}}} \right| = \left| \frac{W}{\omega} \right| \tag{8.1.16}$$

W_{final} is the energy that goes into the final state and W is the energy exchanged between the remote field and the electron undergoing a change of state. Since the energy-frequency ratio of Eq. 8.1.16 is independent of system parameters, and therefore of system details, the ratio is constant. For eigenstates that constant is Planck's constant, $\hbar$. With the upper and lower signs indicating that ω_{initial} is respectively greater than or less than ω_{final}:

$$W_{\text{final}} = -\hbar\omega_{\text{final}}; \quad W_{\text{initial}} = \hbar\omega_{\text{initial}}; \quad W = \pm\hbar\omega \tag{8.1.17}$$

The Manley Rowe results show that if the energy in the initial eigenstate is quantized into energy units proportional to frequency the energy exchanged among the radiation field and the initial and final energy eigenstates all have the same energy-to-frequency ratio.

Although Eq. 8.1.1 through 8.1.14 apply both to linear and nonlinear systems the magnitude of the energy flow at the sum or difference frequencies is zero for linear systems and Eq. 8.1.15 through 8.1.17 are significant for frequency determination only in the nonlinear case.

8.2 Regenerative Feedback

Feedback is a common feature of the physical world. For example, superregenerative feedback creates the monotonic screech of a loudspeaker. A microphone generates an electrical output that is amplified and the signal fed to a loudspeaker that, in turn, emits the same sound, now amplified, and adds to the originally detected sound. If the feedback is large enough the process becomes superregenerative and the sound becomes a monotonic whistle. Some external feature, not part of the analysis, ultimately determines the magnitude of the whistle.

In this section we apply Maxwell's assumption of continuity of path, described in Sec. 5.1, to an electron modeled as an ensemble of charge and current densities; conserved items, such as angular momentum and magnetic moment restrict otherwise possible ensemble distributions. Allowable charge and current density distributions exist in proportion to the number of ways each occurs. All distributions that satisfy the conservation laws will occur, and each will exist for a time proportional to the number of ways in which it can be achieved. Some of the distributions are amenable to a superregenerative buildup and some are not. In the following we detail an expected sequence of events that an electron will undergo if, at the onset, the distribution is amenable to a buildup triggered by an externally applied field.

Let V_0 represent a driving force that creates an initial multipole and let V_1 represent the supplementary force created by all other multipoles. The sum is the final driving force V_F:

$$V_F = \left(V_0 + V_1\right) \tag{8.2.1}$$

Next introduce unknown multiplying coefficient κ as a non-dimensional numeric measure of force augmentation created by other multipoles and defined by the equation:

$$V_1 = \kappa V_F \tag{8.2.2}$$

Combining the two equations gives:

$$V_F = V_0 / (1 - \kappa) \tag{8.2.3}$$

The process is illustrated in Fig. 8.2.1. The applied signal is V_0, V_1 returns to augment the applied signal, and the total transforms into input signal V_F. The arrows of Fig. 8.2.1 indicate energy flows.

A regenerative response occurs if κ is a small, positive number and, for that case, feedback increases the effective field but the output remains linearly dependent upon the applied field. The process is defined to be superregenerative if magnitude κ approaches unity and, as it does, the multipolar magnitudes become limitlessly large. As with a whistling loudspeaker, there is nothing in the analysis that limits signal growth; it is ultimately stopped by something external to the analysis. The input is a trigger of subsequent events; it determines the frequency, phasing, and polarization of the output but it does not affect the magnitude.

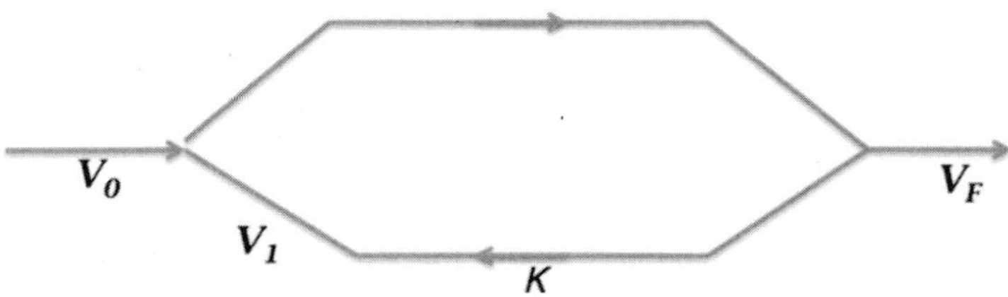

Figure 8.2.1. Input signal V_0 is transmitted through the upper branch to output V_F; portion κ of the output returns through the lower branch, creating input V_1 and forming new input (V_0+V_1) that, in turn, creates new output V_F.

Consider an atom with one electron in an occupied state and a corresponding empty state. The state symmetries permit a transition between them. If the electron's internal structure is correct to support oscillations between the states and if the system is somehow disturbed electromagnetic oscillations begin between the two eigenstates. Oscillations produce dipolar moments that initiate the process and appropriate feedback mechanisms are in place to drive the superregenerative creation of ever-higher order modes. The process continues until stopped by something external to this argument.

8.3 Synergistic Interior Energy Coupling

In this section we detail near-field interactions among collocated radiating multipoles. We saw in Sec. 7.4 that near-field reactive energies are so large that radiation produced by electrically small, high order modes have nearly googol-sized radiation Qs; the Qs are high enough to prevent emission. In this section we show the reactive energy also synergistically drives ever-higher modal orders. If the system has sufficient flexibility to support appropriate charge and current density distributions and if there are no resistive losses then the system creates and supports ever-higher modal orders.

Direct Multipolar Coupling

Electric multipolar moments within a charge distribution are discussed in Sec. A.22. An integral expression for the linear, z-directed electric multipolar moment, p_ℓ, created by distributed electric charges is given by Eq. A.22.7 and repeated here:

$$p_\ell = \int \rho(r) z_1 z_2 ... z_\ell dV \tag{8.3.1}$$

Although a specific charge distribution may be continuous on one scale of dimensions, in the following we drop to a dimensional scale where it is not, but rather granular and capable of responding to local forces by forming an array of granular multipolar moments; we noted in Sec. 6.11 that the Exclusion Principle could be a result of charge granularity. A multipolar moment of order ℓ consists of 2^ℓ equal magnitude, discrete charges, half of which are of each sign. If such a distribution supports dipole moment p_ℓ it may or may not also support higher order moments with the same symmetry, but it does not support lower order ones. There are several means by which multipoles are created within a single contiguous region. Consider the following:

A source of multipoles with boundaries as an inherent constraint is illustrated in Fig. 8.3.1. Arrows indicate driven current densities within a closed volume. Current {1} is the source of a basic electric dipole, and charge continuity requires current {2} as well. Together they form an electric octupole. Circulatory octupolar currents {3} form a magnetic quadrupole. By continuation, this creates all odd numbered TM field sources and all even number TE field sources. A dual magnetic source similarly drives all even numbered TM field sources and all odd numbered TE field sources.

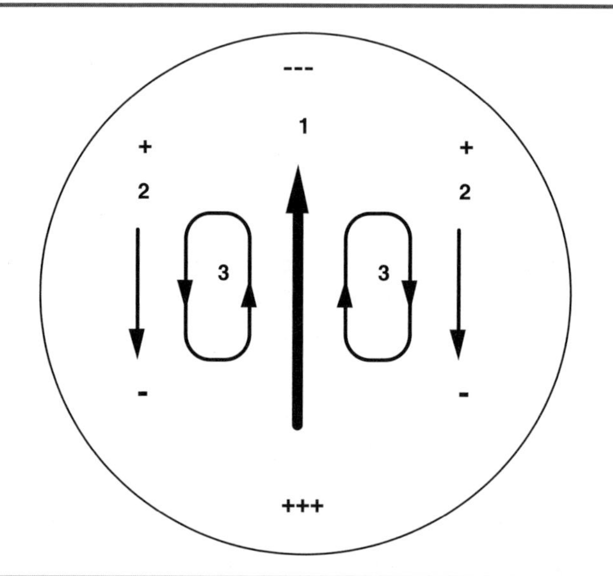

Figure 8.3.1. Source modes resulting from an electric dipole formed within a closed, charged system.

Generally speaking, most specific charge distributions create more than one multipole. As an important example, consider a discrete electric dipole with one positive and one negative charge, $\pm q$, placed respectively at positions $\pm z_0/2$. The dipole moment calculated using Eq. 8.3.1 is $p_1 = q z_0$. Using the same source equation the co-generated multipolar moments of odd orders are $p_\ell = q z_0^\ell / 4^{\ell-1}$. By symmetry, all even order moments are zero. Also by that equation a discrete quadrupole must have at least four charges, two each positive and negative. With linear quadrupoles, the positive charges are at $\pm z_0$ and both negative charges at zero; the quadrupole moment is $p_2 = 2qz_0^2$. Accompanying multipolar moments of even order are $p_\ell = 2qz_0^\ell$ but, by symmetry, all odd order moments are zero. Octupoles must have at least four positive and four negative charges. A discrete, linear octupole has charges $\pm q$ respectively at positions $\pm 3z_0/2$ and charges $\pm 3q$ respectively at $\pm z_0/2$. By Eq. 8.3.1 the octupole moment is $p_3 = 6qz_0^3$; the configuration also supports odd multipolar moments of order five, $p_5 = 211qz_0^5/32$, and greater but the dipole moment is zero. Symmetry requires all even order moments to vanish. The series continues in this way to limitlessly large orders. These co-created multipoles do not exist individually: the presence of one requires the presence of the others.

Is it possible for a radiating electron to support the multipoles necessary to create the field set of Eq. 7.4.1? Consider the following scenario. An applied wave travels parallel with the electron's angular momentum, which we choose as the z-axis. The electric field drives electric oscillations perpendicular to the z-axis and the magnetic field instigates precession of the magnetic moment about that axis; the moments maintain spatial orthogonality as they rotate at the frequency of the applied wave and that orthogonality assures phase addition of the distant field on the $+z$-axis and cancellation on the $-z$-axis.

Overlapping Intermodal Energies

Since we lack detailed information about electrons, we turn to and apply what information we do have and that information includes source symmetries. Just as we did with a receiving-biconical antenna, we consider the interaction of an electron with an applied field as a scattering problem. The interior fields of a spherical scatterer within an applied field are shown by Eq. 2.10.1, where symbols δ_ℓ and γ_ℓ represent, respectfully, the coefficients of interior TE and TM modes. Since the exterior emitter fields have equal values of TE and TM modes, it follows from reciprocity that the internal field coefficients δ_ℓ and γ_ℓ have equal magnitudes, just as they did with a spherical scatterer with equal values of relative permeability and permittivity. The interior fields are:

$$\tilde{\boldsymbol{E}}_{\ell}^{in} = i^{-\ell} \frac{(2\ell+1)}{\ell(\ell+1)}$$

$$\times \left\{ \gamma_{\ell} \left[\begin{array}{l} i\ell(\ell+1)\dfrac{j_{\ell}(\sigma)}{\sigma} P_{\ell}^{1}(\cos\theta)\hat{r} + ij_{\ell}^{\bullet}(\sigma)\dfrac{d}{d\theta}P_{\ell}^{1}(\cos\theta)\hat{\theta} \\[2ex] + j_{\ell}^{\bullet}(\sigma)\dfrac{P_{\ell}^{1}(\cos\theta)}{\sin\theta}\hat{\phi} \end{array} \right] + \delta_{\ell}\left[j_{\ell}(\sigma)\dfrac{P_{\ell}^{1}(\cos\theta)}{\sin\theta}\hat{\theta} - ij_{\ell}(\sigma)\dfrac{d}{d\theta}P_{\ell}^{1}(\cos\theta)\hat{\phi}\right] \right\} e^{i(\omega t-\phi)} \quad (8.3.2)$$

$$\eta'\tilde{\boldsymbol{H}}_{\ell}^{in} = i^{1-\ell} \frac{(2\ell+1)}{\ell(\ell+1)}$$

$$\times \left\{ \delta_{\ell} \left[\begin{array}{l} i\ell(\ell+1)\dfrac{j_{\ell}(\sigma)}{\sigma} P_{\ell}^{1}(\cos\theta)\hat{r} + ij_{\ell}^{\bullet}(\sigma)\dfrac{d}{d\theta}P_{\ell}^{1}(\cos\theta)\hat{\theta} \\[2ex] + j_{\ell}^{\bullet}(\sigma)\dfrac{P_{\ell}^{1}(\cos\theta)}{\sin\theta}\hat{\phi} \end{array} \right] + \gamma_{\ell}\left[j_{\ell}(\sigma)\dfrac{P_{\ell}^{1}(\cos\theta)}{\sin\theta}\hat{\theta} - ij_{\ell}(\sigma)\dfrac{d}{d\theta}P_{\ell}^{1}(\cos\theta)\hat{\phi}\right] \right\} e^{i(\omega t-\phi)} \quad (8.3.3)$$

Collocated multipoles are enmeshed in each other's near fields. We seek to determine if a mechanism exists whereby the sources and local fields interact synergistically. For example, if a driving field induces both an electric dipole and a magnetic quadrupole do the local fields of the newly induced multipoles synergistically drive each other? Do the fields also combine to drive higher order, collocated multipoles? To address the issue we seek to determine and contrast the relative magnitudes of intra- and inter-modal energies and, for this purpose, we use the interior fields of Eq. 8.3.2 and 8.3.3.

Although the equations apply only to regions with no net charge density, we apply them within a region where charge distributions will surely exist. Since we calculate only energy ratios we believe the approach is appropriate. For example in an interior region composed of a magneto-dielectric the effect of the interior distribution of charge and current densities is described by relative values of permeability and permittivity and these values cancel from the intermodal-to-intramodal energy ratios. For the special case of an actively emitting or absorbing electron whatever internal charge and current densities are present the calculated ratios are surely closer to physical reality than either energy by itself.

Using Eq. 8.3.2 and 8.3.3 the modal field energy of order ℓ between radii r and $r + dr$ is:

$$\frac{dW_T}{d\sigma} = \frac{2\pi\varepsilon}{k^3}\gamma_\ell^2\left(2\ell+1\right)\sigma^2\left\{\ell(\ell+1)\frac{j_\ell^2(\sigma)}{\sigma^2}+\left(j_\ell^2(\sigma)+j_\ell^{\bullet 2}(\sigma)\right)\right\} \quad (8.3.4)$$

We test for intra-TE, intra-TM, and inter-(TE+TM) overlap present and associated with z-directed fields, namely the fields:

$$E_{z\ell} = E_{r\ell}\cos\theta - E_{\theta\ell}\sin\theta; \quad H_{z\ell} = H_{r\ell}\cos\theta - H_{\theta\ell}\sin\theta \quad (8.3.5)$$

The convolution integral of z-directed field components yields the coupling energy; with the fields of Eq. 8.3.2 and 8.3.3 the electric magnetic field energies are the same. The independent convolution integrals of interest are:

$$\frac{d}{d\sigma}<\tilde{E}_{z,\ell}^{TM},\tilde{E}_{z,\ell+n}^{TE}>=\gamma_\ell\delta_{\ell+n}\frac{\varepsilon\sigma^2}{4k^3}\int_0^{2\pi}d\phi\int_0^{\pi}\sin\theta d\theta\left(\tilde{E}_{z,\ell}^{TM}\right)\left(\tilde{E}_{z,\ell+n}^{TE}\right)$$

$$\frac{d}{d\sigma}<\tilde{E}_{z,\ell}^{TM},\tilde{E}_{z,\ell+n}^{TM}>=\gamma_\ell\gamma_{\ell+n}\frac{\varepsilon\sigma^2}{4k^3}\int_0^{2\pi}d\phi\int_0^{\pi}\sin\theta d\theta\left(\tilde{E}_{z,\ell}^{TM}\right)\left(\tilde{E}_{z,\ell+n}^{TM}\right) \quad (8.3.6)$$

$$\frac{d}{d\sigma}<\tilde{E}_{z,\ell}^{TE},\tilde{E}_{z,\ell+n}^{TE}>=\delta_\ell\delta_{\ell+n}\frac{\varepsilon\sigma^2}{4k^3}\int_0^{2\pi}d\phi\int_0^{\pi}\sin\theta d\theta\left(\tilde{E}_{z,\ell}^{TE}\right)\left(\tilde{E}_{z,\ell+n}^{TE}\right)$$

Although δ_ℓ and γ_ℓ have equal magnitude, we retain the notation to help track energy sources. Using identities from Table A.21.1 and the above equations we evaluate all cases in which the convolution integrals differ from zero:

- Case 1: TM_ℓ and $TE_{\ell+1}$

$$\frac{d}{d\sigma}<\tilde{E}_{z,\ell}^{TM},\tilde{E}_{z,\ell+1}^{TE}>=\frac{2\pi\varepsilon\sigma^2}{4k^3}\gamma_\ell\delta_{\ell+1}\frac{(2\ell+1)(2\ell+3)}{\ell(\ell+1)^2(\ell+2)}\int_0^\pi \sin\theta d\theta$$

$$\times\left\{\left[\ell(\ell+1)\frac{j_\ell}{\sigma'}P_\ell^1\cos\theta-j_\ell^\bullet\frac{dP_\ell^1}{d\theta}\sin\theta\right]j_{\ell+1}P_{\ell+1}^1\right\} \qquad (8.3.7)$$

$$\frac{d}{d\sigma}<\tilde{E}_{z,\ell}^{TM},\tilde{E}_{z,\ell+1}^{TE}>=\frac{\pi\varepsilon\sigma^2}{k^3}\gamma_\ell\delta_{\ell+1}\frac{\ell}{(\ell+1)}\left[j_{\ell+1}(\sigma')\right]^2$$

This is the overlap energy associated with the z-component of the TM part of mode 'ℓ' and the TE part of mode '$\ell + 1$'. The convolution-to-total energy ratio, Eq. 8.3.7-to-Eq. 8.3.4, is:

$$2\frac{\dfrac{d}{d\sigma}<\tilde{E}_{z,\ell}^{TM},\tilde{E}_{z,\ell+1}^{TE}>}{\dfrac{d}{d\sigma}(W_\ell+W_{\ell+1})}\sim\frac{\ell}{(\ell+1)}\left(\frac{4\gamma_\ell\delta_{\ell+1}}{\gamma_\ell^2(2\ell+1)+\gamma_{\ell+1}^2(2\ell+3)}\right) \qquad (8.3.8)$$

This energy ratio is on the order of one at low numbered modes and large enough for the modes to strongly influence each other; the ratio decreases parabolically with increasing modal number.

- Case 2: $TM_{\ell+1}$ and TE_ℓ

$$\frac{d}{d\sigma}<\tilde{E}_{z,\ell+1}^{TM},\tilde{E}_{z,\ell}^{TE}>=-\frac{2\pi\varepsilon\sigma^2}{4k^3}\beta_{\ell+1}\gamma_\ell$$

$$\times\frac{d}{d\sigma}\int_0^\pi\sin\theta d\theta\left\{\frac{(2\ell+3)}{(\ell+1)(\ell+2)}\left[\begin{array}{c}(\ell+1)(\ell+2)\dfrac{j_{\ell+1}}{\sigma}P_{\ell+1}^1\cos\theta\\[2mm]-j_{\ell+1}^\bullet\dfrac{dP_{\ell+1}^1}{d\theta}\sin\theta\end{array}\right]\frac{(2\ell+1)}{\ell(\ell+1)}j_\ell P_\ell^1\right\}$$

$$\frac{d}{d\sigma}<\tilde{E}_{z,\ell+1}^{TM},\tilde{E}_{z,\ell}^{TE}>=-\frac{\pi\varepsilon\sigma^2}{k^3}\beta_{\ell+1}\gamma_\ell\frac{(\ell+2)}{(\ell+1)}\frac{d}{d\sigma}\left[j_\ell(\sigma)\right]^2$$

$$(8.3.9)$$

This is the overlap energy associated with the z-component of the TM part of mode '$\ell+1$' and the TE part of mode 'ℓ'. The convolution-to-total energy ratio, Eq. 8.3.9-to-Eq. 8.3.4, is:

$$\frac{2\dfrac{d}{d\sigma}\langle \tilde{E}_{z,\ell+1}^{TM}, \tilde{E}_{z,\ell}^{TE}\rangle}{\dfrac{d}{d\sigma}\left(W_\ell + W_{\ell+1}\right)} \sim \frac{(\ell+2)}{(\ell+1)}\left(\frac{4\gamma_\ell\gamma_{\ell+1}}{\gamma_\ell^2(2\ell+1)+\gamma_{\ell+1}^2(2\ell+3)}\right) \qquad (8.3.10)$$

This energy ratio is on the order of one at low numbered modes and large enough for the modes to strongly influence each other; as with Eq. 8.3.8 the ratio decreases parabolically with increasing modal number.

- Case 3: TE_ℓ and $TE_{\ell+2}$

$$\frac{d}{d\sigma}\langle \tilde{E}_{z,\ell}^{TE}, \tilde{E}_{z,\ell+2}^{TE}\rangle$$

$$= -\frac{2\pi\varepsilon\sigma^2}{4k^3}\gamma_\ell\gamma_{\ell+2}\frac{(2\ell+1)(2\ell+5)}{\ell(\ell+1)(\ell+2)(\ell+3)}\frac{d}{d\sigma}\int_0^\pi \sin\theta d\theta\left\{ j_{\ell+2}^* j_\ell P_{\ell+2}^1 P_\ell^1\right\} = 0$$

This is the overlap energy associated with z-component of the TE part of mode 'ℓ' and the TE part of mode '$\ell+2$'; there is no energy overlap.

- Case 4: TM_ℓ and $TM_{\ell+2}$

$$\langle \tilde{E}_{z,\ell+2}^{TE}, \tilde{E}_{z,\ell}^{TE}\rangle = -\frac{2\pi\varepsilon}{4}\gamma_\ell\gamma_{\ell+2}\frac{(2\ell+1)(2\ell+5)}{\ell(\ell+1)(\ell+2)(\ell+3)}$$

$$\times\left[\ell(\ell+1)\frac{j_\ell}{\sigma}P_\ell^1\cos\theta - j_\ell^\bullet\frac{dP_\ell^1}{d\theta}\sin\theta\right]\left[(\ell+2)(\ell+3)\frac{j_{\ell+2}}{\sigma}P_{\ell+2}^1\cos\theta - j_{\ell+2}^\bullet\frac{dP_{\ell+2}^1}{d\theta}\sin\theta\right]$$

$$\langle \tilde{E}_{z,\ell+2}^{TE}, \tilde{E}_{z,\ell}^{TE}\rangle = -\pi\varepsilon\gamma_\ell\gamma_{\ell+2}\frac{\ell(\ell+3)}{(2\ell+3)}\left[j_{\ell+1}\right]^2$$

$$(8.3.11)$$

This is the overlap energy associated with the z-component of the TM part of mode 'ℓ' and the TM part of mode '$\ell+2$'. The convolution-to-total energy ratio, Eq. 8.3.11-to-Eq. 8.3.4, is:

$$\frac{2\dfrac{d}{d\sigma}\langle \tilde{E}_{z,\ell+2}^{TE}, \tilde{E}_{z,\ell}^{TE}\rangle}{\dfrac{d}{d\sigma}(W_\ell + W_{\ell+1})} \sim \frac{\ell(\ell+3)}{(2\ell+3)}\left(\frac{4\gamma_\ell\gamma_{\ell+2}}{\gamma_\ell^2(2\ell+1)+\gamma_{\ell+2}^2(2\ell+5)}\right) \qquad (8.3.12)$$

This energy ratio is on the order of one and large enough for the modes to strongly influence each other; the ratio is essentially constant through all orders.

The intra-to-inter modal energy ratios of Eq. 8.3.8, 8.3.10, and 8.3.12 show energy ratios on the order of one, and hence significant coupling. Significant overlap occurs between intertype modes with $\Delta\ell=\pm1$ and intratype modal types with $\Delta\ell=\pm2$. The $\Delta\ell=\pm1$ interactions are strong at low modal orders and the $\Delta\ell=\pm2$ interactions are strong at all modal orders. Nearest-intertype coupling is important for starting a modal buildup and the next-nearest intratype coupling is important at all modal orders. Together they connect all multipoles of the same symmetry into a single unit.

In summary, an induced dipole drives a continuous chain of higher order multipoles; we detail the chain beginning with an electric dipole. Using modal numbers as subscripts, an induced electric dipole creates a TM_1 field that (1) supplements the drive of a magnetic quadrupole and its TE_2 field and (2) supplements the drive of an electric octupole with its TM_3 field. The multipoles have one displacement in the basal plane and $(\ell-1)$ in the z-direction, which may imply a concurrent expansion of the source in the z-direction. The TE_2 field (a) supplements the drive of the original electric dipole, (b) supplements the drive of the electric octupole, and (c) supplements the drive of a 16-pole magnetic multipole with its accompanying TE_4 field. The electric octupole (A) supplements the drive of the electric dipole, (B) supplements the drive of the magnetic quadrupole, (C) supplements the drive of the 16-pole magnetic multipole, and (D) supplements the drive of a 32-pole electric multipole with its accompanying TE_5 field. The process continues in this way to ever-larger order multipoles until stopped by unknown internal properties of the source. At that point the radiating object is a continuum of coherently oscillating multipoles acting as a single unit.

A dual series of multipoles begins with a magnetic dipole. Although no convolution energy connects the two series once one series is established the other may be formed without additional reactive energy.

Although this method of transmodal coupling is less in magnitude than that discussed in Sec. 8.4, it continues to be effective as the modal orders increase and the recursion relationship of Eq. 7.4.1 is approached.

8.4 Synergistic Exterior Reactive Energy Coupling

This section contains a discussion of transmodal coupling by means of reactive power. Although the power expressions are orthogonal they still dominate transmodal interactions, and since the initial reactive power magnitude is a factor Q larger than the real power, it is dominant. However, as the maximum modal number increases and the recursion relationship approaches that of Eq. 7.4.1 the drives begin to cancel and, indeed, vanish when the ultimate objective is reached. The drives are, therefore, large at the onset but decrease as the ultimate distribution is approached. There are the following types of external synergistic reactive drives: (1) equal parity TM-TE (2) equal parity TM-TM (3) opposite parity TM-TE, plus duals. Collocation of the multipoles that create Eq. 7.4.1 within an electrically small region assures that each is enmeshed in the superimposed near fields of all. This includes the reactive energy that flows throughout the region. To examine effects of these energy flows we analyze three special cases. Two cases involve multipoles with the same angular symmetry and the third two multipoles with opposite symmetry. Although there is mutual, synergistic coupling in all three cases, it is first order in cases with the same parity and second order in the case with opposite parity.

Equal Parity TM-TE Coupling

Consider coupling between different modal types where the modal orders differ by an odd number. In particular, consider a superimposed electric multipole of order ℓ and magnetic multipole of order $\ell + 1$ where each has the orientation and phasing of Eq. 7.4.1.

The radial component of the Poynting vector is the superposition of two independent polarizations: products $E_\theta H_\phi$ and $E_\phi H_\theta$; for simplicity and without loss of rigor we confine our attention to the $E_\theta H_\phi$ product. For ease in handling the algebra to come we make the following definitions:

$$F_\ell = F_0 \frac{(2\ell+1)}{\ell(\ell+1)}; \quad i^{1-\ell}F_0 = 1$$

$$\tag{8.4.1}$$

$$S_\ell^1 = \frac{P_\ell^1}{\sin\theta}; \quad T_\ell^1 = \frac{dP_\ell^1}{d\theta}; \quad y = \omega t_R - \phi$$

With these definitions, using letter notation for the rational portion of the spherical Hankel functions, and introducing 'b' as a small dimensionless number proportional to the magnitude of the higher order term, the angular portions of the time-dependent fields are:

$$E_\ell = \frac{1}{\sigma}\left\{ \begin{bmatrix} -F_\ell\left(C_\ell\sin y - D_\ell\cos y\right)T_\ell^1 \\ +bF_{\ell+1}\left(A_{\ell+1}\sin y - B_{\ell+1}\cos y\right)S_{\ell+1}^1 \end{bmatrix}\hat\theta \\ + \begin{bmatrix} F_\ell\left(C_\ell\cos y + D_\ell\sin y\right)S_\ell^1 \\ -bF_{\ell+1}\left(A_{\ell+1}\cos y + B_{\ell+1}\sin y\right)T_{\ell+1}^1 \end{bmatrix}\hat\phi \right\}$$

$$\tag{8.4.2}$$

$$H_{\ell+1} = \frac{1}{\eta\sigma}\left\{ -\begin{bmatrix} F_\ell\left(A_\ell\sin y - B_\ell\cos y\right)S_\ell^1 \\ +bF_{\ell+1}\left(C_{\ell+1}\sin y - D_{\ell+1}\cos y\right)T_{\ell+1}^1 \end{bmatrix}\hat\theta \\ + \begin{bmatrix} F_\ell\left(A_\ell\cos y + B_\ell\sin y\right)T_\ell^1 \\ +bF_{\ell+1}\left(C_{\ell+1}\cos y + D_{\ell+1}\sin y\right)S_{\ell+1}^1 \end{bmatrix}\hat\phi \right\}$$

The $E_\theta H_\phi$ polarization of the radial Poynting vector is:

$$N_r\big|_{\theta\phi} = \frac{1}{2\eta\sigma^2} \times$$

$$\left\{ [F_\ell]^2 \begin{bmatrix} (A_\ell D_\ell - B_\ell C_\ell) + (A_\ell D_\ell + B_\ell C_\ell)\cos 2y \\ -(A_\ell C_\ell - B_\ell D_\ell)\sin 2y \end{bmatrix} [T_\ell^1]^2 \right.$$

$$+ bF_\ell F_{\ell+1} \begin{bmatrix} (-A_\ell B_{\ell+1} + A_{\ell+1}B_\ell - C_\ell D_{\ell+1} + C_{\ell+1}D_\ell) \\ -(A_\ell B_{\ell+1} + A_{\ell+1}B_\ell - C_\ell D_{\ell+1} - C_{\ell+1}D_\ell)\cos 2y \\ +(A_\ell A_{\ell+1} - B_\ell B_{\ell+1} - C_\ell C_{\ell+1} + D_\ell D_{\ell+1})\sin 2y \end{bmatrix} S_{\ell+1}^1 T_\ell^1$$

$$+ b^2 [F_{\ell+1}]^2 \begin{bmatrix} (A_{\ell+1}D_{\ell+1} - B_{\ell+1}C_{\ell+1}) \\ -(A_{\ell+1}D_{\ell+1} + B_{\ell+1}C_{\ell+1})\cos 2y \\ +(A_{\ell+1}C_{\ell+1} - B_{\ell+1}D_{\ell+1})\sin 2y \end{bmatrix} \left[S_{\ell+1}^1 \right]^2 \right\} \tag{8.4.3}$$

Consider the special case of an applied field driving an induced electric dipole and a much smaller magnetic quadrupole, the TM$_1$-TE$_2$ case. Integrating over the surface of a virtual, circumscribing sphere of radius σ/k:

$$P\big|_{\theta\phi} = \frac{2\pi F_0^2}{2\eta k^2} \left\{ \frac{3}{2}\left[1 - \left(1 - \frac{2}{\sigma^2}\right)\cos 2y - \left(\frac{2}{\sigma} - \frac{1}{\sigma^3}\right)\sin 2y \right] \right.$$

$$+ b\frac{5}{2} \begin{bmatrix} \left(1 - \frac{2}{\sigma^2}\right) - \left(1 - \frac{8}{\sigma^2} + \frac{6}{\sigma^4}\right)\cos 2y \\ -\left(\frac{4}{\sigma} - \frac{8}{\sigma^3} + \frac{3}{\sigma^5}\right)\sin 2y \end{bmatrix}$$

$$+ b^2 \frac{25}{6}\left[1 - \left(1 - \frac{18}{\sigma^2} + \frac{36}{\sigma^4}\right)\cos 2y - 3\left(\frac{2}{\sigma} - \frac{11}{\sigma^3} + \frac{6}{\sigma^5}\right)\sin 2y \right] \right\} \tag{8.4.4}$$

The ratio of the reactive power that oscillates twice each field cycle between field and source to real power that travels permanently outbound from the source is:

$$\frac{Power}{Ratio} = -\frac{3}{17}\left\{\frac{3}{2}\left[\left(1-\frac{2}{\sigma^2}\right)\cos 2y + \left(\frac{2}{\sigma}-\frac{1}{\sigma^3}\right)\sin 2y\right]\right.$$

$$+b\frac{5}{2}\left[\begin{array}{c}\left(1-\dfrac{8}{\sigma^2}+\dfrac{6}{\sigma^4}\right)\cos 2y \\[2mm] +\left(\dfrac{4}{\sigma}-\dfrac{8}{\sigma^3}+\dfrac{3}{\sigma^5}\right)\sin 2y\end{array}\right]$$

$$\left.+b^2\frac{25}{6}\left[\left(1-\frac{18}{\sigma^2}+\frac{36}{\sigma^4}\right)\cos 2y + 3\left(\frac{2}{\sigma}-\frac{11}{\sigma^3}+\frac{6}{\sigma^5}\right)\sin 2y\right]\right\} \qquad (8.4.5)$$

The three rows of Eq. 8.4.5 respectively arise from dipole, mixed, and quadrupole sources. All have the same time dependence and the same sign for each power of σ. This constitutes, therefore, a positive feedback and a synergistic drive for both sources. Initially the dipole field has the larger magnitude and b^2 is a numeric factor much less than one. Although the magnitude of the quadrupole is second order in b the magnitude of the mixed terms are first order, and hence significantly larger and drive both the dipole and the quadrupole. The field builds in this way until stopped by some factor beyond this discussion.

The $E_\phi H_\theta$ polarization provides a parallel set of conditions with identical positive feedback conditions.

Equal Parity, TM-TM Coupling

Consider next modes of the same type with orders that differ by an even number; we use electric multipoles of order ℓ and $\ell + 2$. Each multipole has the orientation and phasing of Eq. 7.4.1. Again, for simplicity and without loss of rigor, we confine our attention to the $E_\theta H_\phi$ polarization of the Poynting vector. For the important special case of an electric dipole and octupole pair, and with b again representing the magnitude of the higher order field, the angular components of actual fields on the surface of a virtual circumscribing sphere of radius σ/k are:

$$
E = \frac{1}{\sigma}\left\{ \begin{array}{l} \left[F_1\left(D_1\cos y - C_1\sin y\right)T_1^1 - bF_3\left(D_3\cos y - C_3\sin y\right)T_3^1 \right]\hat{\theta} \\ + \left[F_1\left(C_1\cos y + D_1\sin y\right)S_1^1 - bF_3\left(C_3\cos y + D_3\sin y\right)S_3^1 \right]\hat{\phi} \end{array} \right\}
$$

$$
H = \frac{1}{\mu\sigma}\left\{ \begin{array}{l} \left[F_1\left(B_1\cos y - A_1\sin y\right)S_1^1 - bF_3\left(B_3\cos y - A_3\sin y\right)S_3^1 \right]\hat{\theta} \\ + \left[F_1\left(A_1\cos y + B_1\sin y\right)T_1^1 - bF_3\left(A_3\cos y + B_3\sin y\right)T_3^1 \right]\hat{\phi} \end{array} \right\}
$$

$$\tag{8.4.6}$$

The $E_\theta H_\phi$ polarized radial Poynting vector is:

$$
N_r\big|_{\theta\phi} = \frac{1}{2\eta\sigma^2}\times
$$

$$
\left\{ \begin{array}{l} F_1 F_1 \left[\begin{array}{l} \left(A_1 D_1 - B_1 C_1\right) \\ + \left(A_1 D_1 + B_1 C_1\right)\cos 2y - \left(A_1 C_1 - B_1 D_1\right)\sin 2y \end{array} \right] T_1^1 T_1^1 \\[2em] - bF_1 F_3 \left[\begin{array}{l} \left(A_1 D_3 + A_3 D_1 - B_1 C_3 - B_3 C_1\right) \\ + \left(A_3 D_1 + A_1 D_3 + B_1 C_3 + B_3 C_1\right)\cos 2y \\ + \left(A_1 C_3 + A_3 C_1 A_1 C_3 - B_3 D_1\right)\sin 2y \end{array} \right] T_1^1 T_3^1 \\[3em] + b^2 F_3 F_3 \left[\begin{array}{l} \left(A_3 D_3 - B_3 C_3\right) \\ + \left(A_3 D_3 + B_3 C_3\right)\cos 2y - \left(A_3 C_3 - B_3 D_3\right)\sin 2y \end{array} \right] T_3^1 T_3^1 \end{array} \right\}
$$

$$\tag{8.4.7}$$

The surface integral on a virtual circumscribing sphere of radius σ/k gives the surface power:

$$
P\big|_{\theta\phi} = \frac{4\pi}{2\eta k^2}\left\langle \frac{3}{4}\left[1 - \left(1 - \frac{2}{\sigma^2}\right)\cos 2y - \left(\frac{2}{\sigma} - \frac{1}{\sigma^3}\right)\sin 2y \right]\right.
$$

$$
- b\frac{7}{4}\left[\left(2 - \frac{39}{\sigma^2} + \frac{120}{\sigma^4}\right)\cos 2y + \left(\frac{14}{\sigma} - \frac{102}{\sigma^3} + \frac{60}{\sigma^5}\right)\sin 2y \right]
$$

$$
+ b^2 \frac{119}{24}\left[1 - \left(1 - \frac{72}{\sigma^2} + \frac{720}{\sigma^4} - \frac{1350}{\sigma^6}\right)\cos 2y \right.
$$

$$
\left.\left. - \left(\frac{12}{\sigma} - \frac{276}{\sigma^3} + \frac{1280}{\sigma^5} - \frac{675}{\sigma^7}\right)\sin 2y \right] \right\rangle
$$

$$\tag{8.4.8}$$

The reactive-to-real power ratio is:

$$\frac{Power}{Ratio} = -\frac{24}{227}\left\{\frac{3}{4}\left[\left(1-\frac{2}{\sigma^2}\right)\cos 2y + \left(\frac{2}{\sigma}-\frac{1}{\sigma^3}\right)\sin 2y\right]\right.$$

$$+b\frac{7}{4}\left[\left(2-\frac{39}{\sigma^2}+\frac{120}{\sigma^4}\right)\cos 2y + \left(\frac{14}{\sigma}-\frac{102}{\sigma^3}+\frac{60}{\sigma^5}\right)\sin 2y\right]$$

$$+b^2\frac{119}{24}\left[1-\left(1-\frac{72}{\sigma^2}+\frac{720}{\sigma^4}-\frac{1350}{\sigma^6}\right)\cos 2y\right.$$

$$\left.\left.-\left(\frac{12}{\sigma}-\frac{276}{\sigma^3}+\frac{1280}{\sigma^5}-\frac{675}{\sigma^7}\right)\sin 2y\right]\right\} \qquad (8.4.9)$$

As with TM_1-TE_2 coupling, there is positive feedback since all three rows have the same time dependence for each power of σ, the initial octupolar field is proportional to the small numeric factor b^2 but the initial positive feedback is proportional to b. The onsets of synergistic drives are therefore proportional to the first power of b and only after the buildup process is underway does b become large enough for significant feedback directly between the source fields.

Opposite Parity, TM-TE Coupling

Consider a superimposition of electric and magnetic dipoles arrayed as in Eq. 7.4.1, the TM_1-TE_1 case. We retain notation F_1 and G_1 respectively for the coefficients of the TM and TE fields, and again b is a small numeric factor. The angular portion of the fields is:

Angular Fields

$$\tilde{\mathbf{E}}_1 = \frac{12}{\sigma}\left\{\begin{array}{l}\left[F_1\left(D_1\cos y - C_1\sin y\right)T_1^1 + bG_1\left(A_1\cos y + B_1\sin y\right)S_1^1\right]\hat{\theta} \\ +\left[F_1\left(D_1\sin y + C_1\cos y\right)S_1^1 + bG_1\left(A_1\sin y - B_1\cos y\right)T_1^1\right]\hat{\phi}\end{array}\right\}$$

$$\tilde{\mathbf{H}}_1 = \frac{F_1}{\eta\sigma}\left\{\begin{array}{l}-\left[F_1\left(A_1\sin y - B_1\cos y\right)S_1^1 + bG_1\left(D_1\sin y + C_1\cos y\right)T_1^1\right]\hat{\theta} \\ +\left[F_1\left(A_1\cos y + B_1\sin y\right)T_1^1 + bG_1\left(D_1\cos y - C_1\sin y\right)S_1^1\right]\hat{\phi}\end{array}\right\}$$

$$(8.4.10)$$

The $E_\theta H_\phi$ polarization of the radial component of the Poynting vector is:

$$
N_r\big|_{\theta\phi} = \frac{1}{\eta\sigma^2}\left\{\left[\left(A_1 D_1 \cos^2 y - B_1 C_1 \sin^2 y - \left(A_1 C_1 - B_1 D_1\right)\sin y \cos y\right)\right]\right.
$$

$$
\left. \times\left(\left[F_1 T_1^1\right]^2 + b^2\left[G_1 S_1^1\right]^2\right) + b\left[F_1 G_1\right]\left[\begin{array}{c}\left(D_1 \cos y - C_1 \sin y\right)^2 \\ +\left(A_1 \cos y + B_1 \sin y\right)^2\end{array}\right]\left[S_1^1 T_1^1\right]\right\}
$$

$$\tag{8.4.11}$$

The cross product term is an odd function of θ and hence vanishes upon integrating over the surface of a circumscribing sphere, leaving:

$$
P\big|_{\theta\phi} = \frac{9}{2\eta\sigma^2}\left[1 - \left(1 - \frac{2}{\sigma^2}\right)\cos 2y - \left(\frac{2}{\sigma} - \frac{1}{\sigma^3}\right)\sin y \cos y\right]\left(\frac{1}{3} + b^2\right) \tag{8.4.12}
$$

In this case, therefore, the initial synergistic feedback is proportional to b^2. Hence for significant feedback to occur the initial value of the TE$_1$ field must be significantly larger than is necessary with multipoles of the same parity. We conclude that although a buildup can occur with this field symmetry, it is less probable than in the cases of equal parity.

8.5 Spontaneous Emission

Consistent with Maxwell's 'assumption of continuity of path' within a statistical ensemble, as described in Sec. 5.1, we envisage an electron as a stable but dynamic system that undergoes continuous evolution. Although such a system would otherwise be expected to radiate from time to time, Chu's work on radiation from electrically small objects shows that emission is prevented by googol-sized atomic Q-traps. This working electron model, a stable but dynamic system that undergoes continuous evolution, is also consistent with Tolman's observation of extensive agreement between statistical mechanical results and experimental results, Dirac's observation of the extraordinary stability of atomic systems, and our earlier conclusion about the magnitudes of reactive radiation reaction forces engendered by local fields. By Maxwell's argument every charge and current density distribution that satisfies the conservation laws is equally likely and, furthermore, over time the system traverses through them all. This includes instances where accelerating charge and current densities produce radiation, but that radiation is confined by a Q-trap and reactive radiation reaction forces drive the charge configuration until the radiation is confined. This suggests that a stable atom might be modeled as an enclosed, spherical cavity with perfectly conducting walls.

Radiation emanating from a confined source requires fields proportional to the sum of spherical Bessel and Neumann functions. The self-consistent field analysis of Sec. 7.6 shows the spherical Bessel functions form an exact radiative solution and, therefore, its field expressions are unchanged by iterations. The spherical Neumann function terms are inexact at large range and hence continued iterations produce ever-larger negative powers of σ; Table 7.6.5 and 7.6.6 contain examples. The tables show the Neumann function terms describe a potential source that supports a standing energy field instead, and a field that can be altered to a traveling energy field by the simple expedient of adding field terms proportional to spherical Bessel function terms if that could be done in a way that avoids the atomic Q-trap.

A special case of considerable interest is an atomic system with eigenstates W_H and W_L. In equilibrium, eigenstate W_H has more energy than W_L. Let the symmetry of the two eigenstates satisfy the radiation selection rules of Eq. 6.10.15 and let the energy difference be $W_H - W_L = \hbar\omega_0$. With one state occupied and the other empty imagine the occupying electron accelerates while transforming between states and, in so doing, generates a broad spectrum of frequencies, all of which are Q-trapped and remain local to the atom. Preferred frequency $\omega = \omega_0$ affects a resonance between the two energy levels large enough in magnitude to be nonlinear and, thereby, satisfy the Manley Rowe relationships. Consider effects at frequency $\omega_0 + \delta\omega$, where $\delta\omega$ is much less than ω_0. These interact to form frequencies ω_0 and $\delta\omega$. But for every frequency $\omega_0 + \delta\omega$ there is a similar frequency $\omega_0 - \delta\omega$ that interacts with frequency $\delta\omega$ to form frequency ω_0. This is the process of frequency pulling; it changes an initial spread of frequencies within a nonlinear medium to a monochromatic line.

What happens next depends upon details of the evolution of the source configuration. An electron, while supporting monochromatic internal oscillations, continues to evolve in accordance with Maxwell's 'assumption of continuity of path'. If it evolves to a configuration that does not support oscillations the energy is re-absorbed and the electron returns to its original state but if it evolves to a configuration that supports oscillations the result is much more interesting.

Consider the following scenario: the electron passes through a configuration that supports a z-directed electric dipole oscillating at frequency ω_0. The z-directed portion of its electric field intensity, as shown in Sec. 8.3, has the appropriate symmetry and therefore drives a collocated magnetic quadrupole and electric octupole. That is, the interior z-directed field of mode ℓ drives other multipoles of orders $\ell - 2$, $\ell - 1$, $\ell + 1$ and $\ell + 2$ of the same parity and degree one. As shown in Sec. 8.4, the exterior oscillating energy also drives all multipoles of the same parity. Our calculations show the magnitudes of intermodal driving forces are large enough for both feedback mechanisms, internal fields and external reactive power, to create a regenerative feedback that drives all intra-parity multipolar moments of degree one.

The exterior reactive energy also provides inter-parity field buildup. As shown in Sec. 8.4 the initial energy of the reactive intra-parity feedback between modes is proportional to small multiplicative factor b and the initial feedback force between inter-parity modes is proportional to b^2. Therefore if b is much less than one the inter-parity drive is not significant. However, if it approaches unity the two feedbacks become equivalent, inter-parity forces become similarly sized with intra-parity ones, and both parities build.

With both parities present the radiation fields are quite different from those with only one. With a dual parity buildup the emission fields of each modal order add constructively on the positive z-axis and destructively elsewhere. The result is a circular polarized, fully $+z$-directed beam with the angular power distribution of Dirac delta function $\delta(\theta,0)$. Since the regenerative field buildup has a single frequency, only that frequency escapes through porthole at $\theta = 0$; it is a monochromatic filter.

By Eq. 7.11.4 there is no reactive power, no reactive energy, and $Q = 0$. Since the angular width of the radiation density asymptotically approaches zero there is a modal value at which radiation is emitted. Upper limit L is determined by the comparative widths of the zero-Q, z-axis porthole and the angular width of the power density distribution. When that value of L is reached the electromagnetic energy is emitted *en masse*. With regard to Einstein's 1917 objection that spontaneous emission leaves the time and direction of spontaneous emission to chance, by the above argument the time and direction of emission is controlled by the unknown evolution of the ensemble of charge and current densities that is an electron. It is therefore unknown and unknowable.

Classical electromagnetic field theory shows there is no other means for radiation exchange between a radiation field and an electrically small source!

The Zipper

These discussions imply an analogy between a zippered jacket and radiating multipoles. The zipper becomes engaged if and only if the jacket sides are brought together and the end pieces meshed until the tab is pulled; a superregenerative response begins if and only if an excited electron with an appropriate inner structure forms and remains until a field is applied with the appropriate phase, frequency and magnitude. Once the zipper is started a slight pull on the tab unites paired zipper teeth, one link at a time, as it moves up the jacket; once a transition is started power goes first into the pair of dipoles and then into ever-higher order pairs of multipoles. The modal near-to-far field energy ratio is equal to the radiative Q of Table 7.4.1. The requirement is that the multipoles be precisely paired and driven with the proper phases and magnitudes. Under these conditions there is nothing to stop growth to higher orders and hence the 'zipper', and the matched pairs of Sec. 7.4 continue to extend to ever-higher modes until terminated by emission.

NOTE: The logical thread of the book continues in Chap. 9. The remainder of Chap. 8 is devoted to special mathematical properties that enable the superregenerative feedback that, in turn, produces Eq. 7.4.1.

8.6 Evaluation of S_{12} on the Axes

Unique properties of Eq. 7.4.1 on the coordinate axes are detailed in Sec. 8.10. Although those fields uniquely meet all required conditions, the question remains why and how that particular field is created. As an aside to answering the question we compute exact values of the sums on the coordinate axes. Values are detailed in Sec. 8.6 to 8.10 and the role of the recursion relationship of Eq. 7.4.1 is noted. If and only if that recursion relationship exists can power escape from the atomic Q-trap; no other relationship provides the properly ordered structure and no other relationship provides the zero-Q porthole that permits emission. The recursion relationship requires the multipolar pairs be precisely matched.

We obtain precise values for the field expressions at all radii on the coordinate axes. We already have precise values over the portion proportional to spherical Bessel functions and seek similarly precise values over spherical Neumann functions.

We begin with S_{12} where, by definition

$$S_{12}(\sigma,\theta) = \sum_{\ell=1}^{\infty} i^{1-\ell}(2\ell+1)y_{\ell}(\sigma)P_{\ell}^{1}(\cos\theta) \tag{8.6.1}$$

Legendre function $P_{\ell}^{1}(\cos\theta)$ is proportional to $\sin\theta$ and hence vanishes on the z-axis:

$$S_{12}(\sigma,0) = S_{12}(\sigma,\pi) = 0 \tag{8.6.2}$$

In the basal plane the Legendre polynomial is:

$$P_{\ell}^{1}(0) = (-1)^{(\ell-1)}\frac{(\ell)!!}{(\ell-1)!!}\delta(\ell,2q+1) \tag{8.6.3}$$

Symbol δ represents the Kronecker delta function and q represents an integer equal to or greater than zero; the function vanishes for even values of q.

The series form of the spherical Neumann functions is given by A.18.10 and repeated here:

$$y_\ell(\sigma) = -\sum_{s=0}^{\ell-1} \frac{(2\ell-2s-1)!!}{(2s)!!}\sigma^{-\ell-1+2s} - \sum_{s=0}^{\infty} \frac{(-1)^s}{(2s-1)!!}\frac{\sigma^{\ell-1+2s}}{(2\ell+2s)!!} \tag{8.6.4}$$

Substituting Eq. 8.6.3 and 8.6.4 into 8.6.1 gives:

$$S_{12}\left(\sigma,\frac{\pi}{2}\right) = -\sum_{\ell o;1}^{\infty}(2\ell+1)\frac{(\ell)!!}{(\ell-1)!!}\times$$
$$\left\{\sum_{s=0}^{\ell-1}\frac{(2\ell-2s-1)!!}{(2s)!!}\sigma^{-\ell-1+2s} + \sum_{s=0}^{\infty}\frac{(-1)^s}{(2s-1)!!}\frac{\sigma^{\ell-1+2s}}{(2\ell+2s)!!}\right\} \tag{8.6.5}$$

Lower limit '$\ell o;1$' indicates the sum is over only odd values of ℓ and the series begins with $\ell = 1$. Since only odd orders contribute to the sum it follows from inspection of Eq. 8.6.5 that only even powers of σ are present in the sum. Therefore the sum has the form:

$$S_{12}(\sigma,\pi/2) = \sum_{ne;-(\ell+1)}^{\infty} A_n\sigma^n \tag{8.6.6}$$

Lower limit '$ne;-(\ell+1)$' indicates the series contains only even values of n and begins at $n = -(\ell+1)$.

Combining Eq. 8.6.5 and 8.6.6 shows:

$$A_n = \left\{\begin{array}{l} -(-1)^{n/2}\sum_{\ell o;1}^{n+1}\frac{(2\ell+1)(\ell)!!(-1)^{(\ell-1)/2}}{(\ell+1)!!(\ell+n+1)!!(n-\ell)!!} \\[2ex] -\lim_{L\to\infty}\sum_{\ell o;n+3}^{L}\frac{(2\ell+1)(\ell)!!(\ell-n-2)!!}{(\ell-1)!!(\ell+n+1)!!} \end{array}\right\} \tag{8.6.7}$$

L is the largest modal number present in the field.

Coefficients versus Powers of σ

Consider as an example the special case $n = +2$; for this case the series describing A_2 follows from Eq. 8.6.7 and is:

$$\left\{ \frac{3}{8} - \frac{7}{32} - \frac{55}{1024} - \frac{105}{4096} - \frac{1995}{131,078} - \frac{5313}{524,288} - \frac{243,243}{33,554,452} - ... \right\} = 0$$

$$(8.6.8)$$

The first two terms come from the first sum of Eq. 8.6.7 and the rest from the second sum. Although the series converges rapidly the sum vanishes only at a limitlessly large value of L. The coefficients of all other positive powers of σ follow similarly and are equal to zero.

- The coefficients of all other positive powers of σ follow similarly and are equal to zero. This ordered null result is a unique property of the recursion relationsip of Eq. 7.4.1

A similar analysis for negative powers of σ shows the total is limitlessly large and hence the series can only represent physical reality if it terminates. The series is:

$$S_{12}\left(\sigma,\pi/2\right) = -\lim_{L\to\infty} \sum_{lo;n-1}^{L} \left(2\ell+1\right)\frac{\left(\ell\right)!!}{\left(\ell-1\right)!!}\frac{\left(\ell+n-2\right)!!}{\left(\ell-n+1\right)!!\sigma^{n}} \qquad (8.6.9)$$

Remembering the lower limit on ℓ is one or greater, for the special case $n = 0$ the series of Eq. 8.6.9 is:

$$-\left(\frac{3}{2} + \frac{21}{16} + \frac{165}{128} + \frac{2625}{2048} + ...\right)$$

$$= -\left(1.5 + 1.3125 + 1.2891 + 1.2817 + ..\right) = -A_0 \qquad (8.6.10)$$

Since each term approaches unity as the modal number increases, for L much larger than one the approximate value is:

$$A_0 \cong 5L/8 \qquad (8.6.11)$$

For the special case $n = -2$, the series is:

$$-\left(3 + \frac{63}{4} + \frac{2475}{64} + \frac{165,375}{2304} + \frac{16,967,475}{147,456} + ...\right) = -A_2 \qquad (8.6.12)$$

The series is proportional to L^3. Similar term-by-term expansions show that $A_n \approx L^{n+1}$ and hence:

$$S_{12}(\sigma, \pi/2) = -\sum_{n=0;e}^{\infty} \frac{A_n}{\sigma^n}$$

$$A_n = -\sum_{\ell o;n-1}^{L} \frac{(2\ell+1)\ell!!\,(\ell+n-2)!!}{(\ell-1)!!\,(\ell-n+1)!!} \approx L^{n+1}$$

$$(8.6.13)$$

Each coefficient A_n contains contributions from all modal orders. It follows from the z-axis values and from Table 7.6.3, Eq. 7.6.3 and 8.6.13 that the axial electric field intensities are:

$$E_r(\sigma, 0, \phi) = E_r(\sigma, \pi, \phi) = 0$$

$$E_r\left(\sigma, \frac{\pi}{2}, \phi\right) = F_0\left\{1 + i\sum_{ne;0}^{\infty} \frac{A_n}{\sigma^{n+1}}\right\} e^{i(\omega t - \phi)}$$

$$(8.6.14)$$

The leading term in the basal plane is sinusoidally time varying and with a magnitude that is independent of range, and it results from summing over spherical Bessel functions. The sum over inverse powers of σ shows the radial field energy is concentrated near the source.

8.7 Evaluation of S_{22} and S_{32} on the Polar Axes

The spherical Neumann portion of sums S_2 and S_3 are:

$$S_{22}(\sigma, \theta) = \sum_{\ell=1}^{\infty} i^{-\ell} \frac{(2\ell+1)}{\ell(\ell+1)} y_\ell(\sigma) \frac{dP_\ell^1(\cos\theta)}{d\theta}$$

$$(8.7.1)$$

$$S_{32}(\sigma, \theta) = \sum_{\ell=1}^{\infty} i^{-\ell} \frac{(2\ell+1)}{\ell(\ell+1)} y_\ell(\sigma) \frac{P_\ell^1(\cos\theta)}{\sin\theta}$$

Associated Legendre polynomial at $\theta = 0$ and π, see Table A.12.1, are equal to:

$$\frac{dP_\ell^1}{d\theta}\bigg|_{\theta=0} = \frac{P_\ell^1}{\sin\theta}\bigg|_{\theta=0} = \frac{\ell(\ell+1)}{2}; \quad \frac{dP_\ell^1}{d\theta}\bigg|_{\theta=\pi} = -\frac{P_\ell^1}{\sin\theta}\bigg|_{\theta=\pi} = (-1)^\ell\frac{\ell(\ell+1)}{2} \quad (8.7.2)$$

Since sums S_2 and S_3 have equal magnitudes on the z-axes we need solve for only one of them. Although the desired sum is over modal orders between one and infinity it is convenient to retain the $\ell = 0$ term as part of the sum and subtract it later in the analysis. We therefore evaluate:

$$S_{32}(\sigma,0) = \frac{\cos\sigma}{\sigma} - \frac{1}{2}\sum_{\ell=0}^{\infty} i^{-\ell}(2\ell+1)\times$$

$$\left\{\sum_{s=0}^{\infty}\frac{(-1)^s}{(2s-1)!!(2\ell+2s)!!}\sigma^{\ell-1+2s} + \sum_{s=0}^{\ell-1}\frac{(2\ell-2s-1)!!}{(2s)!!}\sigma^{-\ell-1+2s}\right\} \quad (8.7.3)$$

Even values of ℓ produce odd powers of σ and *vice versa*. Within the curly brackets the first term contains positive powers of σ from $(+\ell-1)$ to ∞; the second term contains powers from $(-\ell-1)$ to $(+\ell-3)$.

To evaluate the sum, let n be a positive integer, determine the coefficient of σ^n, and then separate results into terms with even and values of n:

$$S_{32} = \left\{\begin{array}{l} i\dfrac{(-1)^{n/2}}{2}\sum_{lo;1}^{n+1}\dfrac{(2\ell+1)\sigma^n}{(n-\ell)!!(n+\ell+1)!!} \\[4mm] +\dfrac{i}{2}\sum_{lo;1,n+3}^{\infty}(-1)^{(\ell-1)/2}\dfrac{(2\ell+1)(\ell-n-2)!!}{(n+\ell+1)!!}\sigma^n \end{array}\right\}$$

$$\pm\left\{\begin{array}{l} \dfrac{\cos\sigma}{\sigma} + \dfrac{(-1)^{[\ell/2-1]}}{2}\sum_{\ell e;0}^{n+1}\dfrac{(2\ell+1)\sigma^n}{(n-\ell)!!(n+\ell+1)!!} \\[4mm] -\dfrac{1}{2}\sum_{\ell e;n-3}^{\infty}(-1)^{\ell/2}\sum_{\ell e;0}^{n+1}\dfrac{(2\ell+1)(\ell-n-2)!!}{(n+\ell+1)!!}\sigma^n \end{array}\right\} \quad (8.7.4)$$

Term '$\ell o;1,n+3$' indicates ℓ begins at the larger of 1 or n+3. The upper $\pm$ sign in Eq. 8.7.4 is for $\theta = 0$ and the lower for $\theta = \pi$. Defining S'_{32} to equal the top row of Eq. 8.7.4 and expanding the series gives:

n even, ℓ odd.

$$S'_{32} = \frac{i(-1)^{n/2}}{2}\left[\begin{array}{l}\left\{\dfrac{3}{(n-1)!!(n+2)!!}+\dfrac{7}{(n-3)!!(n+4)!!}+...+\dfrac{(2n+3)}{(2n+2)!!}\right\} \\[2ex] -\left\{\dfrac{(2n+7)(1)!!}{(2n+4)!!}-\dfrac{(2n+11)(3)!!}{(2n+6)!!}+\dfrac{(2n+15)(5)!!}{(2n+8)!!}-...\right\}\end{array}\right]$$

$$(8.7.5)$$

The bottom row of Eq. 8.7.5 may be evaluated by writing $(2\ell+1) = (\ell+n+1) + (\ell-n)$ and regrouping the terms as:

$$\left\{\begin{array}{l}(2n+4)\dfrac{(1)!!}{(2n+4)!!}+\dfrac{(1)!!}{(2n+4)!!}\left(3-(2n+6)\dfrac{3}{(2n+6)}\right) \\[2ex] -\dfrac{(3)!!}{(2n+6)!!}\left(5-(2n+8)\dfrac{5}{(2n+8)}\right)+...\end{array}\right\}=\dfrac{1}{(2n+2)!!}$$

$$(8.7.6)$$

Inserting Eq. 8.7.6 back into Eq. 8.7.5 gives:

n even, ℓ odd.

$$S'_{32} = \frac{i(-1)^{n/2}}{2}\left\{\begin{array}{l}\dfrac{3}{(n-1)!!(n+2)!!}+\dfrac{7}{(n-3)!!(n+4)!!}+ \\[2ex] ...+\dfrac{(2n+3)}{(2n+2)!!}-\dfrac{1}{(2n+2)!!}\end{array}\right\} \qquad (8.7.7)$$

Evaluating Eq. 8.7.7 for the special case of $n = 0$ gives:

$$\frac{i}{2}\left(\frac{3}{2}-\frac{1}{2}\right)=\frac{i}{2}$$

Repeating the process for $n = 2$ gives:

$$-\frac{i}{2}\left(\frac{3}{8}+\frac{7}{48}-\frac{1}{48}\right)=-\frac{i}{4}$$

Repeating the process for all even values of n equal zero or more gives:

n even, ℓ odd

$$S_{32} = \frac{i}{2}\cos\sigma \tag{8.7.8}$$

Repeating the process for the opposite parity results in:

n odd, ℓ even.

$$S_{32} = \pm\frac{1}{2}\sin\sigma \tag{8.7.9}$$

Combining results for powers of σ equal to and greater than zero shows the sums are:

$$S_{32}(\sigma,0) = \frac{1}{2}\left(ie^{-i\sigma} + \frac{\cos\sigma}{\sigma}\right); \ S_{32}(\sigma,\pi) = \frac{1}{2}\left(ie^{i\sigma} - \frac{\cos\sigma}{\sigma}\right)$$

$$S_{22}(\sigma,0) = \frac{1}{2}\left(ie^{-i\sigma} + \frac{\cos\sigma}{\sigma}\right); \ S_{22}(\sigma,\pi) = -\frac{1}{2}\left(ie^{i\sigma} - \frac{\cos\sigma}{\sigma}\right)$$

$$\tag{8.7.10}$$

$$\mathcal{S}_{32}(\sigma,0) = \frac{1}{2}\left(ie^{-i\sigma} - \frac{\cos\sigma}{\sigma}\right); \ \mathcal{S}_{32}(\sigma,\pi) = -\frac{1}{2}\left(ie^{i\sigma} + \frac{\cos\sigma}{\sigma}\right)$$

$$\mathcal{S}_{22}(\sigma,0) = \frac{1}{2}\left(ie^{-i\sigma} - \frac{\cos\sigma}{\sigma}\right); \ \mathcal{S}_{22}(\sigma,\pi) = \frac{1}{2}\left(ie^{i\sigma} + \frac{\cos\sigma}{\sigma}\right)$$

- Uniquelly to the recursion relationsip of Eq. 7.4.1 the magnitude and phase of each mode is the exact value needed to produce the transcendental functions of Eq. 8.7.10. As with the spherical Bessel function terms, the magnitude of the first term of each sum is independent of distance.

Negative powers of σ are entirely contained in the second sum of Eq. 8.7.3. With positive values of n representing inverse powers of σ, the sum is:

$$-\frac{1}{2}\sum_{\ell=n-1}^{L+1}i^{-\ell}\frac{(2\ell+1)(\ell+n-2)!!}{(\ell-n+1)!!\sigma^{n}}\delta(\ell+n,2q+1) \tag{8.7.11}$$

The series is also evaluated by writing $(2\ell+1)=(\ell-n+1)+(\ell+n)$ and then expanding and regrouping. The result is:

$$\left\{\begin{array}{l}\dfrac{(2n-3)!!}{(0)!!}\left((2n-1)-(2)\dfrac{(2n-1)}{(2)}\right)-\dfrac{(2n-1)!!}{(2)!!}\left((2n+1)-(4)\dfrac{(2n+1)}{(4)}\right)\\[3mm]+\dfrac{(2n+1)!!}{(4)!!}\left((2n+3)-(6)\dfrac{(2n+3)}{(6)}\right)+\cdots+\dfrac{(-1)^{L}(L+n)!!}{(L-n-1)!!}\end{array}\right\}$$
$$\times\left[-\frac{i(-1)^{1-n}}{2\sigma^{n}}\right]$$
$$\tag{8.7.12}$$

- Uniquelly to the recursion relationship of Eq. 7.4.1, Eq. 8.7.12 vanishes, term by term. A deviation from that relationship creates a local field and it, in turn, produces a radiation reaction force that acts to maintain the relationship.

Results for powers of σ greater than or equal to zero are shown in Table 8.7.1; the residual contains only the highest existing modal number

Table 8.7.1. Field sums over spherical Neumann functions on positive and negative z-axes.

$$S_{22}(\sigma,0)=S_{32}(\sigma,0)=\frac{1}{2}\left(ie^{-i\sigma}+\frac{\cos\sigma}{\sigma}\right)+\Xi$$

$$\$_{22}(\sigma,0)=\$_{32}(\sigma,0)=\frac{1}{2}\left(ie^{-i\sigma}-\frac{\cos\sigma}{\sigma}\right)+\frac{i}{\sigma}\frac{\partial}{\partial\sigma}(\sigma\Xi)$$

Ξ is unknown residual with negative powers of σ

$$S_{22}(\sigma,\pi)=-S_{32}(\sigma,\pi)=-\frac{1}{2}\left(ie^{i\sigma}-\frac{\cos\sigma}{\sigma}\right)+(\Xi*-2)$$

$$\$_{22}(\sigma,\pi)=-\$_{32}(\sigma,\pi)=\frac{1}{2}\left(ie^{i\sigma}+\frac{\cos\sigma}{\sigma}\right)+\frac{i}{\sigma}\frac{\partial}{\partial\sigma}(\sigma\Xi*-2\sigma)$$

8.8 Evaluation of S_{32} in the Basal Plane

By definition sum S_{32} is:

$$S_{32}(\sigma,\theta) = \sum_{\ell=1}^{\infty} i^{-\ell} \frac{(2\ell+1)}{\ell(\ell+1)} y_\ell(\sigma) \frac{P_\ell^1(\cos\theta)}{\sin\theta} \tag{8.8.1}$$

In the basal plane the value of angular function, see Table A.12.1, is:

$$P_\ell^1(0) = i^{1-\ell} \frac{(\ell)!!}{(\ell-1)!!} \delta(\ell, 2q+1) \tag{8.8.2}$$

Substituting Eq. 8.8.2 and 8.6.4 into Eq. 8.8.1 gives:

$$S_{32}\left(\sigma, \frac{\pi}{2}\right) = i \sum_{\ell o;1}^{\infty} \frac{(2\ell+1)(\ell-2)!!}{(\ell+1)!!}$$
$$\times \left\{ \sum_{s=0}^{\ell-1} \frac{(2\ell-2s-1)!!}{(2s)!!} \sigma^{-\ell-1+2s} + \sum_{s=0}^{\infty} \frac{(-1)^s}{(2s-1)!!} \frac{\sigma^{\ell-1+2s}}{(2\ell+2s)!!} \right\} \tag{8.8.3}$$

Since there are only odd values of ℓ, only even powers of σ are present in the sum and the sum over the positive powers of σ has the form:

$$S_{32}\left(\sigma, \frac{\pi}{2}\right) = i \sum_{ne}^{\infty} B_n'' \sigma^n \tag{8.8.4}$$

Combining Eq. 8.8.3 and 8.8.4 gives:

$$B_n'' = \left\{ \begin{array}{l} (-1)^{n/2} \sum\limits_{\ell o;1}^{n+1} \dfrac{(2\ell+1)(\ell-2)!!(-1)^{(\ell-1)/2}}{(\ell+1)!!(n-\ell)!!(\ell+n+1)!!} \\[2em] +i \sum\limits_{\ell o;n+3}^{\infty} \dfrac{(2\ell+1)(\ell-2)!!(\ell-n-2)!!}{(\ell+1)!!(\ell+n+1)!!} \end{array} \right\} \tag{8.8.5}$$

Taking coefficients B_0'' and B_2'' as special cases and writing Eq. 8.8.5 out term-by-term gives:

$$B_0'' = \left(\frac{3}{2\times 2} + \frac{7}{4!!\times 4!!} + \frac{11\times 3^2}{6!!\times 6!!} + ... \right) = 1$$

$$\tag{8.8.6}$$

$$B_2'' = \left(-\frac{3}{2\times 4!!} + \frac{7}{4!!\times 6!!} + \frac{11\times 3^2}{6!!\times 8!!} + ... \right) = -\frac{1}{6}$$

Extending the evaluation to all positive values of n and then summing:

$$S_{32}\left(\sigma, \frac{\pi}{2} \right) = i\frac{\sin\sigma}{\sigma} \tag{8.8.7}$$

- Uniquely to the recursion relationship of Eq. 7.4.1 each mode contributes the exact amount required for the positive powers of σ to equal this transcendental function.

To evaluate the coefficients of negative powers of σ, define a new set of coefficients B_n and write the expansion as:

$$S_{32}\left(\sigma, \frac{\pi}{2} \right)_{n<0} = i\frac{B_n}{\sigma^n} \tag{8.8.8}$$

Combining Eq. 8.8.5 and 8.8.8 gives:

$$B_n = \sum_{\ell o;n+1}^{L} \frac{(2\ell+1)(\ell-1)!!}{(\ell+1)!!} \frac{(\ell+n-2)!!}{(\ell-n+1)!!} \tag{8.8.9}$$

The coefficients of the first two terms are:

$$B_2 = i(1.3125 + 1.28906 + 1.28174 + 1.27853 + ...) \approx iL/2$$

$$\tag{8.8.10}$$

$$B_4 = i(36.09375 + 69.21387 + 106.58936 + 165.99072 + ...) \approx iL^3$$

As n increases by one the power of L increases by two.
 The sum of Eq. 8.8.7 and 8.8.8 gives:

$$S_{32}\left(\sigma,\frac{\pi}{2}\right)=\frac{i}{\sigma}\sin\sigma+i\sum_{ne;2}^{L}\frac{B_n}{\sigma^n};\quad \$_{32}\left(\sigma,\frac{\pi}{2}\right)=-\frac{1}{\sigma}\cos\sigma-i\sum_{ne;2}^{L}\frac{(n-1)B_n}{\sigma^{n+1}}$$

$$(8.8.11)$$

8.9 Evaluation of S_{22} in the Basal Plane

By definition sum S_{22} is:

$$S_{22}(\sigma,\theta)=\sum_{\ell=1}^{\infty}i^{-\ell}\frac{(2\ell+1)}{\ell(\ell+1)}y_\ell(\sigma)\frac{dP_\ell^1(\cos\theta)}{d\theta}\tag{8.9.1}$$

In the basal plane the value of angular function, see Table A.12.1, is:

$$\frac{dP_\ell^1(0)}{d\theta}=i^\ell\frac{(\ell+1)!!}{(\ell-2)!!}\delta(\ell,2q)\tag{8.9.2}$$

Since angular derivatives of odd-order Legendre functions vanish at $\theta=\pi/2$, only even values of ℓ appear in the sum. Substituting Eq. 8.6.4 and 8.9.2 into Eq. 8.9.1 gives the basal plane expression:

$$S_{22}\left(\sigma,\frac{\pi}{2}\right)=\frac{\cos\sigma}{\sigma}-\sum_{\ell e;0}^{\infty}\frac{(2\ell+1)(\ell-1)!!}{(\ell)!!}$$

$$\times\left\{\sum_{s=0}^{\ell-1}\frac{(2\ell-2s-1)!!}{(2s)!!}\sigma^{-\ell-1+2s}+\sum_{s=0}^{\infty}\frac{(-1)^s}{(2s-1)!!(2\ell+2s)!!}\sigma^{\ell-1+2s}\right\}\tag{8.9.3}$$

It is helpful in what follows to add and subtract the $\ell=0$ term. Since only even orders of ℓ occur there are only odd powers of σ:

$$S_{22}\left(\sigma,\frac{\pi}{2}\right)=\sum_{no;1}^{\infty}C_n''\sigma^n+\frac{\cos\sigma}{\sigma}\tag{8.9.4}$$

Comparison of Eq. 8.9.4 and 8.9.3 shows:

$$C_n'' = \left\{ \begin{array}{l} (-1)^{(n-1)/2} \displaystyle\sum_{\ell e;0}^{n+1} \frac{(2\ell+1)(\ell-1)!!}{(\ell)!!} \frac{(-1)^{\ell/2}}{(n-\ell)!!(\ell+n+1)!!} \\[3mm] - \displaystyle\sum_{\ell e;n+3}^{\infty} \frac{(2\ell+1)(\ell-1)!!(\ell-n-2)!!}{(\ell)!!} \frac{}{(\ell+n+1)!!} \end{array} \right\} \tag{8.9.5}$$

Evaluation of the special cases $n = 1$ and $n = 3$ shows:

$$C_1'' = -\left\{ -\frac{1}{2} + \frac{5}{2!!\times 4!!} + \frac{9\times 3!!\times 1!!}{4!!\times 6!!} + \ldots \right\} = 0$$

$$C_3'' = -\left\{ \frac{1}{3!!\times 4!!} - \frac{5\times 1!!}{2!!\times 6!!} + \frac{9\times 3!!}{4!!\times 8!!} + \frac{13\times 5!!\times 1!!}{6!!\times 10!!} + \ldots \right\} = 0 \tag{8.9.6}$$

Extending the evaluation to all positive values of n and then summing gives, for ℓ equal to or greater than zero and positive powers of σ, gives:

$$S_{22}\left(\sigma, \frac{\pi}{2}\right) = \frac{\cos\sigma}{\sigma} \tag{8.9.7}$$

- The reduction of Eq. 8.9.3 to this transcendental form is a unique property of the recursion relationship.

To examine coefficients of negative powers of σ define new coefficients:

$$S_{22}\left(\sigma, \pi/2\right)_{n<0} = \sum_{no;1}^{L} \frac{C_n}{\sigma^n} \tag{8.9.8}$$

Combining Eq. 8.9.8 with Eq. 8.9.3 gives:

$$C_n = -\sum_{\ell e;n-1}^{L} \frac{(2\ell+1)(\ell-1)!!(\ell+n-2)!!}{(\ell)!!} \frac{}{(\ell-n+1)!!} \tag{8.9.9}$$

Evaluation of the $n = 1$ term gives:

$$C_1 = 1 + 1.25 + 1.256 + \cdots \cong 5L/8 \tag{8.9.10}$$

The lower limit on ℓ is zero since the zeroth term is added from Eq. 8.9.7 and subtracted from Eq. 8.9.10. With each succeeding increase in n the power of L in the approximate equality increases by two.

Combining Eq. 8.9.8 and 8.9.9 gives:

$$S_{22}\left(\sigma,\frac{\pi}{2}\right) = \frac{\cos\sigma}{\sigma} - \sum_{no;1}^{\infty} \frac{C_n}{\sigma^n}; \quad \$_{22}\left(\sigma,\frac{\pi}{2}\right) = -i\frac{\sin\sigma}{\sigma} + i\sum_{no;1}^{\infty} \frac{(n-1)C_n}{\sigma^{n+1}}$$

$$(8.9.11)$$

8.10 Summary of Axial Fields

Sums of spherical Hankel functions on the $+z$-axis follow from the values for spherical Bessel and Neumann functions obtained in earlier sections. Combining gives the values:

$$S_1(\sigma,0) = 0$$

$$S_2(\sigma,0) = e^{-i\sigma}\left(1 - \frac{i}{2\sigma}\right) + \Xi$$

$$S_3(\sigma,0) = S_2(\sigma,0); \quad \$_3(\sigma,0) = \$_2(\sigma,0)$$

$$\$_2(\sigma,0) = e^{-i\sigma}\left(1 + \frac{i}{2\sigma}\right) + \frac{i}{\sigma}\frac{\partial}{\partial\sigma}(\sigma\Xi)$$

$$(8.10.1)$$

Ignoring the residual and putting the rest into field forms gives the electric field intensity on the positive z-axis:

$$E_r = 0; \quad E_\theta = 2F_0 e^{i(\omega t - \sigma - \phi)}; \quad E_\phi = -i2F_0 e^{i(\omega t - \sigma - \phi)}$$

$$(8.10.2)$$

As seen earlier, these are circularly polarized fields from which the time average Poynting vector is:

$$N = \frac{1}{2}\text{Re}\left(\tilde{E}\times\tilde{H}^*\right) = -\frac{1}{2\eta}\text{Re}\left(i\tilde{E}\times\tilde{E}^*\right)$$

$$(8.10.3)$$

$$N = \hat{z}N_z = \frac{4}{\eta}\hat{z}$$

This axial power density is valid at intermediate ranges and is independent of distance from the source and fully directed with a path parallel to the $+z$-axis.

Each mode contributes a proportionate share to the residual sums in the equatorial plane. The sums are:

$$S_1\left(\sigma,\frac{\pi}{2}\right) = \sigma\left[1+i\sum_{ne;0}^{L}\frac{A_n}{\sigma^{n+1}}\right]; \quad S_2\left(\sigma,\frac{\pi}{2}\right) = \frac{1}{\sigma}\left[\sigma - ie^{-i\sigma}\right] + i\sum_{no;1}^{L}\frac{C_n}{\sigma^n}$$

$$\$_2\left(\sigma,\frac{\pi}{2}\right) = \frac{i}{\sigma}\left[1-e^{-i\sigma}\right] + \sum_{no;3}^{L}\frac{(n-1)C_n}{\sigma^{n+1}}$$

$$S_3\left(\sigma,\frac{\pi}{2}\right) = -\frac{i}{\sigma}\left[1-e^{-i\sigma}\right] + \sum_{ne;2}^{L}\frac{B_n}{\sigma^n}; \quad \$_3\left(\sigma,\frac{\pi}{2}\right) = \frac{i}{\sigma}e^{-i\sigma} - i\sum_{ne;2}^{L}\frac{(n-1)B_n}{\sigma^{n+1}}$$

$$(8.10.4)$$

Letter functions representing the sum coefficients are:

$$A_n = \sum_{\ell o;n-1}^{L}\frac{(2\ell+1)\ell!!\,(\ell+n-2)!!}{(\ell-1)!!\,\,(\ell-n+1)!!}$$

$$B_n = \sum_{\ell o;n+1}^{L}\frac{(2\ell+1)(\ell-2)!!\,(\ell+n-2)!!}{(\ell+1)!!\,\,(\ell-n+1)!!} \qquad (8.10.5)$$

$$C_n = \sum_{\ell e;n-1}^{L}\frac{(2\ell+1)(\ell-1)!!\,(\ell+n-2)!!}{(\ell)!!\,\,(\ell-n+1)!!}$$

Substituting into the field forms gives the electric field in the basal plane:

$$\tilde{E}\left(\sigma,\frac{\pi}{2}\right) = F_0 \left\{ \begin{array}{l} \left[1-i\sum_{ne;0}^{\infty}\frac{A_n}{\sigma^{n+1}}\right]\hat{r} + \left\{\sum_{ne;2}^{\infty}\frac{B_n}{\sigma^n} + \sum_{no;3}^{\infty}\frac{(n-1)C_n}{\sigma^{n+1}}\right\}\hat{\theta} \\[4mm] -i\left\{1+i\sum_{no;1}^{\infty}\frac{C_n}{\sigma^n} - i\sum_{ne;2}^{\infty}\frac{(n-1)B_n}{\sigma^{n+1}}\right\}\hat{\phi} \end{array} \right\} e^{i(\omega t-\phi)}$$

$$(8.10.6)$$

With 'O' representing order, these fields produce the Poynting vector:

$$N\left(\sigma,\pi/2\right)=\frac{F_0^2}{\eta}\left\{O\left(\frac{1}{\sigma^2}\right)\hat{r}+\left[-1+O\left(\frac{1}{\sigma^2}\right)\right]\hat{\theta}-i\left[1+O\left(\frac{1}{\sigma^2}\right)\right]\hat{\phi}\right\}$$

$$(8.10.7)$$

The first term of the zenith-angle power density is $+z$-directed and contributes to energy flow from the lower to the upper hemisphere.

The sums on the negative z-axis are:

$$S_2\left(\sigma,\pi\right)=-\frac{i}{2\sigma}e^{i\left(\omega t+\sigma-\phi\right)};\ \ S_3\left(\sigma,\pi\right)=\frac{i}{2\sigma}e^{i\left(\omega t+\sigma-\phi\right)}$$

$$\mathbb{S}_2\left(\sigma,\pi\right)=-\frac{i}{2\sigma}e^{i\left(\omega t+\sigma-\phi\right)};\ \ \mathbb{S}_3\left(\sigma,\pi\right)=\frac{i}{2\sigma}e^{i\left(\omega t+\sigma-\phi\right)}$$

$$(8.10.8)$$

Substituting the sums into the expression for fields shows the electric field intensity and the Poynting vector both vanish:

$$\tilde{E}\left(\sigma,\pi\right)=0 \tag{8.10.9}$$

The axial sums show the fields produce a power density of $4/\eta$ on the positive z-axis and a z-directed power density of $1/\eta$ in the basal plane. There is no equivalent power on the negative z-axis.

References

R. Adler, "A study of locking phenomena in oscillators," Proc. IRE, Vol. 34, 351–357 (1946)

G. Auletta, *Foundations and Interpretation of Quantum Mechanics*, World Scientific Publishing (2001)

E.H. Armstrong, "Some Recent Developments of Regenerative Circuits," Proc. IRE, Vol. 10, 244 (1922)

M. Bennett, M.F. Shatz, H. Rockwood, K. Wiesenfeld, "Huygens's Clocks," Proc. Royal Soc. London A, Vol. 458, pp. 563–579 (2002)

H.W. Bode, *Network Analysis and Feedback Amplifier Design*, Van Nostrand, 31–35 (1945)

D. Bohm, *Quantum Theory*, Prentice Hall, Princeton NJ (1951)

D. Cox, "Linear amplification with nonlinear components," Communications, IEEE Transactions, Vol. 22, 1942–1945 (1974)

V. Damgov, *Nonlinear and Parametric Phenomena*, World Scientific (2004)

P.A.M. Dirac, *The Principles of Quantum Mechanics*, 4th ed., Oxford (1958)

G. Greenstein, A.G. Zajonc, *The Quantum Challenge: Modern Research on the Foundations of Quantum Mechanics*, Jones and Bartlett Publishers (1997)

L.D. Landau, E.M. Lifshitz, *Quantum Mechanics: Non-Relativistic Theory*, Addison-Wesley, Reading MA (1958)

T.H. Lee, *The Design of CMOS Radio-Frequency Integrated Circuits*, Cambridge (2004)

B. Razavi, "A Study of Injection Locking and Pulling an Oscillators," IEEE J. Solid-State Circuits, Vol. 39 1415–1424 (2004)

D.H. Wolaver, *Phase-Locked Loop Circuit Design*, Prentice Hall (1991)

M.T. Weiss, "Quantum Derivation of Energy Relations Analogous to those for Nonlinear Reactances," Proc. IRE Vol. 45, p. 1012 (1957)

Chapter 9

Absorption, Emission, Entanglements

9.1 Photon Model

An historic photon dichotomy about light is that it diffracts and reflects as a wave but is created and annihilated with particle's kinematics. Planck asked: "What happens to a light-quantum after its emission? Does it pass outwards in all directions continually increasing in volume and tending towards infinite dilution? Alternatively, does it fly like a projectile in one direction only?" In an attempt to address Planck's question we note the explanation of spontaneous emission described in Sec. 8.5 produces a photon that is fully directed, but if and only if the highest modal order is limitlessly large. The creation of a wave train of length 'l' with photon properties requires the conditions expressed by the inequality $l/\lambda \ll L/2\pi$. By Eq. 7.8.1 the product of the upper modal limit times the angular spread during emission is $L\Delta\theta \sim 5.13$. The earth's distance from the sun is about 1.5×10^8 km and the radius of the region illuminated by a solar-emitted photon at that radius, if $L = 1000$, is about 8×10^5 km. Absorption, therefore, requires either that *in situ* energy is absorbed or that it is instantly transferred from throughout the extended region.

Decades of cosmic ray observations confirm that pair production occurs at the top of the atmosphere; the requirements are an external absorber of momentum and a radiation frequency equal to or greater than $\omega_1 = 2m_0c^2/h$. At the top of the atmosphere there is no obvious source of *en situ* energy other than the photon itself. We conclude that a photon and its energy are locally confined. Pair production demonstrates that the extreme fields that exist within high-energy photons affect local vacuum states.

Since frequencies greater than ω_1 so strongly affect local vacuum states, we suggest that lesser frequencies will also affect them but in less obvious ways. Although the total photon field energy is $\hbar\omega$, and is independent of L, its distribution within the source is not. The length of the source is determined by the size of an eigenstate electron and that, in turn, by the size of a host atom; a typical atomic size is on the order of 10^{-10} m. Source width is determined by the physical size of the source singularities and as the order of the singularity increases the size of the region in which the field is concentrated decreases; and the decreased size supports an increased value of maximum field intensity.

Although the wavelength of optical radiation exceeds that required for pair production by several orders of magnitude, the effect establishes a precedent. On the basis of that precedent we hypothesize there exists an axial field intensity above which local vacuum states create limitlessly large values of the constitutive relations. Since the largest fields are on the z-axis we suggest such a filamentary thread lies along the central axis of a photon. Such a thread is sufficient to permanently bind the radiation into a particle of energy. As it travels the filament is continually created at the front of the pulse and, after it passes, space returns to normal. Creation energy, if there is any, is returned upon photon annihilation.

To examine the resultant fields, construct a cylindrical coordinate system with coordinates (ρ,ϕ,z). The origin lies at the center of the source and the z-axis is coincident with that of the original source. By Eq. 7.10.7, the photon field is TEM and a potential expressed in circular coordinates is a complete source of TEM fields. Since the angular variation is as $\exp(-i\phi)$ the order is one and the transverse fields decrease as the inverse square of distance ρ from the filament. The resulting descriptive equation, therefore, is:

$$\mathbf{E} = \frac{\Psi_0}{\rho^2}\{\hat{\rho} - i\hat{\varphi}\}\,e^{i(\omega t - \varphi - kz)}; \quad \tilde{\mathbf{H}} = \frac{\Psi_0}{\eta\rho^2}\{i\hat{\rho} + \hat{\varphi}\}\,e^{i(\omega t - \varphi - kz)} \tag{9.1.1}$$

A photon consists of a fixed length of this radiation field. Unlike antenna radiation, it is confined by non-radiative means and therefore it possesses particle-like kinematic and scattering properties. Its energy is equal to the volume integral of Eq. 9.1.1:

$$W = \hbar\omega = 2\pi X_0^2 \frac{l}{d^2} \tag{9.1.2}$$

Constant X_0 depends upon L and F_0 as shown in Eq. 7.10.7, d is the radius of the filamentary thread, and l is the photon length. The angular momentum is manifest in the rotation of this pseudo-particle, with its intrinsic mass density rotating about the z-axis at the rate that satisfies Eq. 7.7.10.

9.2 Binding by Standing Waves

An applied radiation field creates a push or pull on a scatterer. It is calculable by substituting field values just off the surface into the electromagnetic stress tensor Eq. 1.6.8. Note that Fig. 2.19.1, 2.10.2 and 2.10.3 are examples of the effect on spherical scatterers. In this section we examine the force between two identical, radiating electric dipoles within a z-directed string of identical ones.

Each dipole has moment p and produces the fields:

$$\tilde{E} = \frac{pk^3}{4\pi\varepsilon}\left\{\hat{r}\left[\frac{2}{\sigma^3}+\frac{2i}{\sigma^2}\right]\cos\theta + \hat{\theta}\left[\frac{1}{\sigma}-\frac{i}{\sigma^2}-\frac{1}{\sigma^3}\right]\sin\theta\right\}e^{i\omega t_R}$$

$$\eta\tilde{H} = \frac{pk^3}{4\pi\varepsilon}\hat{\phi}\left[\frac{1}{\sigma}-\frac{i}{\sigma^2}\right]\sin\theta\, e^{i\omega t_R} \tag{9.2.1}$$

The two dipoles are separated by distance $2z_0$ along the z-axis. We solve for the force between them by using the fields on the virtual xy plane that passes through z_0 to evaluate the stress tensor, and then integrate the result over the virtual plane to obtain the total force. For that purpose we shift to rectangular coordinates. The fields of the lower dipole on the virtual plane are:

$$\frac{\tilde{E}_x}{x} = \frac{\tilde{E}_y}{y} = \frac{p_1 k^3}{4\pi\varepsilon}\left[\frac{1}{\sigma}+\frac{i}{\sigma^2}+\frac{1}{\sigma^3}\right]\frac{z_0}{r^2}e^{i\omega t_R}$$

$$\tilde{E}_z = \frac{p_1 k^3}{4\pi\varepsilon}\left\{\left[\frac{2}{\sigma^3}+\frac{2i}{\sigma^2}\right]\frac{z_0^2}{r^2}-\left[\frac{1}{\sigma}-\frac{i}{\sigma^2}-\frac{1}{\sigma^3}\right]\frac{\rho^2}{r^2}\right\}e^{i\omega t_R} \tag{9.2.2}$$

$$\eta \frac{\tilde{H}_y}{x} = -\eta \frac{\tilde{H}_x}{y} = \frac{p_1 k^3}{4\pi\varepsilon}\left[\frac{1}{\sigma} - \frac{i}{\sigma^2}\right]\frac{1}{r} e^{i\omega t_R}$$

$$\rho^2 = x^2 + y^2$$

The fields of the upper dipole on the virtual plane follow from Eq. 9.2.2 by replacing z_0 by $-z_0$. The net field is the sum of the two:

$$\tilde{E}_x = \tilde{E}_y = 0$$

$$\tilde{E}_z = \frac{2p_1 k^3}{4\pi\varepsilon}\left\{\left[\frac{2}{\sigma^3} + \frac{2i}{\sigma^2}\right]\frac{z_0^2}{r^2} - \left[\frac{1}{\sigma} - \frac{i}{\sigma^2} - \frac{1}{\sigma^3}\right]\frac{\rho^2}{r^2}\right\} e^{i\omega t_R} \tag{9.2.3}$$

$$\eta \frac{\tilde{H}_y}{x} = -\eta \frac{\tilde{H}_x}{y} = \frac{p_1 k^3}{4\pi\varepsilon}\left[\frac{1}{\sigma} - \frac{i}{\sigma^2}\right]\frac{1}{r} e^{i\omega t_R}$$

The stress tensor on the virtual plane is:

$$T_{zz} = \frac{\varepsilon}{4}\left(\frac{2p_1 k^3}{4\pi\varepsilon}\right)^2 \left\{ \begin{array}{l} (kz_0)^4\left[\dfrac{4}{\sigma^8} + \dfrac{4}{\sigma^{10}}\right] + \left[\dfrac{1}{\sigma^6} - \dfrac{1}{\sigma^8} + \dfrac{1}{\sigma^{10}}\right](k\rho)^4 \\[2ex] + \dfrac{4}{\sigma^{10}}(k\rho)^2 (kz_0)^2 - \left[\dfrac{1}{\sigma^4} + \dfrac{1}{\sigma^6}\right](k\rho)^2 \end{array}\right\} \tag{9.2.4}$$

Positive and negative values indicate, respectfully, attractive and repulsive pressures. Integrating over the plane yields the total force:

$$\mathcal{F} = \frac{\pi\varepsilon}{4}\left(\frac{2p_1 k^2}{4\pi\varepsilon}\right)^2 \left\{-\frac{1}{2} + \frac{1}{2(kz_0)^2} + \frac{5}{4(kz_0)^4}\right\} \tag{9.2.5}$$

The lead term on the right is repulsive and independent of distance from the dipoles. In an infinite string of such objects all exert equal forces and with equal numbers of sources above and below, the net force is zero. The other terms are distance dependent and attractive. The result is, therefore, a net attractive force with a resultant bunching of the oscillators.

Although the particle-like fields of Eq. 9.1.1 do not produce interparticle forces, it is only the steady-state portion of the solution. Since photons are characterized by a single frequency, we are assured it is also the dominant effect. Transient fields, however, are essential for the buildup and decay of steady state fields and, for the sake of argument, we take the bunching result of Eq. 9.2.5 as a basis for locally attractive forces between neighboring transient fields.

9.3 Photon Creation and Annihilation

Let an atom possess eigenstates with inherent frequencies ω_H and ω_L where $\omega_H - \omega_L = \omega_0$. One eigenstate is occupied and state symmetries are those necessary to support an electron transition. Since eigenfunction electrons respond nonlinearly to a driving force, frequencies ω_H, ω_L, and ω_0 are all present. Let a photon with frequency ω_0 and modeled as described in Sec. 9.1 approach the atom. As the leading edge of the photonic and atomic fields overlap a standing wave pattern forms and, by the arguments of Sec. 9.2, creates an attractive force between them. Although the effect on the relatively massive atom is negligible, it is significant on the zero rest mass photon; under the influence of the attractive force it adjusts course and approaches the atom along the negative z-axis. Its field structure has the exact symmetry necessary for an active electron to create or annihilate a photon.

Photon Absorption

The approaching photon imposes its own template on the active, low-energy state electron. The template has the polarization, phase, and recursion relationship of the incoming photon and forms the exact electronic structure necessary for energy $\hbar\omega_0$, and hence momentum $\hbar\omega_0/c$, to flow from the photon to the nonlinearly reacting electron. The Manley Rowe equations show it is accompanied by energy $\hbar\omega_L$, which also flows to the high-energy state. The energy transfer process annihilates the photon and its energy together with the low-state energy; together they change the electronic configuration to that of the high-energy eigenstate.

Photon Emission

The approaching photon imposes its own template on the active, high-energy state electron. The template has the precise configuration of the incoming photon, including polarization, phase, recursion relationship, and field distribution necessary to create a daughter photon. Therefore, and even as the mother photon continues on its way, energy $\hbar\omega_0$ and momentum $\hbar\omega_0/c$ flow from the high-energy state to a daughter photon. The Manley Rowe equations show the transfer is simultaneous with energy $\hbar\omega_L$ flowing into the low-energy state. Upon completion the electron has adopted the configuration of the low-energy eigenstate and the daughter photon exits the region along the positive z-axis.

Since the onset of energy transfer does not depend upon which eigenstate is initially occupied the probabilities of induced absorption and emission are the same.

The processes are illustrated in Fig. 9.3.1.

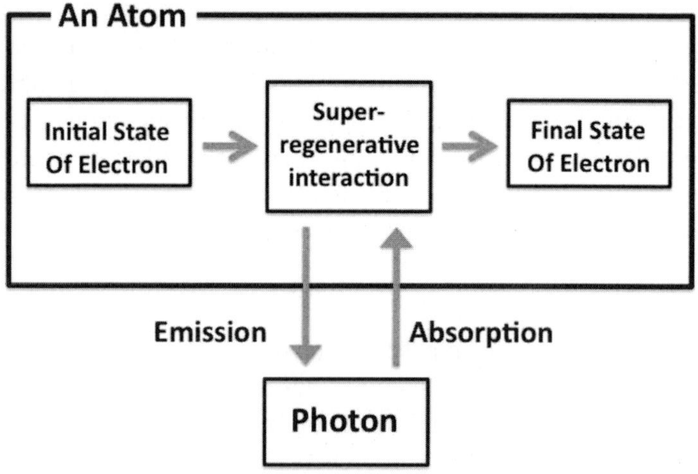

Figure 9.3.1. Diagram illustrating power flows in a superregenerative source.

Entanglement

Two entangled objects are described by a single wave function. After emission there is no direct means of breaking the entanglement and hence it remains. Mother and daughter photons have the same frequency, polarization, and phase and are parts of a single, entangled wave function.

9.4 Electrons

Electron structure has been the topic of continued discussions ever since Thompson discovered them in 1897, and a word of warning before beginning a discussion may be in order. We suggest reading Saxe's poem about elephants, see Sec. A.23. Like the wise men's observations of an elephant, our information about an electron is limited and subject to interpretation. A strange side problem is the temptation to believe smaller things are simpler things. However, we ask: is an acorn simpler than an oak tree?

We can infer much about electrons from photon characteristics. Our results are consistent with an electron as a continuously evolving configuration of charge and current densities. The evolution creates the statistical character of quantum mechanics and is part of the answer to Einstein's question of why spontaneous emission isn't either immediate or nonexistent. An essential axiom for this model is what Maxwell, in quite a different context, called 'the assumption of continuity of path': every electric and magnetic charge density distribution that satisfies the conservation laws is equally likely and over time the system traverses through all possibilities.

The proposed model responds nonlinearly to applied forces and the case for nonlinearity seems overwhelming. Two sinusoidally time varying fields generate sum and/or difference frequencies and each frequency supports equi-energy trigonometric pulses. This is predicted by the Manley Rowe equations, but not for linear systems. Within nonlinear systems oscillations exhibit frequency pulling that results in a monochromatic frequency line and, with sufficient feedback, a superregenerative response: all these characterize active electrons.

As discussed in Chap. 7 and 8, a trapped electron supports oscillations between eigenstates and the oscillations form electric and magnetic multipoles. An electric multipole of order ℓ requires a minimum of $2^{\ell-1}$ charges and hence the intra-electron charge distribution must be sufficiently fine to contain at least that many charges. For the case $L = 1000$ this implies 10^{299} iotas of charges. Yet electrons satisfy the exclusion principle and it is a derivable result of electrostatics if and only if the charge is granular. Therefore L must be large enough to provide a highly directional output, yet it cannot be limitlessly large.

We note that knowledge of the multipolar moments required for radiative energy exchange provides an opportunity to re-create the charge and current distributions and hence to detail electronic structure during energy exchanges.

Reference

J.D. Jackson, *Classical Electrodynamics*, 2nd ed., John Wiley (1975) pp. 339–342

Chapter 10

Epilogue

10.1 The Radiation Scenario

Exchanges of Radiation Energy

In his classic 1917 paper Einstein expressed regret that his formulation of quantized radiation was inconsistent with results calculated using Maxwell's continuum equations. In his 1919 Nobel address Planck inquired if particles of radiation exist. To examine if photon particles are consistent with continuum electromagnetism we begin with the fundamental theorem that each and every electromagnetic field may be analytically described by a multipolar expansion. Therefore if a continuum electromagnetism description of photons is possible it is expressible as a multipolar expansion. An electric multipole of order ℓ consists of an ℓth order discontinuity that contains, on an appropriate dimensional scale, 2^ℓ separated charges of opposite sign; magnetic multipoles are the electromagnetic dual of electric ones. Each electric multipolar moment contains ℓ incremental spacings between separated charges and, since that spacing is incrementally small, the moment of the lowest modal order dominates. Since each magnetic multipolar moment due to an electric current contains $\ell + 1$ incremental spacings between separated charges, the magnitude of each magnetic moment is equivalent to the one higher order electric moment. Therefore, electrically small antennas that respond linearly to an applied field are dominated by electric dipole radiation.

The transmitting biconical antenna analyzed by Schelkunoff had perfectly conducting antenna arms. So did the receiving biconical antenna analyzed by Grimes, and with a sink at the terminals that responded linearly to an applied field. These are examples of operational transmitting and receiving antennas each of which sits in a reactive

305

energy field of its own making. Emitted radiation separates into a part that travels forever outward, the far field, and a part that remains attached to the antenna, the near field; the near field supports oscillating reactive energy and the far field supports outwardly traveling energy. Chu showed the characteristics of an electrically small antenna is dominated by its geometry and on the surface of a circumscribing virtual sphere fit snugly about the source the ratio of oscillating to traveling energy is at least as large as $(1/ka)^3$; for an atom producing mid-optical radiation that ratio is about 10^9. Schrödinger emphasized that a radiating point charge within an atom should produce a spectrum of frequencies, not a narrow bandwidth, and that frequency should change with the source energy.

We start our examination of radiation generated by electrically small objects with the unique multipolar expansion that describes a z-directed plane wave, as expressed in spherical coordinates. For a plane wave traveling from $z = -\infty$ to $+\infty$ the expression is a sum over modal orders from one to infinity of the product of a Legendre polynomial of unit degree times a spherical Bessel function of the same order, and the product is multiplied by a unique modal recursion relationship. The result of replacing spherical Bessel functions with spherical Hankel functions produces another field that still satisfies the Maxwell equations but dramatically changes the field pattern: the new field describes a multipolar source at the origin of coordinates from which fully z-directed radiation emanates and travels to $+\infty$ in the upper half plane; it vanishes in the lower half. We find that each differential length of that radiation supports the kinematic ratios Einstein showed is necessary for a photon. By the electromagnetic uniqueness theorem this field expression is the unique continuum electromagnetic field that describes photons; there can be no other. There is, however, an immediate difficulty: the mathematics of the event requires output power to increase with increasing modal number and the physics of the event requires output power to decrease with increasing modal number.

The melding of the mathematical and physical requirements begins by forming the radial component of the Poynting vector produced by the fields of Eq. 7.4.1. It is an orthogonal sum over all modal products and consists of vector products between angular electric and magnetic field components, $E_\theta H_\phi$ and $E_\phi H_\theta$. We show in Chap. 7 that on the positive z-

axis that the Poynting vector supports no reactive power. This is a dramatic difference from antennas. In the absence of reactive power there is no reactance and the multimodal radiation field *en toto*, not just the far field, moves away from the source at speed c. Source requirements for such emission is that each multipolar order support equal phases and magnitudes of TE and TM fields and that modal magnitudes increase slowly with number in accordance with the recursion relationships of Eq. 7.4.1. Radiation with these properties is exchanged between the field and a source of arbitrary radius with no emission barrier. Since this conclusion (1) is based upon the continuum Maxwell equations and since (2) the calculated particle-like properties of the radiation match observed properties it fulfills Einstein's objective of bringing together quantum and continuum radiation.

10.2 Nonlinear Media

Simply cataloging properties of the equation set as discussed in Sec. 10.1 leaves open the question of how to create the required modal coefficients. There is no way to construct the necessary modal coefficients within a linear medium. Neither are there ways within a linear medium to create either the power-frequency proportionality of quantized radiation or the narrow bandwidth of atomic absorption-emission lines. The only possible explanation that is based upon classical physics requires an eigenstate electron to act nonlinearly: when not transitioning it responds regeneratively to internal oscillations and when transitioning it responds superregeneratively. We note there is no *a priori* reason to expect an electron to be a simple object and hence to have only linear internal responses; complex natural phenomena typically respond linearly to driving forces over only restricted ranges. We are driven to regard an electron as a statistical ensemble of charge and current distributions that are both situation- and time-dependent while operating within the constraints of conservation laws. Maxwell's 'assumption of continuity of path' is an essential postulate: namely, every non-radiative distribution of electric and magnetic charge densities that satisfies the conservation laws is equally likely and furthermore, over time, the system traverses through them all.

The case for an eigenstate electron to respond nonlinearly to applied forces seems undeniable. The presence of additive frequencies, those Dirac called '… an unexpected connexion …', is predicted in nonlinear systems by the Manley Rowe relationship, since two independent frequencies create both their sum and difference frequencies. Furthermore, there is a long history of such frequencies within the nonlinear element of superheterodyne radio receivers. The relationship also predicts equal energies in each trigonometric energy pulse; hence output power is proportional to the frequency at which it occurs: the mere fact of nonlinearity is enough to predict a constant frequency-to-power ratio, although it is silent about what the ratio is. Nonlinearity also predicts that an initial spread of frequencies will undergo frequency pulling to a narrow bandwidth, as was the case for Huygens's mechanical clocks.

To understand spontaneous emission we begin with a host atom with eigenstates of energies W_H and W_L, $W_H > W_L$, and of course inherent frequencies ω_H and ω_L. The state symmetries satisfy the radiation selection rules of Chap. 6 and $W_H - W_L = \hbar\omega_0$. Originally the upper state is occupied and the lower one is empty. If the system responds nonlinearly to a disturbance an anticipated response is electric dipoles that oscillate at frequency ω_0. The creation of an oscillating electric dipole begins a chain of events that drives all other electric and magnetic multipoles of the same parity. In addition, if the original dipole is large enough it crosses parity lines and drives a magnetic dipole. The magnetic dipoles, in turn, drive a dual chain of multipolar moments with that parity. The buildup becomes ordered and hence superregenerative if and only if the multimodal coefficients of Eq. 7.4.1 are created. This process satisfies the physics of the multimodal buildup. When the upper modal limit reaches a critical value the radiation field moves *en toto* away from the source at speed c. The superregenerative system response acts as a narrow band filter for exchanged radiation: only a monochromatic frequency survives the superregenerative buildup.

The newly created photon is centered on the z-axis and, using cylindrical coordinates, both electric and magnetic fields are radially directed outward from the axis. Reasoning by analogy with pair production, we postulate the intense local fields on the z-axis rend a filamentary region of vacuum states into limitlessly large effective values of the constitutive relations. This simple structure is sufficient to permanently bind the field into a single entity; it also supports radial fields that extend outward, decreasing in magnitude as the inverse square of the radius.

Photon absorption begins with the upper and lower atomic states respectively empty and occupied. If a photon's radially extended field intercepts the local field of an oscillating atom the superimposed fields form standing waves that, in turn, create an attractive force between the two objects. It forces the zero mass photon towards the atom where the photon's electromagnetic structure is the precise template needed for energy transfer with no reactive energy penalty. Since the system is nonlinear, energy flows at frequency ω_L, the low energy state, and frequency ω_0, the photon, into the upper state at frequency ω_H. In the process the photon is annihilated and the electron transitions to the upper level.

Stimulated photon emission begins with the upper atomic state occupied and the lower one empty. A photon passes near enough to the atom so the radially extended field intercepts the local field of the atom, and the superimposed fields form a standing wave pattern that, in turn, creates an attractive force between the two objects. That force brings the zero rest mass photon to the atom where the photon's electromagnetic structure is the precise template needed for energy exchange without a reactive energy penalty. Since the system is nonlinear, energy flows from the upper state into both the lower state and a new photon. In the process a new photon is created and the electron transitions to the lower level. The mother photon continues on its way even as the daughter is freed from its source; since a single Hamiltonian and wave function describe both during the birth process they remain entangled after emission.

10.3 Selected Historical Developments

Quantum Theory

Particles have played a critical role in the analysis of physical events at least since Newton examined them in the latter part of the seventeenth century. For several centuries after Newton, studies involving combinations of elastic spheres dominated physics. Perhaps as a result, electrical problems were first interpreted by analyzing the behavior of charged particles; Maxwell used the force between "two very small bodies" to discuss implications of Coulomb's law. Although Maxwell's equations showed that fields are essential to explain occurrences that particles alone cannot, still the lore of particles permeated physics at the end of the nineteenth century. Therefore after Thompson discovered and measured particle-like free electrons the idea of electrons as particles was widely and quickly accepted, along with the Lorentz particle-electron and, briefly later, the Bohr atomic model. The finely honed and widespread skills of classical mechanics were carried over to quantum effects; even the name "quantum mechanics" is indicative of such an origin.

Building upon his earlier work with particles, Newton compared the propagation of light with that of projectiles. According to him, luminous bodies eject projectiles of light that continue in flight until acted upon by other objects. Quite differently, Huygens compared the propagation of light with the propagation of sound through air and waves on water. According to him, light was not a thing but a disturbance that propagated through space. Light is emitted over a spread of angles by a luminous object. It bounces off objects and the total of all bounces off all objects that it intercepts form an observer's field of view. After Huygens more than a century passed before Young in England and, in an arguably unparalleled technical outpouring, Fresnel in France confirmed that light is propagated as a wave with transverse vibrations and, therefore, two polarizations.

The modern theoretical basis for the wave theory of light is based upon Maxwell's continuum electromagnetic field equations. Hertz discovered the photoelectric effect by showing that ultraviolet light increases current emission from a cathode; he also constructed a transmitting-receiving pair of electric dipoles and confirmed that electromagnetic waves transport energy through space. His dipole radiation has rotational symmetry about its axis, a symmetry that Einstein later referred to as "spherical" symmetry.

In 1900 Planck showed that quantizing electromagnetic energy in units of $W = \hbar\omega$ accounts for otherwise significant discrepancies between observations and calculated laws that describe equilibrium conditions between radiation and matter. He wrote that a most suitable body for energy exchange seemed to be Hertz's dipole with its "spherical" waves. A few years later, 1916, Planck showed that his 1900 expression for radiation density was consistent with thermodynamic reciprocity only if he added a zero-point energy term. That term is temperature independent and exists throughout all space.

In 1905 Einstein used field energy quantization to explain the photoelectric effect. In 1917 he extended and simplified Planck's derivation of the radiation law; he also showed that the laws of statistical mechanics require quantized radiant energy exchanges to be accompanied by quantized momentum exchanges of equal value upon emission or absorption. This, in turn, is satisfied if all of each unit of radiated energy travels in the same direction. Einstein wrote: "… (Atomic) emission in spherical waves does not occur; the molecule suffers recoil of magnitude $\hbar\omega/c$. This seems to make a quantum theory of radiation almost unavoidable."

These events showed quite conclusively that electromagnetic energy is exchanged between atoms and radiation fields in quantized units, and this result, in turn, led to a fundamental difficulty. Einstein wrote that quite differently from results of the Maxwell wave equation "monochromatic radiation … behaves in thermodynamic theoretical relationships as though it consists of distinct independent energy quanta of magnitude $W = \hbar\omega$."

Planck wrote, "There is one particular question the answer to which will, in my opinion, lead to an extensive elucidation of the entire problem. What happens to the energy of a light-quantum after its emission? Does it pass outwards in all directions, according to Huygens' wave theory, continually increasing in volume and tending towards infinite dilution? Alternatively, does it, as in Newton's emanation theory fly like a projectile in one direction only? In the former case the quantum would never again be in a position to concentrate its energy at a spot strongly enough to detach an electron from its atom." These are the primary reasons it is commonly accepted that light propagates as if it consists of waves and exchanges energy as if it consists of particles. In 1923 Compton reported the electron scattering effect known by his name, together with a supporting argument based upon light as a particle and in 1926 Goudsmit and Uhlenbeck reported an electron's 'quantized momentum'.

A fundamental change occurred in the research methods of modern physics during the 1920s. Instead of continuing to build upon the base of classical physics, in the manner of Planck and Einstein, efforts became directed towards pure mathematics. Arguably the first major step in this direction was publication by Schrödinger in 1926 of the equation that bears his name. It was followed a few years later by Dirac's publication of his equations. Both the Schrödinger and the Dirac equations correctly describe the behavior of electrons in equilibrium but the Schrödinger equation is restricted to non-relativistic electron speeds and the Dirac equations is valid at all speeds; electron spin must be added in an *ad hoc* way to the Schrödinger equation but is inherent to the Dirac equations. Both equations yield the probability that an electron will enter a transition and, when it does, the correct input and output energy-frequency relationships are predicted. Both equations treat electrons as waves and yet an electron also exhibits particle properties.

Schrödinger developed his equation by using the de Broglie wavelength and an analogy with a known relationship between classical mechanics and geometric optics. According to Mehra, in the search for his equations Dirac: "started playing with equations rather than trying to introduce the right physical idea. A great deal of (the) work is just playing with the equations and seeing what they give." The negative corollary of this method is the model used to derive an equation determines its interpretation and neither Schrödinger's nor Dirac's equation came with a model; ergo neither provides an obvious physical interpretation of its results. Therefore, although both equations give statistically correct time-average values of measurable quantities the question of how to interpret them remains.

In an attempt to combine quantum and classical theories, an immediate difficulty is that accelerating charges radiate energy. Stability requires closed current loops and/or a spherically symmetric region of charge that pulsates radially, or combinations thereof. Yet the Bohr orbit is some 20,000 larger than Lorentz's estimated electron size, point charges must accelerate to maintain an orbital position, and atoms are stable. Neither Schrödinger nor Dirac successfully addressed this issue.

Schrödinger had severe reservations about the interpretation accorded his equation; he preferred to explain electron stability by ascribing physical significance to electron waves. For this reason he rejected the idea that wave functions represent a probability of occupation by a point electron and preferred a mass density created by standing electron waves. He was deeply interested in how electrons transition between eigenstates. The electromagnetic equations require a transient solution, transient solutions support a continuous spectrum of emitted radiation, and yet a continuous spectrum is not observed. His linear equations do not describe energy jumps, and yet such jumps occur. Schrödinger commented to Bohr "If we have to put up with these damned jumps I'm sorry I got involved." Bohr answered that although Schrödinger's arguments were correct, since quantum jumps occur it must be that the pictorial concepts used to describe the classical physics of such events are insufficient. Schrödinger's reaction is evidenced by later comments about the "quaint basic assumption" of a discontinuity between states.

Both Dirac and Schrödinger were concerned about the frequencies of eigenstate radiation. Dirac wrote, "One would expect to be able to include the various frequencies in a scheme comprising certain fundamental frequencies and their harmonics. This is not observed to be the case. Instead, there is observed a new and unexpected connexion between the frequencies." He went on to say that this result is "quite unintelligible from the classical standpoint."

In stark contrast with quantum theory and as shown in Chap. 1, electromagnetic field theory rests on only a few, very general axioms. Quantum theory requires that classical electromagnetic laws apply partially, but not fully, within atoms. To some, it seems incongruous that nature should require such disparate and seemingly conflicting bases for such strongly overlapping sciences. To this end, Einstein wrote that: "I am, in fact, firmly convinced that the essentially statistical character of contemporary quantum theory is solely to be ascribed to the fact that this theory operates with an incomplete description of physical systems." He also said he had devoted more time to thinking about this subject than any other. Although he believed that the mathematics of quantum theory was uniquely correct, he was bothered by the statistical nature of radiation onset from an atom that is initially in a high-energy state. He argued that either an atom is stable or it is unstable. If it is stable it will not spontaneously decay, and if it is unstable it will begin the decay process without a time delay. Yet an atom is stable until it spontaneously undergoes a discontinuous energy drop and emits a pulse of radiation. He concluded that the wave function description of this event is incomplete. Einstein wrote: "Assuming the success of efforts to accomplish a complete physics description, the statistical quantum theory would, within the frame-work of future physics, take an approximately analogous position to statistical mechanics within the framework of classical mechanics. I am rather firmly convinced that the development of theoretical physics will be of this type; but the path will be lengthy and difficult."

Classical Theory

While attention was focused on mathematical developments in quantum theory, developments continued in electrical circuitry and electromagnetic radiation. Centuries earlier Huygens described the synchronization of two mechanical clocks by small but periodic nonlinear interactions. In 1901, the year after Planck's first paper requiring quantized radiation, Fessenden made the first heterodyne radio receiver and in 1907 Mie combined the classical wave theory of light with spherical functions to analyze scattering of light by electrically small metallic particles. In 1918, the year after Einstein's quantum radiation paper, Armstrong, then a Major in the U.S. Army in France, developed a superheterodyne radio receiver. His receiver required two nonlinear mixers, one to establish an intermediate frequency and one to extract an audio output from it. Inputs to the first mixer were the incoming radio signal and a local oscillator; these were passed through a nonlinear element from which the intermediate frequency was extracted. In the early 1920s he led a continuing discussion of nonlinearities, regeneration, and superregeneration in electronic circuits. The superheterodyne frequency mix is the same as Dirac's atomic frequency mix was, twelve years later, that he described as an 'unexpected connexion between frequencies' that is 'quite unintelligible from a classical standpoint'. Major technical advances in both nonlinear circuitry and radiation analysis occurred during World War II. Shortly afterward works of Bode, in 1945, and Adler, in 1946, utilized nonlinear responses of electronic circuits to driving forces, including frequency pulling and both regenerative and superregenerative feedback.

Chu, in 1947, published a quantitative analysis of radiation Q as a function of the antenna size-to-wavelength ratio. About the same time Schelkunoff, 1948, made the first multimodal solution of a full-sized transmitting antenna that included all fields near and far. Manley and Rowe, in 1956, published the power-frequency relationships applicable to nonlinear systems. In 1960 Harrington showed that the maximum possible gain of a single radiated mode is $(2\ell+1)$, where ℓ is the modal number. In 1973 Beers, ignoring the pattern difference between dipoles and photons, showed that after scaling a dipole to an electron's size-to-wavelength ratio an equivalent power exchange would require reactive power equivalent to one-tenth the solar power received by the entire earth! Yet it happens. Grimes, in 1982, extended Shelkunoff's solution for a transmitting antenna to a full solution of a receiving antenna, including an expression for the received momentum.

Combining Quantum and Classical

Although Einstein, Schrödinger, and many others continued to press for a more physical explanation of quantum phenomena, there was no obvious and concerted attempt to incorporate lessons learned from electronics or the additional knowledge and techniques developed with electromagnetic fields. The radiation frequencies, the line widths, and the energy-frequency relationships are directly predictable from the Manley Rowe equations applied to a nonlinear system. Both the potential for nonlinearities and Einstein's points about the time and direction of spontaneous emission follow from our electron model. Also, using it, fully directed radiation pulses result from the continuum theory when combined with the processes discussed in the three preceding chapters.

Our derivation of the Schrödinger equation in Chap. 4 conserves an eigenstate electron's energy, charge, and angular momentum. With the axiom that an electron, *per se*, is a statistical ensemble of electric and magnetic charge densities only the probabilities of measureable results are predictable because the state of the electronic ensemble at any instant is unknown and, presumably, unknowable. The model provides a physical explanation of photon creation and annihilation and it satisfies Einstein's requirement that quantum mechanics be in an 'approximately analogous position to statistical mechanics within the framework of classical mechanics'.

References

R. Adler, "A study of locking phenomena in oscillators," Proc. IRE, vol. 34, 351–357 (1946)

E.H. Armstrong, "Some Recent Developments of Regenerative Circuits," Proc. IRE, vol. 10, 244 (1922)

Y. Beers, 'The supergain atom', *Am. J. Phys.* vol. 41, pp. 275–278 (1973)

M. Bennett, M.F. Shatz, H. Rockwood, K. Wiesenfeld, "Huygens's Clocks," Proc. Royal Soc. London A, vol. 458, pp. 563–579 (2002)

H.W. Bode, *Network Analysis and Feedback Amplifier Design*, Van Nostrand, 31–35 (1945)

N. Bohr, "On the Constitution of Atoms and Molecules," *Phil. Mag.* vol. 26, 1–19 (1913), Also in *The World of the Atom*, H. A. Boorse and L. Motz, eds., Basic Books (1966) pp. 751–765

M. Born, "Max Karl Ludwig Planck" *Obituary Notice of Fellows of the Royal Society*, vol. 6, pp. 161–180 (1948), also in *The World of the Atom*, H. A. Boorse and L. Motz, eds., Basic Books (1966) p. 475

M. Born, E. Wolf, *Principles of Optics Electromagnetic Theory of Propagation, Interference and Diffraction of Light*, 3rd ed. Pergamon Press (1965)

L.J. Chu, "Physical Limitations of Omni-Directional Antennas," *J. Appl. Phys.*, vol. 19, pp. 1163–1175 (1948)

A.H. Compton, "The Spectrum of Scattered X-Rays," *Phys. Rev.* vol. 21, pp. 409–413 (1923); also in in *The World of the Atom*, H. A. Boorse and L. Motz, eds., Basic Books (1966) p. 925–929

A.H. Compton, "A Quantum theory of the Scattering of X-Rays by Light Elements," *Phys. Rev.* vol. 21, pp. 483–522 (1923); also in in *The World of the Atom*, H. A. Boorse and L. Motz, eds., Basic Books (1966) p. 911–924

A.M. Dirac, *The Principles of Quantum Mechanics*, 4th ed., Oxford Press (1958) pp. 1–2

A. Einstein, "Concerning a Heuristic Point of View About the Creation and Transformation of Light," *Ann. Phys.* vol. 17, 132–148 (1905); also in *The*

World of the Atom, H. A. Boorse and L. Motz, eds., Basic Books (1966) p. 544–557

A. Einstein, "The Quantum Theory of Radiation," *Phys. Z.*, vol. 18, 121–128 (1917); also in *The World of the Atom*, H. A. Boorse and L. Motz, eds., Basic Books (1966) p. 888–901

A. Einstein, "Reply to Criticisms" in *Albert Einstein, Philosopher and Scientist* (P.A. Schilpp, editor) Harper Torchbooks Science Library (1959) pp. 665–688

R.A. Fessenden, *U. S. Patent* No. 706,740 (1902).

S.A. Goudsmit, G.E. Uhlenbeck "Spinning Electrons and the Structure of Spectra," *Nature* vol. 117, p. 264 (1926); also in in *The World of the Atom*, H. A. Boorse and L. Motz, eds., Basic Books (1966) p. 947–951

R.F. Harrington, "Effect of Antenna Size on Gain, Bandwidth, and Efficiency," *J. Res. National Bureau of Standards* vol. 64D, 1–12 (1960)

H. Hertz *Electric Waves: Researches on the Propagation of Electric Action with Finite Velocity Through Space* (1893) Translated by D. E. Jones, Dover Publications (1962)

C. Huygens, "Treatise on Light," (1678) trans. by S.P. Thompson, Dover Publication (1912); also *The World of the Atom*, H. A. Boorse and L. Motz, eds., Basic Books (1966) pp. 69–85

H.A. Lorentz, *The Theory of Electrons*, (1909) also Dover Publications 2nd ed., (1952); also Pais, A., "The Early History of the theory of the Electron: 1897-1947," in *Aspects of Quantum Theory*, A. Salam, P. Wigner, eds., Cambridge Press (1972) pp. 79–93

J.M. Manley, H.E. Rowe, "Some general properties of nonlinear elements, Part I: General energy relations," *Proc. IRE*, vol. 44, pp. 904–913 (1956), and "General energy relations," *Proc. IRE* vol. 47, pp. 2115–2116 (1959)

J.C. Maxwell, *A Treatise on Electricity and Magnetism*, 3rd ed., (1891) reprint Dover Publications (1954)

J. Mehra, "'The Golden Age of Theoretical Physics': P. A. M. Dirac's Scientific Work from 1924-1933," in *Aspects of Quantum Theory*, Cambridge Press (1972) p. 45

G. Mie, "A Contribution to Optical Extinction by Metallic Colloidal Suspensions," *Ann. Physik.* Vol. 25, p. 377 (1908)

W.J. Moore, *Schrödinger, Life and Thought*, Cambridge Press, 1989, pp. 227–228

I. Newton, *Mathematical Principles of Natural Philosophy*, (1687) Translated by I.B. Cohen, A. Whitman, Univ. of California Press (1999)

I. Newton, *Optiks*, (1704); Dover Publications (1952)

R. Penrose, *The Emperor's New Mind Concerning Computers, Minds, and the Laws of Physics*, Penguin Books (1991)

M. Planck, "The Origin and Development of the Quantum Theory," Nobel Prize in Physics Address, 1919, in *The World of the Atom*, H. A. Boorse and L. Motz, eds., Basic Books (1966) p. 491–501

G.A. Schott, *Electromagnetic Radiation and the Mechnical Reactions Arising From It*, Cambridge Universtiy Press (1912)

E. Schrödinger, *Ann. Phys.* Vol. 79, p. 489 (1926); also "Wave Mechanics," in *The World of the Atom*, H. A. Boorse and L. Motz, eds., Basic Books (1966) pp. 1060–1076

Appendices

A.1 Introduction to Tensors

The application of field concepts to classical physics is made easier by the use of tensors. Tensor notation simplifies what would otherwise be tedious notational bookkeeping. The simplest and lowest rank tensor is a scalar, the next higher ranking tensor is a vector, and higher order tensors are referred to simply as tensors:

Table A.1.1. Properties of tensors.

Rank	
$r = 0$	Scalar
$r = 1$	Vector
$r = 2$	Tensor

The number of numbers that it takes to construct a tensor, N_0, depends upon the rank of the tensor and the number of dimensions. If N and r are, respectively, the number of dimensions and the rank, the number of numbers is:

$$N_0 = N^r \tag{A.1.1}$$

Independently of the number of dimensions, a scalar is fully described by a single number. Examples are the speed of light, c, and electron charge, q. Scalars have the same value in all inertial frames.

It takes as many numbers as there are dimensions to describe a vector. Example vectors are electric field intensity, velocity, and position. Let vector A be known in three dimensions. The three numbers represent components along each of the three orthogonal coordinate axes, (x_1, x_2, x_3). If the same vector is determined using a set of axes rotated to new coordinate positions (x'_1, x'_2, x'_3) the result is the new vector components:

$$A'_1 = A_1 \cos(x'_1, x_1) + A_2 \cos(x'_2, x_1) + A_3 \cos(x'_3, x_1)$$

$$A'_2 = A_1 \cos(x'_1, x_2) + A_2 \cos(x'_2, x_2) + A_3 \cos(x'_3, x_2) \qquad \text{(A.1.2)}$$

$$A'_3 = A_1 \cos(x'_1, x_3) + A_2 \cos(x'_2, x_3) + A_3 \cos(x'_3, x_3)$$

The directional cosine of the angle between axes 'i' in the prime coordinates and axis 'j' in the unprimed coordinates is signified by $\cos(x'_i, x_j)$. Making the definition that the direction cosine $c_{jk} = \cos(x'_j, x_k)$ Eq. A.1.2 takes the more compact form:

$$A'_j = \sum_{k=1}^{3} c_{ik} A_k \qquad \text{(A.1.3)}$$

It follows that:

$$A_r = \sum_{k=1}^{3} c_{kr} A'_k \qquad \text{(A.1.4)}$$

An example of a second rank tensor is the stress tensor in crystals. Such a tensor transforms between coordinate systems as:

$$T'_{rs} = \sum_{i=1}^{3} \sum_{j=1}^{3} c_{ri} c_{sj} T_{ij} \qquad \text{(A.1.5)}$$

The number of direction cosines for a transformation between coordinates systems is the same as the rank of the tensor.

Like all other vectors, a position vector transforms between coordinate frames as:

$$x'_1 = c_{11}x_1 + c_{12}x_2 + c_{13}x_3$$
$$x'_2 = c_{21}x_1 + c_{22}x_2 + c_{23}x_3 \qquad (A.1.6)$$
$$x'_3 = c_{31}x_1 + c_{32}x_2 + c_{33}x_3$$

By definition the rotation matrix is:

$$\left(c_{ij} \right) = \begin{pmatrix} c_{11} & c_{12} & c_{13} \\ c_{21} & c_{22} & c_{23} \\ c_{31} & c_{32} & c_{33} \end{pmatrix} \qquad (A.1.7)$$

The length of a differential vector in three dimensions is:

$$(\Delta r)^2 \equiv (\Delta x'_1)^2 + (\Delta x'_2)^2 + (\Delta x'_3)^2 \equiv (\Delta x_1)^2 + (\Delta x_2)^2 + (\Delta x_3)^2 \qquad (A.1.8)$$

The transformation equalities derived from Eq. A.1.8 are:

$$c_{11}^2 + c_{21}^2 + c_{31}^2 = 1 = c_{12}^2 + c_{22}^2 + c_{32}^2 = c_{13}^2 + c_{23}^2 + c_{33}^2 \qquad (A.1.9)$$

This may be written as

$$\sum_{i=1}^{3} c_{ij}c_{ij} = 1 \ \text{ and } \ \sum_{i=1}^{3} c_{ij}c_{ik} = 0; \ j \neq k$$

The Kronecker delta function is defined by the relationship:

$$\delta_{jk} = 1 \ \text{ if } j = k; \quad \delta_{jk} = 0 \ \text{ if } j \neq k \qquad (A.1.10)$$

Using this definition, the condition on directional cosines may be written more compactly as:

$$\sum_{i=1}^{3} c_{ij}c_{ik} = \delta_{jk} \qquad (A.1.11)$$

A useful exercise is to show the determinant is normalized:

$$\det \left| c_{ij} \right| = 1 \qquad (A.1.12)$$

Solution: Let the volume of cube $x_1 x_2 x_3$ equal one. The volume in the transformed coordinates is unchanged by describing it in another frame, so it too is equal to one. The volume is given by:

$$V = x'_1 \cdot \left(x'_2 \times x'_3 \right) = 1$$

Writing out cross products in terms of directional cosines gives:

$$x'_2 \times x'_3 = x_1 \left(c_{22}c_{33} - c_{23}c_{32} \right) + x_2 \left(c_{33}c_{11} - c_{31}c_{13} \right) + x_3 \left(c_{11}c_{22} - c_{12}c_{21} \right)$$

From which it follows that:

$$x'_1 \cdot \left(x'_2 \times x'_3 \right) = \left[\begin{array}{l} c_{11} \left(c_{22}c_{33} - c_{23}c_{32} \right) + c_{12} \left(c_{23}c_{31} - c_{21}c_{33} \right) \\ + c_{13} \left(c_{21}c_{32} - c_{22}c_{31} \right) \end{array} \right]$$

The determinant of c_{ij} is:

$$\left| c_{ij} \right| = c_{11} \left(c_{22}c_{33} - c_{23}c_{32} \right) + c_{12} \left(c_{23}c_{31} - c_{21}c_{33} \right) + c_{13} \left(c_{21}c_{32} - c_{22}c_{31} \right)$$

Comparing the above equations gives:

$$\left| c_{ij} \right| = 1$$

A.2 Tensor Operations

A common summation convention that reduces the number of symbols otherwise be required is that if an index occurs twice is summed over all possible values. That is:

$$c_{ij} c_{ik} = \delta_{jk} \tag{A.2.1}$$

Consider some arithmetic operations on tensor fields. Tensor addition is defined only for tensors of equal rank; for example, addition of a scalar and a vector is not defined. Addition of tensors of equal rank is by:

$$C_{ij} = A_{ij} + B_{ij} \qquad \text{(A.2.2)}$$

Proof consists of showing that the sum obeys the coordinate rotation properties of a second rank tensor. In Eq. A.2.2 indices 'i' and 'j' appear only once in each term and, therefore, are running indices. Equation A.2.2 consists of nine separate summations.

Subtraction is accomplished by multiplying B_{ij} by minus one and adding; multiplication is defined between tensors of arbitrary rank. By definition,

$$C_{i..jr..s} = A_{i..j} B_{r..s} \qquad \text{(A.2.3)}$$

The rank of C is the sum of the ranks of A and B. For example the product of vector A_i and scalar a is aA_i, another vector. The product between two vectors is a second rank tensor, for example the product of A_i and B_j is $C_{ij} = A_i B_j$, where indices 'i' and 'j' are both running indices; C_{ij} represents nine numbers.

Division by tensors other than rank zero is not defined.

In addition to these scalar-like arithmetic operations there are operations confined to tensors. Tensor contraction is accomplished by equating two indices. Equal indices signify a summation and summation results in a tensor reduced in rank by two from the initial one. The process is, therefore, restricted to tensors of rank $r \geq 2$. As an example:

$$A'_{rstu} = c_{ri} c_{sj} c_{tk} c_{u\ell} A_{ijk\ell} \qquad \text{(A.2.4)}$$

After equating 's' and 't' and summing:

$$A'_{rssu} = c_{ri} c_{sj} c_{sk} c_{u\ell} A_{ijkl} \qquad \text{(A.2.5)}$$

Since:

$$c_{sj} c_{sk} = \delta_{jk} \quad \text{and} \quad A'_{rssu} = c_{ri} c_{u\ell} A_{ijj\ell} \qquad \text{(A.2.6)}$$

This is a tensor of rank two less than the starting one.

A common example is the scalar product between two vectors, A_i and B_i. To evaluate, begin with the product:

$$C_{ij} = A_i B_j \tag{A.2.7}$$

Equating indices 'i' and 'j' and summing over the indices gives:

$$C_{ii} = A_i B_i = D \tag{A.2.8}$$

The product forms scalar D.

A.3 Tensor Symmetry

Physically real tensors are either symmetric or antisymmetric. Terms of different symmetry are defined as:

Symmetric tensor $\qquad\qquad A_{rstu} = A_{rtsu}$

Antisymmetric tensor $\qquad\quad A_{rstu} = -A_{rtsu}$
$$\tag{A.3.1}$$

Symmetric and antisymmetric tensors have, respectively, $N(N+1)/2$ and $N(N-1)/2$ terms.

An important special case is a three-dimensional tensor of rank two, say T_{ij}. Such tensors transform as:

$$T'_{ij} = c_{ir}c_{js}T_{rs} \tag{A.3.2}$$

Equation A.3.2 is short hand notation for the nine terms of T'_{ij}, each of which contains nine separate numbers. For example:

$$T'_{12} = c_{1r}c_{2s}T_{rs} = \begin{bmatrix} c_{11}c_{21}T_{11} + c_{11}c_{22}T_{12} + c_{11}c_{23}T_{13} \\ +c_{12}c_{21}T_{21} + c_{12}c_{22}T_{22} + c_{12}c_{23}T_{23} \\ +c_{13}c_{21}T_{31} + c_{13}c_{22}T_{32} + c_{13}c_{23}T_{33} \end{bmatrix} \tag{A.3.3}$$

If T_{ij} is antisymmetric, $T_{ij} = -T_{ji}$ and the transformation simplifies to:

$$T'_{23} = c_{11}T_{23} + c_{12}T_{31} + c_{13}T_{12}$$

$$T'_{31} = c_{21}T_{23} + c_{22}T_{31} + c_{23}T_{12} \tag{A.3.4}$$

$$T'_{12} = c_{31}T_{23} + c_{32}T_{31} + c_{33}T_{12}$$

The proof of Eq. A.3.4 follows by writing out the terms in the form:

$$T'_{12} = \begin{bmatrix} 0 + c_{11}c_{22}T_{12} - c_{11}c_{23}T_{31} \\ -c_{12}c_{21}T_{12} + 0 + c_{12}c_{23}T_{23} \\ +c_{13}c_{21}T_{31} - c_{13}c_{22}T_{23} + 0 \end{bmatrix}$$

$$= \left(c_{11}c_{22} - c_{12}c_{21}\right)T_{12} + \left(c_{12}c_{23} - c_{13}c_{22}\right)T_{23} + \left(c_{13}c_{21} - c_{11}c_{23}\right)T_{31}$$

Similarly:

$$T'_{23} = \left(c_{21}c_{32} - c_{22}c_{31}\right)T_{12} + \left(c_{22}c_{33} - c_{23}c_{32}\right)T_{23} + \left(c_{23}c_{31} - c_{21}c_{33}\right)T_{31}$$

From the determinant:

$$c_{11}\left(c_{22}c_{33} - c_{23}c_{32}\right) + c_{12}\left(c_{23}c_{31} - c_{21}c_{33}\right) + c_{13}\left(c_{21}c_{32} - c_{22}c_{31}\right) = 1$$

Combining this result with $c_{ij}c_{ik} = \delta_{jk}$ results in:

$$c_{11}c_{11} + c_{12}c_{12} + c_{13}c_{13} = 1$$

The latter two equations combine to show that:

$$c_{11} = \left(c_{22}c_{33} - c_{23}c_{32}\right); \quad c_{12} = \left(c_{23}c_{31} - c_{21}c_{33}\right); \quad c_{13} = \left(c_{21}c_{32} - c_{22}c_{31}\right)$$

Substitution of this result back into the expansion results in Eq. A.3.4.

This result shows that an antisymmetric second rank tensor, T_{ij}, transforms like a vector. It is tempting to call it a vector, but if the coordinate system is switched from a right hand system to a left hand system the components change sign. It is therefore a pseudovector.

A.4 Differential Operations on Tensor Fields

The gradient operation increases the rank of a tensor by one. As an example, let $\sigma(r)$ represent a scalar field. Taking the partial differential:

$$\frac{\partial \sigma(r)}{\partial x_i} = V_i \tag{A.4.1}$$

Make the equality:

$$\frac{\partial \sigma(r)}{\partial x_i} = \frac{\partial x'_j}{\partial x_i}\frac{\partial \sigma(r)}{\partial x'_j} = c'_{ij} V_i$$

Combining gives:

$$V_i = c'_{ij} V'_j \tag{A.4.2}$$

Since V_i transforms as a vector, it is a vector, and the divergence operation decreases the rank of a tensor by one. The gradient and divergence operations may be conducted on tensors of any rank. For example:

$$T_{ij..kl} = \frac{\partial R_{ij..k}}{\partial x_\ell} \tag{A.4.3}$$

$$\sigma(r) = \frac{\partial V_i}{\partial x_\ell} \tag{A.4.4}$$

To show that $\sigma(r)$ is a scalar, write it as:

$$\sigma(r) = \frac{\partial x'_k}{\partial x_i}\frac{\partial \left(c'_{ij} V'_j \right)}{\partial x'_k} = c'_{ij} c'_{ik} \frac{\partial V'_j}{\partial x'_k} = \frac{\partial V'_j}{\partial x'_j} = \sigma'(r) \tag{A.4.5}$$

The divergence operation may be conducted on tensors of any rank:

$$T_{i..j} = \frac{\partial R_{i..jk}}{\partial x_k} \tag{A.4.6}$$

Proof follows in the same way as for Eq. A.4.4.

The curl operation begins with the vector differential operation:

Table A.4.1. Table of vector properties.

1.	$A \times (B \times C) = B(A \cdot C) - C(A \cdot B)$
2.	$grad(\phi\psi) = \nabla(\phi\psi) = \phi\nabla\psi + \psi\nabla\phi$
3.	$div(\phi A) = \nabla \cdot (\phi A) = \phi\nabla \cdot A + \nabla\phi \cdot A$
4.	$curl(\phi A) = \nabla \times (\phi A) = \phi\nabla \times A + \nabla\phi \times A$
5.	$\nabla \cdot (A \times B) = B \cdot (\nabla \times A) - A \cdot (\nabla \times B)$
6.	$\nabla \times (A \times B) = A(\nabla \cdot B) - B(\nabla \cdot A) + (B \cdot \nabla)A - (A \cdot \nabla)B$
7.	$\nabla(A \cdot B) = A \times (\nabla \times B) + B \times (\nabla \times A) + (B \cdot \nabla)A + (A \cdot \nabla)B$
8.	$\nabla^2(1/r) = 0,\ if\ r > 0$
9.	$\nabla^2 A = \nabla(\nabla \cdot A) - \nabla \times (\nabla \times A)$

Table A.4.2. Integrals over closed surfaces.

10.	$\oint A \cdot dS = \int (\nabla \cdot A) dV$
11.	$\oint \phi dS = \int (\nabla\phi) dV$
12.	$\oint A \times dS = -\int (\nabla \times A) dV$

Table A.4.3. Integrals over open surfaces.

13.	$\oint \phi d\ell = \int dS \times \nabla\phi$
14.	$\oint A \cdot d\ell = \int (\nabla \times A) \cdot dS$

$$T_{i..j..kn} = \frac{\partial R_{i..j..k}}{\partial x_n} - \frac{\partial R_{i..n..k}}{\partial x_j} \tag{A.4.7}$$

This increases the rank by one. A particularly useful special case is for vectors. Let:

$$T_{in} = \frac{\partial R_i}{\partial x_n} - \frac{\partial R_n}{\partial x_i} \tag{A.4.8}$$

Note that T_{ij} is antisymmetric and has $N(N–1)/2$ independent numbers. In three dimensions, it has three, the same as a vector and we already saw that antisymmetric second rank tensors transform like vectors. Therefore, the curl of a vector changes the vector to an antisymmetric second rank tensor that is pseudovector. The pseudovector acts like a vector in any given coordinate system but changes sign if the systems are changed from a left to right hand system.

A.5 Green's Function

The 4-Laplacian of the electromagnetic potential is defined by Eq. 1.5.4, and repeated here:

$$\frac{\partial^2 A_\nu}{\partial X_\beta \partial X_\beta} = -\mu J_\nu \tag{A.5.1}$$

We seek to integrate that differential equation in order to obtain a general expression for the electromagnetic potential itself. For this purpose, it is helpful to define a similar but simpler function, to integrate that function, then to use the integral to obtain an expression for the electromagnetic potential. The function is Green's function $G(X_\alpha, X'_\alpha)$. By definition it is:

$$\frac{\partial^2 G(X_\alpha, X'_\alpha)}{\partial X_\beta \partial X_\beta} = -\left[\delta(X_\alpha - X'_\alpha)\right]^4 \tag{A.5.2}$$

In Eq. A.5.2, the four-dimensional delta function indicates the Dirac delta. By definition the one-dimensional Dirac delta function satisfies the integral relationship:

$$\int f(x)d(x-x')dx = \left|\begin{array}{c} f(x') \\ 0 \end{array}\right. \tag{A.5.3}$$

The upper or lower solution applies if the range of integration respectively does or does not include x'. The integrand magnitude of a Dirac delta function increases without limit and the width Δx decreases without limit in a way that retains a product value of one.

Construct the equation:

$$\int d\big[g(x)-g(x')\big]dx = \int\left(\frac{d\big[g(x)-g(x')\big]}{dg(x)/dx}\right)dg(x)$$

It follows from the definition of the delta function that:

$$\int f(x)d\big[g(x)-g(x')\big]dx = \frac{f(x)}{dg(x)/dx}\bigg|_{x=x'} \tag{A.5.4}$$

The method used to integrate Eq. A.5.1 is a four-dimensional extension of a common three-dimensional technique. The procedure begins with the quadruple integral:

$$\int\int\int\int\left\{A_\alpha\frac{\partial^2 G}{\partial X_\beta \partial X_\beta} - G\frac{\partial^2 A}{\partial X_\beta \partial X_\beta}\right\}dX_1 dX_2 dX_3 dX_4 = 0 \tag{A.5.5}$$

The equality results since all integrals are evaluated at $\pm\infty$ and the integrand decreases with distance rapidly enough so the integral is zero at the infinite limits. Substituting Eq. A.5.2 into the first term in the integrand and substituting Eq. A.5.1 into the second gives:

$$A_\alpha\big(X_\beta\big) = \mu\int\int\int\int J_\alpha\big(X'_\gamma\big)G\big(X'_\gamma, X_\gamma\big)dX_1 dX_2 dX_3 dX_4 \tag{A.5.6}$$

Since $J_\alpha(X'_\gamma)$ is known but $G(X'_\gamma, X_\gamma)$ is not, it is necessary to solve for $G(X'_\gamma, X_\gamma)$ using Eq. A.5.6 before, in turn, solving for $A_\alpha(X_\beta)$. For this purpose, consider an aside on the four dimensional Fourier transform pair:

$$F\left(X_{\gamma}\right)=\left(\frac{1}{2\pi}\right)^{2}\iiiint H\left(K_{\gamma}\right)e^{-iX_{\gamma}K_{\gamma}}dK_{1}dK_{2}dK_{3}dK_{4}$$

$$H\left(K_{\gamma}\right)=\left(\frac{1}{2\pi}\right)^{2}\iiiint F\left(X_{\gamma}\right)e^{iX_{\gamma}K_{\gamma}}dX_{1}dX_{2}dX_{3}dX_{4}$$

(A.5.7)

K_γ and X_γ are unknown conjugate variables to be determined. Making the definition that:

$$F\left(X_{\gamma}\right)=\left[\delta\left(X_{\gamma}-X'_{\gamma}\right)\right]^{4}$$

(A.5.8)

Combining Eq. A.5.7 and A.5.8 results in:

$$\left[\delta\left(X_{\gamma}-X'_{\gamma}\right)\right]^{4}=\left(\frac{1}{2\pi}\right)^{2}\iiiint e^{-iK_{\gamma}\left(X_{\gamma}-X'_{\gamma}\right)}dK_{1}dK_{2}dK_{3}dK_{4}$$

$$H\left(K_{\gamma}\right)=\left(\frac{1}{2\pi}\right)^{2}e^{iK_{\gamma}K'_{\gamma}}$$

(A.5.9)

It is convenient to introduce an additional function, $g(K_\gamma)$, defined by the equation:

$$G\left(X_{\gamma},X'_{\gamma}\right)=\iiiint g\left(K_{\alpha}\right)e^{-iK_{\gamma}\left(X_{\gamma}-X'_{\gamma}\right)}dK_{1}dK_{2}dK_{3}dK_{4}$$

(A.5.10)

To solve for the function $g(K_\gamma)$ consider, as an example, the conjugate pair x and k_x to be a single dimension of Eq. A.5.2 and A.5.9 with the equalities:

$$\frac{\partial^{2}G\left(x-x'\right)}{\partial x^{2}}=-\delta\left(x-x'\right); \quad \delta\left(x-x'\right)=\left(\frac{1}{2\pi}\right)$$

(A.5.11)

$$G\left(x,x'\right)=\int g\left(k_{x}\right)e^{-ik_{x}\left(x-x'\right)}dk_{x}$$

Differentiating $G(x,x')$ twice with respect to x gives:

$$\delta(x-x') = k_x^2 \int g(k_x) e^{-ik_x(x-x')} dk_x \tag{A.5.12}$$

Combining gives:

$$k_x^2 g(k_x) = \frac{1}{2\pi} \tag{A.5.13}$$

Extension to four dimensions gives:

$$g(K_\gamma) = \left(\frac{1}{2\pi}\right)^2 \frac{1}{K_\alpha K_\alpha} \tag{A.5.14}$$

Substituting Eq. A.5.14 back into Eq. A.5.10 results in:

$$G(X_\gamma, X'_\gamma) = \left(\frac{1}{2\pi}\right)^4 \int\int\int\int \frac{e^{-iK_\gamma(X_\gamma - X'_\gamma)}}{K_\alpha K_\alpha} dK_1 dK_2 dK_3 dK_4 \tag{A.5.15}$$

It is convenient to use three-dimensional notation to evaluate Eq. A.5.15. For this purpose note that the four variable set (x,y,z,ict) is complex and, if the exponentials are to remain oscillating functions, it is necessary that the conjugate variable set K_α also be complex. Writing it as $(k_x, k_y, k_z, i\omega/c)$ and substituting into Eq. A.5.15 gives:

$$G(X_\gamma, X'_\gamma) = \frac{i}{c}\left(\frac{1}{2\pi}\right)^4 \int\int\int\int \frac{e^{-iK_\gamma(X_\gamma - X'_\gamma)}}{k^2 - \omega^2} dk\,d\omega \tag{A.5.16}$$

where:

$$k^2 = k_x^2 + k_y^2 + k_z^2; \quad dk = dk_x dk_y dk_z \tag{A.5.17}$$

Equation A.5.16 is the sum of two Cauchy integrals, integrals that may be evaluated by use of the Cauchy integral identity:

$$2\pi i f(z') = \oint \frac{f(z)}{z - z'} dz \tag{A.5.18}$$

Introducing p as a small, real, positive number used as a construction tool whose value is eventually put equal to zero, Eq. A.5.16 may be written as:

$$G(X_\gamma, X'_\gamma) = \frac{ic}{(2\pi)^4} \iiint dk \oint \frac{d\omega e^{-iK_\gamma(X_\gamma - X'_\gamma)}}{(\omega - ck - ip)(\omega + ck + ip)}$$

Moving the space portion of the exponential out from under the time-dependent integral results in:

$$G(X_\gamma, X'_\gamma) = \frac{ic}{(2\pi)^4} \iiint dk e^{-ik \cdot r} \oint \frac{d\omega e^{-i\omega(t - t')}}{(\omega - ck - ip)(\omega + ck + ip)} \tag{A.5.19}$$

Restating, the problem is that given an electric charge at (r', t') to find the function $G(X_\gamma, X'_\gamma)$. The field is zero before the charge is introduced. That is, with $t'' = t - t'$ all fields are zero for $t'' < 0$. The last integral of Eq. A.5.19 may be evaluated first along the real axis and then back around an infinite, complex ω path. For $t'' < 0$ the return path encompasses the lower half-plane, where no poles are enclosed. For $t'' > 0$ the return path is around the upper half of the complex plane, where two poles are enclosed. Evaluation of the integral gives:

$$\oint \frac{d\omega e^{-i\omega t''}}{(\omega - ck - ip)(\omega + ck + ip)} = 2\pi i \left[\frac{e^{-it''(ck + ip)}}{2ck} - \frac{e^{-it''(-ck + ip)}}{2ck} \right] = \frac{2\pi}{ck} \sin(ckt'') \tag{A.5.20}$$

Combining Eq. A.5.20 and A.5.21 gives:

$$G(X_\gamma, X'_\gamma) = \frac{i}{(2\pi)^3} \iiint \frac{1}{k} dk e^{-ik \cdot r} \sin(ckt'') \tag{A.5.21}$$

Next let $\boldsymbol{R}$ be the space vector from source point $\boldsymbol{r'}$ to field point $\boldsymbol{r}$, and choose it to be in the z-direction. Then $\boldsymbol{k}\cdot\boldsymbol{r} = kR\cos\theta$ where θ is the polar angle. Also, replace $d\boldsymbol{k}$ with $k^2dk\,\sin\theta\,d\theta\,d\phi$:

$$G\left(X_\gamma, X'_\gamma\right) = \frac{i}{(2\pi)^3}\iiint kdk\,\sin\theta\,d\theta\,d\phi\,e^{-ikR\cos\theta}\sin\left(ckt''\right) \qquad (A.5.22)$$

Evaluating the angular integrals over an enclosing sphere gives:

$$G\left(X_\gamma, X'_\gamma\right) = \frac{2i}{(2\pi)^2}\int_0^\infty \frac{dk}{R}\sin\left(kR\right)\sin\left(ckt''\right) \qquad (A.5.23)$$

Since the integral of Eq. A.5.23 is an even function of k, it may, without changing the value of the integral, be replaced by the equation:

$$G\left(X_\gamma, X'_\gamma\right) = \frac{i}{8\pi^2 R}\int_{-\infty}^\infty dk\left[e^{-i\left(\omega t''-kR\right)} - e^{i\left(\omega t''+kR\right)}\right] \qquad (A.5.24)$$

Equation A.5.10 shows that Eq. A.5.24 is the sum of two Dirac delta functions. The second one is evaluated at advanced time $t'' < 0$ when there are no charges, and if causality applies all results from it are equal to zero. Working with the retarded time $t'' > 0$ when there are charges, using Eq. A.5.3 it follows that:

$$\delta\left(R - ct''\right) = -\frac{1}{c}\delta\left(t'' - R/c\right) \qquad (A.5.25)$$

Combining with the first term in the integrand of Eq. A.5.24 gives:

$$G\left(X_\gamma, X'_\gamma\right) = \frac{1}{4i\pi Rc}\delta\left(t'' - R/c\right) \qquad (A.5.26)$$

This completes the derivation of $G(X_\gamma, X'_\gamma)$.

A.6 The Potentials

To obtain potential A_ν of a moving charge density, substitute Eq. A.5.26 into Eq. A.5.5. The result is:

$$A_\nu\left(X_\gamma\right) = \frac{\mu}{4\pi}\iiint dV' \int dt \frac{J_\nu(X'_\alpha)}{R\left(X_\gamma, X'_\gamma\right)}\delta\left(t''- R/c\right)$$

(A.6.1)

Distance $R(X_\gamma, X'_\gamma)$ is the distance between the source and field points. Using Eq. A.5.3 to evaluate Eq. A.6.1 gives:

$$A(r,t) = \frac{\mu}{4\pi}\iiint \frac{J\left(r',t\right)}{\left(\mathrm{R} - \boldsymbol{R}\cdot\boldsymbol{v}/c\right)}dV'; \quad \Phi(r,t) = \frac{\mu c^2}{4\pi}\iiint \frac{\rho\left(r',t\right)}{\left(\mathrm{R} - \boldsymbol{R}\cdot\boldsymbol{v}/c\right)}dV'$$

(A.6.2)

This is the final form for the electromagnetic 4-potential of a moving charge, the Liénard-Wiechert potentials. Distance is from the point of field emission to the field point when the radiation is received.

To obtain the potential A_ν of an oscillating charge density, note that Eq. A.6.2 remains applicable except, for this case, the average velocity of the oscillating charge is zero. The resulting equation is:

$$A_{0\nu}\left(r\right)e^{i\omega t} = \frac{\mu}{4\pi}\int dV' \int dt' \frac{J_{0\nu}\left(r',t'\right)}{R\left(r,r'\right)}\delta\left(t',t - R/c\right)$$

(A.6.3)

Subscripts "0" indicate the value is independent of time t. Applying Eq. A.6.3 to a differential volume in space gives shows that in three dimensions the potentials at position r due to a current density are given by:

$$A(r)e^{i\omega t} = \frac{\mu}{4\pi}\int \frac{J\left(r'\right)e^{i\omega(t-R/c)}}{R(r,r')}dV'$$

$$\Phi(r)e^{i\omega t} = \frac{1}{4\pi\varepsilon}\int \frac{\rho\left(r'\right)e^{i\omega(t-R/c)}}{R(r,r')}dV'$$

(A.6.4)

These are the retarded potentials.

A.7 Equivalent Sources

It is shown in Sec. 1.10 that the force fields satisfy the partial differential equations:

$$\nabla \times (\nabla \times \boldsymbol{E}) + \varepsilon\mu \frac{\partial^2 \boldsymbol{E}}{\partial t^2} = 0 \ \text{ and } \ \nabla \times (\nabla \times \boldsymbol{B}) + \varepsilon\mu \frac{\partial^2 \boldsymbol{B}}{\partial t^2} = 0 \qquad (A.7.1)$$

Other helpful relationships are:

$$\nabla \times (\nabla \times \boldsymbol{E}) = \nabla (\nabla \bullet \boldsymbol{E}) - \nabla^2 \boldsymbol{E} \ \text{ and } \ \varepsilon\mu \frac{\partial^2 \boldsymbol{E}}{\partial t^2} = -k^2 \boldsymbol{E} \qquad (A.7.2)$$

Combining the two equations gives:

$$\nabla^2 \boldsymbol{E} + k^2 \boldsymbol{E} = 0 \ \text{ and } \ \nabla^2 \boldsymbol{B} + k^2 \boldsymbol{B} = 0 \qquad (A.7.3)$$

These are the Helmholtz equations for the field intensities. In rectangular coordinates the form of the vector and scalar Laplacian operators are identical.

The objective is to obtain expressions for the field vectors at any field point, $\boldsymbol{r}(x,y,z)$, and external to a field-generating volume, as a function of field values on the surface of the volume. The development requires three vector integral equations, the divergence theorem and two related ones. Let dS represent a scalar differential area on the surface of the volume and n be a unit vector directed normal to the surface at the same point. At the surface, fields $\boldsymbol{F}$ and ϕ have the continuity properties of electromagnetic fields: they are continuous with continuous first derivatives.

$$\oint d\boldsymbol{S} \bullet \boldsymbol{F} = \int \nabla \bullet \boldsymbol{F} dV; \ \ \oint d\boldsymbol{S} \times \boldsymbol{F} = \int \nabla \times \boldsymbol{F} dV; \ \ \oint \phi d\boldsymbol{S} = \int \nabla \phi dV \qquad (A.7.4)$$

Next let ϕ and ψ each represent scalar fields and construct the function $\phi \nabla \psi$. Substituting the new function into the divergence equation gives:

$$\oint \phi \nabla \psi \bullet d\boldsymbol{S} \equiv \int \nabla \bullet (\phi \nabla \psi) dV = \int \left[\nabla \phi \bullet \nabla \psi + \phi \nabla^2 \psi \right] dV \qquad (A.7.5)$$

Reversing the roles of ϕ and ψ and subtracting the result from Eq. A.7.5 gives:

$$\oint [\phi\nabla\psi - \psi\nabla\phi]\cdot d\mathbf{S} = \int [\phi\nabla^2\psi - \psi\nabla^2\phi]dV \tag{A.7.6}$$

For the special case of an oscillating charge, the defining equation for Green's function in three-dimensional form satisfies an equation similar to that of the Helmholtz wave equation. With point $\mathbf{r}'(x',y',z')$ representing the source position:

$$\nabla^2 G(\mathbf{r},\mathbf{r}') + k^2 G(\mathbf{r},\mathbf{r}') = -\delta(\mathbf{r},\mathbf{r}') \tag{A.7.7}$$

In free space, the solution is:

$$G(\mathbf{r},\mathbf{r}') = -\frac{e^{-i\mathbf{k}\cdot(\mathbf{r}-\mathbf{r}')}}{R(\mathbf{r},\mathbf{r}')} \tag{A.7.8}$$

The objective is to construct a virtual sphere about a source then to calculate the fields at an arbitrary field point, $\mathbf{r}(x,y,z)$, in terms of the fields that exist on the surface of the virtual sphere. In this way, the fields can be obtained without knowledge of the source itself. For this purpose, begin by substituting into Eq. A.7.6 that $\phi = G$ and ψ is equal to one component of the electric field intensity. Then repeat twice with ψ representing the other electric field components and sum over the three equations. The result is the vector form of Eq. A.7.6:

$$\oint \{\mathbf{E}(r')[\mathbf{n}'\cdot\nabla'G(r,r')] - G(r,r')[\mathbf{n}'\cdot\nabla']\mathbf{E}(r')\}dS'$$
$$= \int [\mathbf{E}(r')\nabla^2 G(r,r') - G(r,r')\nabla^2\mathbf{E}(r')]dV' \tag{A.7.9}$$

Next, let the field point be in the vicinity of the source and construct a virtual sphere with a radius just large enough to contain both source and field points. Substituting Eq. A.7.3 and A.7.7 into the volume integrals of Eq. A.7.9 results in:

$$\int \mathbf{E}(r')\delta(r,r')dV' = \mathbf{E}(r) \tag{A.7.10}$$

The second term in the surface integral of Eq. A.7.9 may be written:

$$G\left(n'\boldsymbol{\cdot}\nabla'\right)\boldsymbol{E} = n'\boldsymbol{\cdot}\nabla'\left(G\boldsymbol{E}\right) - \boldsymbol{E}\left(n'\boldsymbol{\cdot}\nabla'G\right) \qquad (A.7.11)$$

Combining shows that:

$$\boldsymbol{E}(r) = \oint \left\{ 2\boldsymbol{E}(r')\left[n'\boldsymbol{\cdot}\nabla'G(r,r')\right] - \left[n'\boldsymbol{\cdot}\nabla'\right]\left[G(r,r')\boldsymbol{E}(r')\right] \right\} dS'$$

$$(A.7.12)$$

This equation expresses the field intensity at the field point in terms of field values on the surface of a virtual sphere surrounding both the source and the field. Although Eq. A.7.12 expresses the electric field intensity at the field point in terms of values on the surface of an external virtual sphere, it is not satisfactory since the divergence operation contains derivatives of the electric field intensity.

Next, consider the field point to be outside the virtual sphere and, for completeness construct a second virtual sphere. It is concentric with the first one and of radius large enough to contain both source and field positions, and it contains no other sources. With no sources, the volume integral of Eq. A.7.9 is equal to zero. The integral similar to Eq. A.7.12, for this case, is equal to zero:

$$\oint \left\{ 2\boldsymbol{E}(\mathbf{r}')\left[n'\boldsymbol{\cdot}\nabla'G(\mathbf{r},\mathbf{r}')\right] - \left[n'\boldsymbol{\cdot}\nabla'\right]\left[G(\mathbf{r},\mathbf{r}')\boldsymbol{E}(\mathbf{r}')\right] \right\} dS' = 0 \qquad (A.7.13)$$

The integral is taken over both the inner and outer spherical surfaces, with the normal direction always extending outward from the field containing volume. Next, let the radius of the outer surface increase without limit, in which case, as discussed below, the integral over the exterior surface is equal to zero. Comparison of Eq. A.7.12 and A.7.13 shows that the surface integral over the inner surface is equal to $-\boldsymbol{E}(\boldsymbol{r})$.

To restate Eq. A.7.12 in a way that involves field vectors only note that the last term of Eq. A.7.13 may be written as a volume integral:

$$\oint n'\boldsymbol{\cdot}\nabla'(G\boldsymbol{E})dS = \int \nabla'^2(G\boldsymbol{E})dV$$

Substitute vector (GE) into the vector identity:

$$\nabla^2 (GE) = \nabla\left[\nabla \cdot (GE)\right] - \nabla\times\left[\nabla\times(GE)\right]$$

After combining and using the second and third integrals of Eq. A.7.4 to return to surface integrals, Eq. A.7.13 goes to:

$$\oint\left\{2E(n'\cdot\nabla'G) - n'\left[\nabla'\cdot(GE)\right] + n'\times\left[\nabla'\times(GE)\right]\right\}dS = 0 \qquad \text{(A.7.14)}$$

Completing both the divergence and curl operations and using the Maxwell equations to substitute for vector operations results in:

$$\oint\left\{2E(n'\cdot\nabla G) - n'(E\cdot\nabla G) - i\omega n'\times B + n'\times(\nabla'G\times E)\right\}dS = 0 \qquad \text{(A.7.15)}$$

Substituting for the triple product and simplifying results in:

$$\oint\left\{-i\omega(n'\times B)G + (n'\cdot E)\nabla'G + (n'\times E)\times\nabla'G\right\}dS = 0 \qquad \text{(A.7.16)}$$

By Eq. A.7.13, the value of the surface integral is equal to $-E(r)$. However, as defined above the normal is from the field-containing region to the source-containing region. Reversing the direction so the normal extends outward gives:

$$E(r) = \oint\left\{\begin{array}{l}-i\omega\left[n'\times B(r')\right]G(r,r') \\ +\left[n'\cdot E(r')\right]\nabla'G(r,r') + \left[n'\times E(r')\right]\times\nabla'G(r,r')\end{array}\right\}dS$$

$$\text{(A.7.17)}$$

The requirements are that the field point is external to the contained region and the source is fully contained by it.

This equation, when combined with Eq. A.7.8, is the exact expression for the exterior electric field intensity in terms of the surface fields on a source-containing region. It is not necessary to know anything about the source other than it created the surface fields. Although both the external electric field intensity and the fields on the surface are unique, the inverse is not true: the source necessary to produce $E(r)$ is not unique.

The corresponding expression for the magnetic flux density follows in a similar way. Carrying out the calculation gives:

$$B(r) = \oint \left\{ \begin{array}{l} i\omega\mu\varepsilon\left[n' \times E(r')\right]G(r,r') \\ + \left[n' \cdot B(r')\right]\nabla'G(r,r') - \left[n' \times B(r')\right] \times \nabla'G(r,r') \end{array} \right\} dS$$

(A.7.18)

With static sources Green's function decreases as the inverse of the radius, the electric field intensity decreases as the inverse square of the radius, and the surface area increases as the square of the radius. Therefore in the limit as the radius becomes infinite the contribution to field intensities $E(r)$ and $B(r)$ due to the outer surface goes to zero. On the other hand, for dynamic sources, Green's function and the electric field intensity both decrease as the inverse of the radius and the surface area increases as the square of the radius. From this point of view, contributions of the outer surface integral to $E(r)$ and $B(r)$ remain constant in the limit of infinite radius. That the null result remains, however, may be seen by application of a dynamic boundary condition: The radius of the outer sphere is greater than the speed of light, c, times whatever time is of interest in the problem. Even with dynamic sources, the outer surface integral has no influence on fields $E(r)$ or $B(r)$.

A.8 Spherical Shell Dipole

Calculations of electromagnetic effects about virtual shells commonly consider exterior effects but ignore interior ones. To establish the approximate magnitude of the resultant error, consider the interior-to-standing exterior energy ratio for an electric dipole. For this purpose, begin with a spherical shell of radius a that supports a surface electric charge density:

$$\rho(\theta,t) = \frac{q}{2\pi a^2}\cos\theta\, e^{i(\omega t - ka)}$$

(A.8.1)

This is the static and time varying charge combination that occurs by driving an originally uniformly charged sphere to its maximum extent as a dipole. The resulting electric dipole moment, see Eq. A.22.7, is:

$$p_z = \frac{2qa}{3} e^{i(\omega t - ka)}$$

(A.8.2)

The time dependence of the change is associated with the surface current density:

$$\mathbf{I}_s = -i\frac{qkc}{2\pi a}\sin\theta e^{i(\omega t - ka)}\hat{\theta}$$

(A.8.3)

The time varying charge and current densities create, respectively, electric and magnetic fields in both interior and exterior regions about the shell. The field forms follow from Eq. 1.13.9; all coefficients are equal to zero except those of order one and degree zero. The field components are:

$$E_{re} = F\frac{2}{\sigma}h_1(\sigma)\cos\theta e^{i(\omega t - \sigma)}; \qquad E_{ri} = G\frac{2}{\sigma}j_1(\sigma)\cos\theta e^{i(\omega t - \sigma)}$$

$$E_{\theta e} = -Fh_1^\bullet(\sigma)\sin\theta e^{i(\omega t - \sigma)}; \qquad E_{\theta i} = -Gj_1^\bullet(\sigma)\sin\theta e^{i(\omega t - \sigma)}$$

$$\eta H_{\phi e} = iFh_1(\sigma)\sin\theta e^{i(\omega t - \sigma)}; \qquad \eta H_{\phi i} = iGj_1(\sigma)\sin\theta e^{i(\omega t - \sigma)}$$

(A.8.4)

F and G are complex field coefficients to be determined. Phase factor $\exp(-ika)$ is added as a notational convenience.

Applying the field boundary conditions of Eq. 1.14.28 and 1.14.30 gives:

$$\frac{qk}{4\pi\varepsilon a}e^{i(\omega t - ka)} = Fh_1(ka) - Gj_1(ka)$$

$$\frac{qk}{2\pi\varepsilon a}e^{i(\omega t - ka)} = Fh_1(ka) + Gj_1(ka)$$

(A.8.5)

Solving for the coefficients gives:

$$F = \frac{3q}{8\pi\varepsilon a^2}\frac{(ka)}{h_1(ka)}e^{i(\omega t - ka)}; \qquad G = \frac{q}{8\pi\varepsilon a^2}\frac{(ka)}{j_1(ka)}e^{i(\omega t - ka)}$$

(A.8.6)

Inserting the coefficients into the fields gives:

Exterior

$$E_e = \frac{3}{2}\left(\frac{q}{4\pi\varepsilon a^2}\right)\left(\frac{ka}{h_1(ka)}\right)\left\{\hat{r}\frac{2}{\sigma}h_1(\sigma)\cos\theta - \hat{\theta}h_1^{\bullet}(\sigma)\sin\theta\right\}e^{i(\omega t-\sigma)} \quad (A.8.7)$$

$$\eta H_e = \frac{3i}{2}\left(\frac{q}{4\pi\varepsilon a^2}\right)\left(\frac{ka}{h_1(ka)}\right)\hat{\phi}h_1(\sigma)\sin\theta e^{i(\omega t-\sigma)}$$

Interior

$$E_i = \frac{1}{2}\left(\frac{q}{4\pi\varepsilon a^2}\right)\left(\frac{ka}{j_1(ka)}\right)\left\{\hat{r}\frac{2}{\sigma}j_1(\sigma)\cos\theta - \hat{\theta}j_1^{\bullet}(\sigma)\sin\theta\right\}e^{i(\omega t-\sigma)} \quad (A.8.8)$$

$$\eta H_i = \frac{i}{2}\left(\frac{q}{4\pi\varepsilon a^2}\right)\left(\frac{ka}{j_1(ka)}\right)\hat{\phi}j_1(\sigma)\sin\theta e^{i(\omega t-\sigma)}$$

The fields just off the surface of an emitter with ka much less than one are:

Exterior

$$E_e = \frac{3}{2}\left(\frac{q}{4\pi\varepsilon a^2}\right)\left\{2\hat{r}\cos\theta + \hat{\theta}\sin\theta\right\}e^{i(\omega t-ka)} \quad (A.8.9)$$

$$\eta H_e = \frac{3}{2}\left(\frac{q}{4\pi\varepsilon a^2}\right)(ka)\hat{\phi}\sin\theta e^{i(\omega t-ka)}$$

Interior

$$E = \left(\frac{q}{4\pi\varepsilon a^2}\right)\left\{\hat{r}\cos\theta - \hat{\theta}\sin\theta\right\}e^{i(\omega t-ka)} = \left(\frac{q}{4\pi\varepsilon a^2}\right)\hat{z}e^{i(\omega t-ka)} \quad (A.8.10)$$

$$\eta H = \frac{i}{2}\left(\frac{kq}{4\pi\varepsilon a}\right)\hat{\phi}\sin\theta e^{i(\omega t-ka)}$$

The time-average, outbound power density follows from Eq. A.8.8:

$$N_r = \frac{9}{8\eta}\left(\frac{q}{4\pi\varepsilon a^2}\right)^2\left|\frac{ka}{h_1(ka)}\right|^2 \text{Re}\left[ih_1^{\bullet}(\sigma)h_1^*(\sigma)\right]\sin^2\theta \cong \frac{9}{8\eta}\left(\frac{q}{4\pi\varepsilon a^2}\right)^2\frac{(ka)^6}{\sigma^2}\sin^2\theta$$

$$(A.8.11)$$

The approximate equality is after incorporating $ka \ll 1$.

The time-average energy density at each exterior point is:

$$w_T = \frac{9}{16\varepsilon}\left(\frac{q}{4\pi a^2}\right)^2 \left|\frac{ka}{h_1(ka)}\right|^2 \left\{\frac{4}{\sigma^2}\left|h_1(\sigma)\right|^2 \cos^2\theta + \left[\left|h_1^{\bullet}(\sigma)\right|^2 + \left|h_1(\sigma)\right|^2\right]\sin^2\theta\right\}$$

$$w_T \cong \frac{9}{16\varepsilon}\left(\frac{q}{4\pi a^2}\right)^2 (ka)^6 \left\{\left[\frac{4}{\sigma^6}+\frac{4}{\sigma^4}\right]\cos^2\theta + \left[\frac{2}{\sigma^2}+\frac{1}{\sigma^6}\right]\sin^2\theta\right\}$$

$$(A.8.12)$$

In this case the source-associated energy density is equal to the energy density minus the energy density of the Poynting vector at each point. Combining Eq. A.8.11 and A.8.12 gives the source-associated energy density:

$$w_S \cong \frac{9}{16\varepsilon}\left(\frac{q}{4\pi a^2}\right)^2 (ka)^6 \left\{\left(\frac{4}{\sigma^6}+\frac{4}{\sigma^4}\right)\cos^2\theta + \frac{1}{\sigma^6}\sin^2\theta\right\} \qquad (A.8.13)$$

Integrating over exterior space gives the total source associated standing energy:

$$W_S = \frac{3q^2}{32\pi\varepsilon a}(1+2ka) \qquad (A.8.14)$$

The time-average interior energy density follows from Eq. A.8.8:

$$w_I = \frac{1}{4\varepsilon}\left[\frac{q}{8\pi a^2}\right]^2 \left\{\left(4-\frac{2\sigma^2}{5}\right)\cos^2\theta + \left(4+\frac{\sigma^2}{5}\right)\sin^2\theta\right\} \qquad (A.8.15)$$

Integrating over interior space gives:

$$W_i = \left(\frac{q^2}{48\pi\varepsilon a}\right)\left[1+\frac{(ka)^2}{10}\right] \qquad (A.8.16)$$

Taking the ratio of Eq. A.8.16 to Eq. A.8.14 shows the ratio of the source-associated energies is:

$$W_I/W_S \cong 2/9 \qquad (A.8.17)$$

For the important case of a radiating dipole shell the often-ignored time-average interior stored energy is about 22% of the exterior source associated energy. So long as the region remains small the ratio is independent both of size and wavelength.

A.9 Gamma Functions

Products in which successive factors differ by one occur frequently in the formation of power series. If ℓ is an integer, such products may be expressed as special products of a certain number of integers, beginning with one. The factorial of integer ℓ is, by definition:

$$\ell! = 1 \cdot 2 \cdot 3 \cdot \ldots \cdot \ell \tag{A.9.1}$$

This may be written in the compact form:

$$\ell! = \ell(\ell-1)! \tag{A.9.2}$$

The same symbolism is useful for noninteger numbers, v. A similar equation is defined as:

$$v! = v(v-1)! \tag{A.9.3}$$

Similarly:

$$v(v-1)(v-2)\ldots(v-m+1) = \frac{v!}{(v-m)!} \tag{A.9.4}$$

Let $f(v)$ be any function that satisfies the condition:

$$f(v) = vf(v-1) \tag{A.9.5}$$

Taking the ratio of Eq. A.9.5 to A.9.3 gives:

$$\phi(v) = \frac{f(v)}{v!} = \frac{f(v-1)}{(v-1)!} = \phi(v-1) \tag{A.9.6}$$

It follows that $\phi(v)$ is a periodic function of period v.

Euler proposed that the definition of a noninteger factorial be:

$$v! = \int_0^\infty t^v e^{-t} dt \tag{A.9.7}$$

Euler's definition is valid over the range:

$$-1 < v \le 0 \tag{A.9.8}$$

Factorial $v!$ can be evaluated for any value of v using Eq. A.9.3. For example, if $-2 < v < -1$ then $v!$ Is:

$$v! = \frac{(v+1)!}{v+1} = \frac{(v+1)!(v+2)!}{(v+1)(v+2)} = \frac{(v+1)!(v+2)!...(v+n)!}{(v+1)(v+2)...(v+n)} \tag{A.9.9}$$

Equation A.9.9 shows a simple pole exists for v equal to a negative integer. Other results of Eq. A.9.7 include:

$$0! = 1 = 1!; \quad (-1/2)! = \sqrt{\pi}; \quad (-v)!(v-1)! = \frac{\pi}{\sin(\pi v)} \tag{A.9.10}$$

The Stirling formula for the approximate value of $v!$, in the limit of large values of v, is:

$$v! = \left(\frac{v}{e}\right)^v \sqrt{2\pi v} \tag{A.9.11}$$

A related and frequently recurring product form has succeeding numbers that differ by two:

$$v(v-2)(v-4)(v-6)...$$

The series is denoted by the double factorial:

$$v!! = v(v-2)(v-4)(v-6)... \tag{A.9.12}$$

It follows that for even and odd integers, respectively, are:

$$(2\ell)!! = 2^\ell(\ell)! \quad \text{and} \quad (2\ell+1)!! = \frac{(2\ell+1)!}{(2\ell)!!} \tag{A.9.13}$$

The left and right equations of Eq. A.9.13 for $\ell = 0$ show that:

$$(0)!! = 1 \quad \text{and} \quad (1)!! = 1 \tag{A.9.14}$$

Although Eq. A.9.15 is in indeterminate form for $\ell = -1$, consider the identity:

$$\frac{(\ell)!}{(\ell)!!} = (\ell - 1)!! \tag{A.9.15}$$

Table A.9.1 contains a listing of useful and selected sums involving factorials.

Table A.9.1. A table of sums over factorials.

$$2\sum_{me;0}^{\ell} \frac{(\ell+m-1)!(\ell-m-1)!!}{(\ell+m)!!(\ell-m)!!} U(m)\delta(\ell+m,2q) = 1$$

$$\sum_{mo;1}^{\ell} \frac{(-1)^{(\ell-m)/2} m(\ell+m-1)!!(\ell-m-1)!!}{(\ell+m)!!(\ell-m)!!} \delta(\ell+m,2q) = \frac{\ell!!}{(\ell-1)!!}$$

$$2\sum_{me;2}^{\ell} \frac{(-1)^{(\ell-m)/2} m^2 (\ell+m-1)!!(\ell-m-1)!!}{\ell(\ell+1)(\ell+m)!!(\ell-m)!!} \delta(\ell+m,2q) = \frac{(\ell-1)!!}{\ell!!}$$

$$4\sum_{m=0}^{\ell-1} \frac{(\ell+m)!!(\ell-m)!!}{\ell(\ell+1)(\ell+m-1)!!(\ell-m-1)!!} U(m)\delta(\ell+m,2q+1) = 1$$

$$4\sum_{m=0}^{\ell} \frac{m^2 (\ell+m-1)!!(\ell-m-1)!!}{\ell(\ell+1)(\ell+m)!!(\ell-m)!!} \delta(\ell+m,2q) = 1$$

$$8\sum_{me;0}^{\ell} \frac{(-1)^{(\ell-m-1)/2} (\ell+m)!!(\ell-m)!!}{\ell(\ell+1)(\ell+m-1)!!(\ell-m-1)!!} \delta(\ell+m,2q+1)U(m) = \frac{\ell!!}{(\ell-1)!!}$$

$$2\sum_{mo;1}^{\ell} \frac{(-1)^{(\ell-m-1)/2} m(\ell+m)!!(\ell-m)!!}{\ell(\ell+1)(\ell+m-1)!!(\ell-m-1)!!} \delta(\ell+m,2q+1) = \frac{(\ell-1)!!}{\ell!!}$$

$$2\sum_{s=0}^{(\ell-m-1)/2} \frac{\ell(-1)^s \delta(\ell+m,2q+1)}{(\ell+m+2s+1)!!(\ell-m-2s-1)!!} = \frac{1}{(\ell+m-1)!!(\ell-m-1)!!}$$

For the special case $\ell = 0$, Eq. A.9.15 gives:

$$(-1)!! = 1 \tag{A.9.16}$$

A.10 Azimuth Angle Trigonometric Functions

Solutions of Eq. 1.12.11 are trigonometric functions:

$$\Phi(\phi) = \sum_m \left[C_m \cos(m\phi) + D_m \sin(m\phi) \right] \tag{A.10.1}$$

An equally satisfactory solution is:

$$\Phi(\phi) = \sum_m \left[\hat{C}_m e^{jm\phi} + \hat{D}_m e^{-jm\phi} \right] \tag{A.10.2}$$

By definition $j^2 = (-1)$. For cases of interest here, the azimuth angle occupies the full range of angle from 0 through 2π. This condition requires the solution to satisfy the relationship:

$$\Phi(\phi) = \Phi(\phi + 2\pi) \tag{A.10.3}$$

Equations A.10.1 through A.10.3 are jointly satisfied only if m represents the full range of positive integers, including zero.

The trigonometric functions form an orthogonal set. Trigonometric identities show that:

$$\int_0^{2\pi} \cos(m\phi)\cos(n\phi)\,d\phi = \frac{1}{2}\int_0^{2\pi} d\phi \left\langle \cos\left[(m-n)\phi\right] + \cos\left[(m+n)\phi\right] \right\rangle \tag{A.10.4}$$

Evaluating the integral on the right gives:

$$\int_0^{2\pi} \cos(m\phi)\cos(n\phi)\,d\phi = \left(\frac{\sin\left[(m-n)\phi\right]}{2(m-n)} + \frac{\sin\left[(m+n)\phi\right]}{2(m+n)} \right)_0^{2\pi} \tag{A.10.5}$$

Since both m and n are positive integers, the second term on the right of Eq. A.10.5 is always zero; the first term is also positive unless $m = n$, for which case the result is indeterminate. Evaluation may be accomplished by either evaluating the indeterminate or by substituting into the integrand the identity:

$$\cos^2(m\phi) \equiv \frac{1}{2}\left[1 + \cos(2m\phi)\right] \tag{A.10.6}$$

Integrating Eq. A.10.6 gives:

$$\int_0^{2\pi} d\phi \cos^2(m\phi) = \frac{1}{2}\int_0^{2\pi} d\phi\left[1 + \cos(2m\phi)\right] = \pi \tag{A.10.7}$$

Combining Eq. A.10.5 through A.10.7:

$$\int_0^{2\pi} \cos(m\phi)\cos(n\phi)\,d\phi = \pi\delta(m,n) \tag{A.10.8}$$

The Kronecker delta function is indicated by $\delta(m,n)$. By definition:

$$\delta(m,n) = 1 \text{ if } m = n$$
$$= 0 \text{ if } m \neq n \tag{A.10.9}$$

It follows in a similar way that:

$$\int_0^{2\pi} \sin(m\phi)\sin(n\phi)\,d\phi = \pi\delta(m,n) \tag{A.10.10}$$

$$\int_0^{2\pi} \sin(m\phi)\cos(n\phi)\,d\phi = 0 \tag{A.10.11}$$

Combining Eq. A.10.8 through A.10.11 gives:

$$\int_0^{2\pi} e^{j(m-n)\phi}d\phi = 2\pi\delta(m,n) \tag{A.10.12}$$

For example, Eq. A.10.12 may be used to evaluate the product function:

$$\int_0^{2\pi} e^{j(m-m')\phi} \sin\phi \, d\phi = \frac{1}{2j}\int_0^{2\pi}\left(e^{j(m'-m+1)\phi} - e^{j(m'-m-1)\phi}\right)d\phi \tag{A.10.13}$$
$$= \pi j\left[\delta(m',m+1)-\delta(m',m-1)\right]$$

It is sometimes more convenient to express functions in terms of the order of trigonometric functions. The formulas for going from power to order follow directly from the geometry; all possible combinations are given by the four sums:

$$\cos^{2\ell}\phi \equiv \frac{1}{2^{2\ell}}\left\{\sum_{k=0}^{\ell-1}\frac{2(2\ell)!}{(2\ell-k)!k!}\cos\left[2(\ell-k)\phi\right]+\frac{(2\ell)!}{(\ell!)^2}\right\} \tag{A.10.14}$$

$$\sin^{2\ell}\phi \equiv \frac{1}{2^{2\ell}}\left\{\sum_{k=0}^{\ell-1}(-1)^{\ell-k}\frac{2(2\ell)!}{(2\ell-k)!k!}\cos\left[2(\ell-k)\phi\right]+\frac{(2\ell)!}{(\ell!)^2}\right\} \tag{A.10.15}$$

$$\cos^{2\ell-1}\phi \equiv \frac{4}{2^{2\ell}}\left\{\sum_{k=0}^{\ell-1}\frac{(2\ell-1)!}{(2\ell-k-1)!k!}\cos\left[(2\ell-2k-1)\phi\right]\right\} \tag{A.10.16}$$

$$\sin^{2\ell-1}\phi \equiv \frac{4}{2^{2\ell}}\left\{\sum_{k=0}^{\ell-1}(-1)^{\ell-k-1}\frac{(2\ell-1)!}{(2\ell-k-1)!k!}\sin\left[(2\ell-k-1)\phi\right]\right\} \tag{A.10.17}$$

An expansion for $1/(\sin\phi)$ is necessary to accomplish needed calculations. To form the expansion, note that $1/(\sin\phi)$ is an odd function of ϕ and therefore expressible as:

$$\frac{1}{\sin\phi} = \sum_{s=1}^{\infty} A_s \sin(2s-1)\phi \tag{A.10.18}$$

To evaluate coefficients A_s, multiply Eq. A.10.18 by $\sin(2p-1)\phi$ and integrate over the full range of the variable:

$$\int_0^{2\pi} d\phi \frac{\sin(2p-1)\phi}{\sin\phi}$$

$$= \sum_{s=1}^{\infty} A_s \int_0^{2\pi} d\phi \sin\left[(2s-1)\phi\right]\sin\left[(2p-1)\phi\right] = \pi A_p \delta(p,s) \quad \text{(A.10.19)}$$

The left side of Eq. A.10.19 is a periodic trigonometric function for all terms except $p = 1$ and periodic terms integrate to zero. This leaves:

$$\int_0^{2\pi} d\phi \frac{\sin(2p-1)\phi}{\sin\phi} = 2\pi \quad \text{(A.10.20)}$$

Substituting back into Eq. A.10.18:

$$\frac{1}{\sin\phi} = 2\sum_{s=1}^{\infty} \sin(2s-1)\phi \quad \text{(A.10.21)}$$

The corresponding sum, where cosines replace sines, is equal to zero:

$$\sum_{s=0}^{\infty} \cos\left[(2s-1)\phi\right] = 0 \quad \text{(A.10.22)}$$

It follows that:

$$\frac{1}{\sin\phi} = 2j\sum_{s=0}^{\infty} e^{-j(2s+1)\phi} \quad \text{(A.10.23)}$$

A.11 Legendre Functions

To obtain the general solution of Eq. 1.12.10 first solve the special case $m = 0$, for which the equation is:

$$\frac{d^2\Theta}{d\theta^2} + \cot\theta \frac{d\Theta}{d\theta} + \nu(\nu+1)\Theta = 0 \qquad\qquad (A.11.1)$$

The character of separation constant ν depends upon the boundary conditions applicable to the region in which the equation is applied. In the case of spherical waves in free space, for example, ν is an integer, and denoted by $\nu = \ell$. For lossless waves in conical structures ν is real and noninteger. If there is loss, ν is imaginary. In this book since only lossless problems are considered ν is real in all cases.

Since Eq. A.11.1 contains a singularity on the polar axes, the character of the functions $\Theta(\theta)$ in the region away from the axes are of special interest. Rather than go directly to a solution, consider first the solution form in the region of interest. A useful substitution is:

$$\Theta = \frac{1}{(\sin\theta)^{1/2}} \tilde{\Theta} \qquad\qquad (A.11.2)$$

Differentiating gives:

$$\frac{d\Theta}{d\theta} = \frac{1}{(\sin\theta)^{1/2}} \frac{d\tilde{\Theta}}{d\theta} - \frac{\cos\theta}{2(\sin\theta)^{3/2}} \tilde{\Theta}$$
$$\qquad\qquad (A.11.3)$$
$$\frac{d^2\Theta}{d\theta^2} = \frac{1}{(\sin\theta)^{1/2}} \frac{d^2\tilde{\Theta}}{d\theta^2} - \frac{\cos\theta}{(\sin\theta)^{3/2}} \frac{d\tilde{\Theta}}{d\theta} + \left[\frac{3\cos^2\theta}{4(\sin\theta)^{5/2}} + \frac{1}{2(\sin\theta)^{1/2}} \right] \tilde{\Theta}$$

Combining Eq. A.11.1 through A.11.3 gives:

$$\frac{d^2\tilde{\Theta}}{d\theta^2} + \left\{ \left(\nu + \frac{1}{2}\right)^2 + \frac{1}{4}\left(1 + \cot^2\theta\right) \right\} \tilde{\Theta} = 0 \qquad\qquad (A.11.4)$$

For θ near $\pi/2$, $\cot^2\theta$ is much less than one and Eq. A.11.4 is nearly equal to:

$$\frac{d^2\tilde{\Theta}}{d\theta^2} + \left\{ \left(\nu + \frac{1}{2}\right)^2 + \frac{1}{4} \right\} \tilde{\Theta} = 0 \qquad\qquad (A.11.5)$$

Solutions of Eq. A.11.5 are:

$$\Theta(\theta) = \frac{1}{(\sin\theta)^{1/2}}\left[A_\nu \cos(\kappa_\nu\theta) + B_\nu \sin(\kappa_\nu\theta)\right] \tag{A.11.6}$$

By definition:

$$\kappa_\nu = \left[\left(\nu + \frac{1}{2}\right)^2 + \frac{1}{4}\right]^{1/2} \tag{A.11.7}$$

Near the equator the zenith angle functions are trigonometric functions of $(\kappa_\nu\theta)$ normalized by the square root of $\sin\theta$. The interval over which Eq. A.11.6 is valid increases with increasing values of ν.

To examine the solution near its singularity, begin near the positive z-axis, where

$$\cot\theta \cong \frac{1}{\theta} \tag{A.11.8}$$

Equation A.11.1 has the form:

$$\frac{d^2\Theta}{d\theta^2} + \frac{1}{\theta}\frac{d\Theta}{d\theta} + \nu(\nu+1)\Theta = 0 \tag{A.11.9}$$

Equation A.11.9 is the cylindrical Bessel equation.

To begin the solution process, make the definition:

$$\beta = \left[\nu(\nu+1)\right]^{1/2} \tag{A.11.10}$$

The two independent solutions of Eq. A.11.9 may be written as:

$$\Theta(\theta) = A_\nu J_0(\beta\theta) + B_\nu Y_0(\beta\theta) \tag{A.11.11}$$

$J_0(\beta\theta)$ and $Y_0(\beta\theta)$ represent, respectively, cylindrical Bessel and Neumann functions of zero order. In the limit as θ goes to zero, $J_0(\beta\theta)$ goes to unity and $Y_0(\beta\theta)$ becomes logarithmically singular:

$$\Theta(0) = \hat{C}_v Y_0(0) \tag{A.11.12}$$

By symmetry, near the negative z-axis the function takes the same form. The local solution is:

$$\Theta(\pi - \theta) = \hat{A}_v Y_0\big(\beta(\pi - \theta)\big) + \hat{B}_v J_0\big(\beta(\pi - \theta)\big) \tag{A.11.13}$$

Combining gives:

$$\Theta(\pi) = \hat{D}_v J_0(0) + \hat{C}_v Y_0(0) \tag{A.11.14}$$

If one solution of Eq. A.11.1 is $P_v(\cos\theta)$, the other is $P_v(-\cos\theta)$. The first solution is regular on the positive z-axis and singular on the negative z-axis and the second solution is singular on the positive z-axis and regular on the negative z-axis. Both are periodic functions, see Eq. A.11.6, in the center region. The full solution is:

$$\Theta(\theta) = A_v P_v(\cos\theta) + B_v P_v(-\cos\theta) \tag{A.11.15}$$

For values of m different from zero it is convenient to rewrite Eq. A.11.1 by introducing the variable χ where:

$$\chi = \frac{1}{2}(1 - \cos\theta) = \sin^2(\theta/2) \tag{A.11.16}$$

Derivatives are:

$$\frac{d\Theta}{d\chi} = \frac{d\Theta}{d(\cos\theta)} \frac{d(\cos\theta)}{d\chi} = -2\frac{d\Theta}{d(\cos\theta)}$$

$$\frac{d^2\Theta}{d\chi^2} = \frac{d^2\Theta}{d(\cos\theta)^2}\left[\frac{d(\cos\theta)}{d\chi}\right]^2 + \frac{d\Theta}{d(\cos\theta)}\frac{d^2(\cos\theta)}{d\chi^2} \tag{A.11.17}$$

Combining Eq. A.11.16 and A.11.17 with Eq. A.11.1 gives:

$$\chi(1-\chi)\frac{d^2\Theta}{d\chi^2}+(1-2\chi)\frac{d\Theta}{d\chi}+\nu(\nu+1)\Theta=0 \tag{A.11.18}$$

A power series expansion results in:

$$\Theta(\chi)=\sum_{j=0}^{\infty}a_j\chi^j;\quad \frac{d}{d\chi}\Theta(\chi)=\sum_{j=0}^{\infty}a_j j\chi^{j-1}$$

$$\frac{d^2}{d\chi^2}\Theta(\chi)=\sum_{j=0}^{\infty}a_j j(j-1)\chi^{j-2} \tag{A.11.19}$$

Combining Eq. A.11.18 and A.11.19 yields:

$$\sum_{j=0}^{\infty}\left[(j+1)^2 a_{j+1}-j(j+1)a_j+\nu(\nu+1)a_j\right]\chi^j=0 \tag{A.11.20}$$

Since Eq. A.11.20 is an identity in χ, the coefficient of each power of χ is separately equal to zero and results in the recursion relationship:

$$\frac{a_{j+1}}{a_j}=\frac{j(j+1)-\nu(\nu+1)}{(j+1)^2} \tag{A.11.21}$$

If ν is an integer, the numerator of Eq. A.11.21 is equal to zero at $\nu=j$ and the series terminates. Substituting Eq. A.11.21 in the first series of Eq. A.11.19 results in the solution:

$$P_\nu(\cos\theta)=\sum_{j=0}^{\infty}\frac{(-1)^j(\nu+j)!}{(j!)^2(\nu-j)!}\sin^{2j}\left(\frac{\theta}{2}\right) \tag{A.11.22}$$

$P_\nu(-\cos\theta)$ is also a solution; changing the sign in Eq. A.11.17 gives:

$$\chi=\frac{1}{2}(1+\cos\theta)=\cos^2(\theta/2) \tag{A.11.23}$$

Combining Eq. A.11.22 and A.11.23 yields the second solution:

$$P_\nu\left(-\cos\theta\right) = \sum_{j=0}^{\infty} \frac{\left(-1\right)^j \left(\nu+j\right)!}{\left(j!\right)^2 \left(\nu-j\right)!} \cos^{2j}\left(\theta/2\right) \tag{A.11.24}$$

Neither Eq. A.11.22 nor Eq. A.11.24 is fully an even or odd function of θ. Since it is convenient to work with equations of definite parity, it is convenient to define the new functions:

$$L_\nu\left(\cos\theta\right) = \frac{1}{2}\left\{P_\nu\left(\cos\theta\right) + P_\nu\left(-\cos\theta\right)\right\}$$

$$\tag{A.11.25}$$

$$M_\nu\left(\cos\theta\right) = \frac{1}{2}\left\{P_\nu\left(\cos\theta\right) - P_\nu\left(-\cos\theta\right)\right\}$$

The functional symmetry is:

$$L_\nu\left(\cos\theta\right) = L_\nu\left(-\cos\theta\right); \qquad M_\nu\left(\cos\theta\right) = -M_\nu\left(-\cos\theta\right) \tag{A.11.26}$$

Since the regions of convergence for $P_\nu(\cos\theta)$ are $-1 < \cos\theta \le 1$ and $0 \le \theta < \pi$, the region of convergence for L_ν and M_ν is $0 < \theta < \pi$.

A.12 Legendre Polynomials

To solve problems where a solution is necessary on the z-axis all functions must remain bounded at the endpoints: $0 \le \theta \le \pi$. This happens only if separation constant n is an integer and, for that case, the series solution of Eq. A.11.21 terminates. Both Eq. A.11.22 and A.11.24 remain bounded on both the $\pm z$-axes, but are not independent. Since the product $\ell(\ell+1)$ is the same for $\ell = n$ as it is for $\ell = -(n+1)$, solutions are the same for the range of integers respectively from 0 to $+\infty$ and from -1 to $-\infty$. Therefore, only positive values of ℓ need be considered.

To characterize Legendre polynomials, it is more convenient to redo the expansion than to work with the solutions of Sec. A.11. For this purpose, rewrite Eq. A.11.1 by replacing noninteger ν by integer ℓ and defining $\chi = \cos\theta$, to obtain:

$$\left(1-\chi^2\right)\frac{d^2\Theta}{d\chi^2}-2\chi\frac{d\Theta}{d\chi}+\ell(\ell+1)\Theta=0 \tag{A.12.1}$$

The power series expansion is:

$$\Theta(\chi)=\sum_{j=0}^{\infty}a_j\chi^j \qquad \frac{d\Theta(\chi)}{d\chi}=\sum_{j=0}^{\infty}a_j j\chi^{j-1}$$

$$\frac{d^2\Theta(\chi)}{d\chi^2}=\sum_{j=0}^{\infty}a_j j(j-1)\chi^{j-1} \tag{A.12.2}$$

Combining Eq. A.12.1 and A.12.2 yields:

$$\sum_{j=0}^{\infty}\left[(j+1)(j+2)a_{j-2}-j(j+1)a_j+\ell(\ell+1)a_j\right]\chi^j=0 \tag{A.12.3}$$

Since Eq. A.12.1 is an identity in χ it follows that:

$$\frac{a_{j+2}}{a_j}=\frac{j(j+1)-\ell(\ell+1)}{(j+1)(j+2)} \tag{A.12.4}$$

Substituting Eq. A.12.4 into the expansion of the first of Eq. A.12.2 gives:

$$\Theta_\ell(\chi)=a_0\sum_{j=0}^{\infty}\frac{(-1)^j\,\chi^{2j}\,(\ell)!!(\ell+2j-1)!!}{(2j)!(\ell-1)!!(\ell-2j)!!} \tag{A.12.5}$$

Values are:

$$\Theta_0(\chi)=a_0;\ \Theta_1(\chi)=a_0\left(1-3\chi^2\right);\ \Theta_2(\chi)=a_0\left(1-10\chi^2+\frac{35}{3}\chi^4\right) \tag{A.12.6}$$

Arbitrary constant a_0 is redefined for each value of ℓ and valued to make $\Theta_\ell(0)=1$. With that definition, functions $\Theta_\ell(\chi)$ are defined to be

Legendre polynomials of the first kind, and indicated by $P_\ell(\cos\theta)$. Therefore $P_\ell(\chi)$ is given by the series:

$$P_\ell(\chi) = \frac{1}{2^\ell} \sum_{s=0}^{[\ell/2]} \frac{(-1)^s}{s!} \frac{(2\ell-2s)!}{(\ell-s)!(\ell-2s)!} \chi^{\ell-2s} \tag{A.12.7}$$

The symbol $[\ell/2]$ indicates the largest integer contained in $\ell/2$. From Eq. A.12.7, it follows that:

$$P_\ell(1) = 1 \ ; \ P_{2\ell+1}(0) = 0 \ ; \ P_{2\ell}(0) = (-1)^\ell \frac{(2\ell-1)!!}{(2\ell)!!} \ ; \ P_\ell(-\chi) = (-1)^\ell P_\ell(\chi)$$

Values of selected functional combinations on the axes are given in Table A.12.1.

Table A.12.1. Values of selected functions on the axes.

	$P_\ell(\cos\theta)$	$\dfrac{dP_\ell^1(\cos\theta)}{d\theta}$	$\dfrac{P_\ell^1(\cos\theta)}{\sin\theta}$
$\theta = 0$	1	$\dfrac{1}{2}\ell(\ell+1)$	$\dfrac{1}{2}\ell(\ell+1)$
$\theta = \pi$	$(-1)^\ell$	$\dfrac{1}{2}\ell(\ell+1)(-1)^\ell$	$\dfrac{1}{2}\ell(\ell+1)(-1)^{\ell+1}$
$\theta = \dfrac{\pi}{2}$	$i^\ell \dfrac{(\ell-1)!!}{\ell!!}\delta(2q,\ell)$	$i^\ell \dfrac{(\ell+1)!!}{(\ell-2)!!}\delta(2q,\ell)$	$i^{\ell-1}\dfrac{(\ell)!!}{(\ell-1)!!}\delta(2q+1,\ell)$

For even values of ℓ, the expansion may be written:

$$P_\ell(\chi) = \frac{1}{2^\ell(\ell)!} \frac{d^\ell}{d\chi^\ell} \sum_{s=0}^{\ell} \frac{(-1)^s \ell!}{s!(\ell-s)!} \chi^{2\ell-2s} \tag{A.12.8}$$

The binomial expansion is:

$$(\chi^2-1)^\ell = \sum_{s=0}^{\ell} \frac{(-1)^s \ell!}{s!(\ell-s)!} \chi^{2\ell-2s} \tag{A.12.9}$$

Combining results gives another expression for Legendre polynomials:

$$P_\ell(\chi) = \frac{1}{2^\ell (\ell)!} \frac{d^\ell}{d\chi^\ell} (\chi^2 - 1)^\ell \qquad (A.12.10)$$

Equation A.12.10 is the Rodriques formula for Legendre polynomials.

Comparison using Eq. A.11.27 and Eq. A.12.9 shows, for integer orders:

$$L_{2\ell}(\chi) = P_{2\ell}(\chi); \quad M_{2\ell+1}(\chi) = P_{2\ell+1}(\chi) \qquad (A.12.11)$$

The second integer-order solution to Eq. A.12.1 is commonly defined as:

$$Q_\nu(\chi) = \frac{\pi}{2\sin(\nu\pi)} \left[P_\nu(\chi)\cos(\nu\pi) - P_\nu(-\chi) \right] \qquad (A.12.12)$$

$Q_\nu(\chi)$ is a solution of the Legendre differential equation and, when ν is equal to integer ℓ, it is in indeterminate form. Differentiating numerator and denominator then letting ν become an integer gives:

$$
\begin{aligned}
Q_\ell(\chi) &= \lim_{\nu \to \ell} \frac{1}{2\cos(\nu\pi)} \left[-\pi\sin(\nu\pi)P_\nu(\chi) + \cos(\nu\pi)\frac{dP_\nu(\chi)}{d\nu} - \frac{dP_\nu(-\chi)}{d\nu} \right] \\
&= \frac{1}{2} \left\{ \frac{dP_\nu(\chi)}{d\nu} - \frac{dP_\nu(-\chi)}{d\nu} \right\}_{\nu=\ell}
\end{aligned}
$$

$$(A.12.13)$$

Using Eq. A.12.13, the functions at the lowest three orders are:

$$Q_0(\chi) = \ln\left[\cot(\theta/2)\right]; \quad Q_1(\chi) = (\cos\theta)\ln\left(\cot\frac{\theta}{2}\right) - 1$$

$$Q_2(\chi) = \frac{1}{2}(3\cos^2\theta - 1)\ln\left(\cot\frac{\theta}{2}\right) - \frac{3}{2}\cos\theta$$

$$(A.12.14)$$

Comparison with the noninteger functions, Eq. A.12.12, shows that:

$$Q_{2\ell}(\chi) = \frac{\partial M_v(\chi)}{\partial v}\bigg|_{v \Rightarrow 2\ell}; \quad Q_{2\ell+1}(\chi) = \frac{\partial L_v(\chi)}{\partial v}\bigg|_{v \Rightarrow 2\ell+1} \qquad (A.12.15)$$

In the remainder of this book we are concerned only with the zero-order function $Q_0(\chi)$.

A.13 Associated Legendre Functions

Associated Legendre functions are solutions for the extended case $m > 0$. The Legendre differential equation, Eq. 1.12.10, may be rewritten as:

$$\left(1-\chi^2\right)\frac{d^2\Theta}{d\chi^2} - 2\chi\frac{d\Theta}{d\chi} + \left[v(v+1) - \frac{m^2}{\left(1-\chi^2\right)}\right]\Theta = 0 \qquad (A.13.1)$$

Solutions are most easily obtained by starting with the $m = 0$ equation and differentiating m times to obtain:

$$\left(1-\chi^2\right)\frac{d^{m+2}\Theta}{d\chi^{m+2}} - 2(m+1)\chi\frac{d^{m+1}\Theta}{d\chi^{m+1}} + \left[v(v+1) - m(m+1)\right]\frac{d^m\Theta}{d\chi^m} = 0 \quad (A.13.2)$$

Introducing construction function $W(\theta)$ and solving for the first two derivatives:

$$\Theta(\theta) = W(\theta)\sin^m\theta$$

$$\frac{d\Theta}{d\chi} = \frac{dW}{d\chi}\sin^m\theta - mW\sin^{m-2}\theta\cos\theta$$

$$(A.13.3)$$

$$\frac{d^2\Theta}{d\chi^2} = \frac{d^2W}{d\chi^2}\sin^m\theta - 2m\frac{dW}{d\chi}\sin^{m-2}\theta\cos\theta$$

$$+ m(m-2)W\sin^{m-4}\theta\cos^2\theta - mW\sin^{m-2}\theta$$

Substituting Eq. A.13.3 into Eq. A.13.1 gives:

$$\frac{d^2W}{d\chi^2}\sin^2\theta - 2(m+1)\frac{dW}{d\chi}\cos\theta + \left[v(v+1)-m(m+1)\right]W = 0 \qquad \text{(A.13.4)}$$

For the special case where v is an integer, comparison of Eq. A.13.2 and A.13.4 shows that W is:

$$W_\ell(\theta) = \frac{d^m P_\ell(\cos\theta)}{d\chi^m} \qquad \text{(A.13.5)}$$

Since Eq. A.13.5 satisfies the associated Legendre differential equation, the solution of that equation is:

$$\Theta(\theta) = P_\ell^m(\cos\theta) = \sin^m\theta\,\frac{d^m P_\ell(\cos\theta)}{d\chi^m} \qquad \text{(A.13.6)}$$

The equality holds for all integer orders, ℓ, and degrees, m. Combining Eq. A.13.6 with Rodriques formula shows the extension to associated Legendre functions is:

$$P_\ell^m(\chi) = \frac{1}{(2\ell)!!}\left(1-\chi^2\right)^{m/2}\frac{d^{\ell+m}}{d\chi^{\ell+m}}\left(\chi^2-1\right)^\ell \qquad \text{(A.13.7)}$$

A.14 Orthogonality

Integral relationships for products of Legendre polynomials follow directly from differential equation Eq. A.13.1 for integer orders. Multiplying the differential equation by another Legendre polynomial of the same degree but unspecified order and then integrating yields:

$$\int_{-1}^{1} P_n^m(\chi)\left\{\frac{d}{d\chi}\left[\left(1-\chi^2\right)\frac{d^2 P_\ell^m}{d\chi^2}\right] + \left[\ell(\ell+1)-\frac{m^2}{\left(1-\chi^2\right)}\right]P_\ell^m\right\}d\chi = 0$$

$$\text{(A.14.1)}$$

To evaluate the left side, integrate the first term once by parts. The result is:

$$\int_{-1}^{1}\left\{\left(1-\chi^{2}\right)\frac{dP_{\ell}^{m}}{d\chi}\frac{dP_{n}^{m}}{d\chi}+\left[\ell(\ell+1)-\frac{m^{2}}{\left(1-\chi^{2}\right)}\right]P_{\ell}^{m}P_{n}^{m}\right\}d\chi=0 \qquad (A.14.2)$$

Next exchange positions of ℓ and n, repeat the process, and subtract the second integral from the first. The result is:

$$\left[\ell(\ell+1)-n(n+1)\right]\int_{-1}^{1}P_{\ell}^{m}(\chi)P_{n}^{m}(\chi)d\chi=0 \qquad (A.14.3)$$

The result shows that the associated Legendre polynomials form an orthogonal set.

To evaluate the integral with $\ell = n$, put I_1 equal to the integral:

$$I_{1}=\int_{-1}^{1}\left[P_{l}^{m}(\theta)\right]^{2}d\theta \qquad (A.14.4)$$

Combining Eq. A.13.7 and A.14.4:

$$I_{1}=\frac{(-1)^{m}}{\left[(2\ell)!!\right]^{2}}\int_{-1}^{1}d\chi\,\frac{d^{\ell+m}}{d\chi^{\ell+m}}\left(\chi^{2}-1\right)^{\ell}\left[\left(\chi^{2}-1\right)^{m}\frac{d^{\ell+m}}{d\chi^{\ell+m}}\left(\chi^{2}-1\right)^{\ell}\right] \qquad (A.14.5)$$

Integrating by parts $\ell + m$ times leaves:

$$I_{1}=\frac{1}{\left[(2\ell)!!\right]^{2}}\int_{-1}^{1}d\chi\left\{(1-\chi^{2})^{\ell}\frac{d^{\ell+m}}{dx^{\ell+m}}\left[(\chi^{2}-1)^{m}\frac{d^{\ell+m}}{dx^{\ell+m}}(\chi^{2}-1)^{\ell}\right]\right\} \qquad (A.14.6)$$

Only the highest power of χ survives the indicated differentiation operations:

$$\frac{d^{\ell+m}}{d\chi^{\ell+m}}\left[(\chi^{2}-1)^{m}\frac{d^{\ell+m}}{d\chi^{\ell+m}}(\chi^{2}-1)^{\ell}\right]=\frac{d^{\ell+m}}{d\chi^{\ell+m}}\left[\chi^{2m}\frac{d^{\ell+m}}{d\chi^{\ell+m}}\chi^{2\ell}\right]$$

$$=\frac{(2\ell)!}{(\ell-m)!}\frac{d^{\ell+m}}{d\chi^{\ell+m}}\chi^{\ell+m}=\frac{(2\ell)!(\ell+m)!}{(\ell-m)!} \qquad (A.14.7)$$

Combining Eq. A.14.6 and A.14.7 leaves:

$$I_1 = \frac{(2\ell)!(\ell+m)!}{\left[(2\ell)!!\right]^2(\ell-m)!}\int_{-1}^{1}d\chi\left(1-\chi^2\right)^{\ell} \tag{A.14.8}$$

Using the binomial expansion, Eq. A.12.9, and integrating gives:

$$I_1 = \frac{2(-1)^{\ell}(2\ell-1)!!(\ell+m)!}{(2\ell)!!(\ell-m)!}\sum_{s=0}^{\ell}\frac{(-1)^s\,\ell!}{s!(\ell-s)!(2\ell-2s+1)} \tag{A.14.9}$$

The sum may be written in closed form as:

$$I_1 = \frac{2}{(2\ell+1)}\frac{(\ell+m)!}{(\ell-m)!} \tag{A.14.10}$$

Combining Eq. A.14.3 and A.14.10 gives the orthogonality relationship for associated Legendre polynomials:

$$I_1 = \int_{-1}^{1}P_{\ell}^{m}(\chi)P_{n}^{m}(\chi)d\chi = \frac{2}{(2\ell+1)}\frac{(\ell+m)!}{(\ell-m)!}\delta(\ell,n) \tag{A.14.11}$$

A similar integral, the value of which follows after a slight extension of the above, is:

$$I_2 = \int_{-1}^{1}P_{\ell}^{m}(\chi)P_{\ell}^{m+2s}(\chi)d\chi = (-1)^s\frac{2}{(2\ell+1)}\frac{(\ell+m)!}{(\ell-m-2s)!} \tag{A.14.12}$$

A.15 Recursion Relationships

It is helpful to compile a table of identities involving associated Legendre polynomials. A convenient starting point for determining the recursion relationships is Eq. A.14.7. For the case that ℓ is odd and the upper limit is $(\ell-1)/2$. Substituting $p = (\ell-1-2s)/2$ yields:

$$P_\ell(\chi) = \frac{\chi(-1)^{(\ell-1)/2}}{2^\ell} \sum_{p=0}^{(\ell-1)/2} \frac{(-1)^p (\ell+1+2p)!}{(2p+1)!\left(\frac{\ell-1-2p}{2}\right)!\left(\frac{\ell+1+2p}{2}\right)!} \chi^{\ell-2s} \qquad (A.15.1)$$

Rewrite the equation as:

$$P_\ell(\chi) = (-1)^{(\ell-1)/2} \frac{(\ell)!!}{(\ell-1)!!} \sum_{s=0}^{(\ell-1)/2} \frac{(-1)^s \chi^{2s+1} (\ell+2s)!!(\ell-1)!!}{(2s+1)!(\ell)!!(\ell-1-2s)!!} \qquad (A.15.2)$$

For the case of ℓ even, the upper limit is $\ell/2$. In a similar way the equation goes to:

$$P_\ell(\chi) = (-1)^{\ell/2} \frac{(\ell-1)!!}{(\ell)!!} \sum_{s=0}^{\ell/2} \frac{(-1)^s \chi^{2s} (\ell-1+2s)!!(\ell)!!}{(2s)!(\ell-1)!!(\ell-2s)!!} \qquad (A.15.3)$$

With Eq. A.15.3, replace ℓ by $(\ell+1)$ and write out the first few terms, then repeat with ℓ replaced by $(\ell-1)$. The resulting series are:

$$P_{\ell+1}(\chi) = (-1)^{(\ell+1)/2} \frac{(\ell)!!}{(\ell+1)!!} \left\{ 1 - \frac{\chi^2}{2!} \frac{(\ell+2)!!}{(\ell)!!} \frac{(\ell+1)!!}{(\ell-1)!!} + \frac{\chi^4}{4!} \frac{(\ell+4)!!}{(\ell)!!} \frac{(\ell+1)!!}{(\ell-3)!!} - \cdots \right\}$$

$$P_{\ell-1}(\chi) = (-1)^{(\ell-1)/2} \frac{(\ell-2)!!}{(\ell-1)!!} \left\{ 1 - \frac{\chi^2}{2!} \frac{(\ell)!!}{(\ell-2)!!} \frac{(\ell-1)!!}{(\ell-3)!!} + \frac{\chi^4}{4!} \frac{(\ell+2)!!}{(\ell-2)!!} \frac{(\ell-1)!!}{(\ell-5)!!} - \cdots \right\}$$

These expressions combine to form the indicated sum:

$$\ell P_{\ell-1}(\chi) + (\ell+1) P_{\ell+1}(\chi) = (-1)^{(\ell-1)/2} \frac{(2\ell+1)(\ell)!!}{(\ell-1)!!)} \left\{ \frac{\chi^2}{1!} - \frac{\chi^4}{3!}(\ell+2)(\ell-1) + \cdots \right\}$$

$$= (-1)^{(\ell-1)/2} \frac{(2\ell+1)(\ell)!!}{(\ell-1)!!)} \sum_{s=0}^{(\ell-1)/2} \frac{(-1)^s \chi^{2s+2} (\ell+2s)!!(\ell-1)!!}{(2s+1)(\ell)!!(\ell-1-2s)!!} \qquad (A.15.4)$$

Combining the above gives:

$$(2\ell+1)\chi P_\ell(\chi) = \ell P_{\ell-1}(\chi) + (\ell+1) P_{\ell+1}(\chi) \qquad (A.15.5)$$

Proofs for even values of ℓ follow in a parallel way and give the same result. Equation A.15.5 is the first recursion relationship.

The same technique with the indicated operations results in the second recursion relationship:

$$(2\ell+1)P_\ell(\chi) = \frac{dP_{\ell+1}(\chi)}{d\chi} - \frac{dP_{\ell-1}(\chi)}{d\chi} \tag{A.15.6}$$

The integral expression of Eq. A.15.7 follows from Eq. A.15.6:

$$\int P_\ell(\chi)d\chi = \left[\frac{P_{\ell+1}(\chi)-P_{\ell-1}(\chi)}{(2\ell+1)}\right] \tag{A.15.7}$$

Differentiating Eq. A.15.5 by χ and adding $\ell \times$ Eq. 15.6 gives:

$$(\ell+1)P_\ell(\chi) = \frac{dP_{\ell+1}(\chi)}{d\chi} - \chi\frac{dP_\ell(\chi)}{d\chi} \tag{A.15.8}$$

Differentiating Eq. A.15.8 m times, multiplying through by $\sin^{m+1}\theta$, and using Eq. A.15.7 gives:

$$P_{\ell+1}^{m+1}(\chi) = \chi P_\ell^{m+1}(\chi) + (\ell+m+1)\sin\theta P_\ell^m(\chi) \tag{A.15.9}$$

A series of identities follows by mixing and matching. A selection is listed in Table A.15.1.

Associated Legendre functions have even or odd parity, respectively, if the sum $(\ell + m)$ is even or odd:

$$P_\ell^m(\chi) = (-1)^{\ell+m} P_\ell^m(-\chi) \tag{A.15.10}$$

Let delta be a Kronecker delta and let q equal any of the field of positive integers, including zero. At $\theta = \pi/2$, the equator, functional values are:

$$P_\ell^m(0) = (-1)^{(\ell-m)/2}\frac{(\ell+m+1)!}{(\ell-m)!}\delta(2q,\ell+m) \tag{A.15.11}$$

$$\frac{d^r}{d\theta^r}P_\ell^m(\cos\theta)\Big|_{\frac{\pi}{2}} = P_\ell^{m+r}(0) \tag{A.15.12}$$

Table A.15.1. A table of identities for Legendre functions.

1.
$$\frac{dP_\ell^m}{d\theta} = \frac{1}{2}[(\ell+m)(\ell-m+1)P_\ell^{m-1} - P_\ell^{m+1}]$$

2.
$$\frac{mP_\ell^m}{\sin\theta} = \frac{1}{2}[(\ell+m)(\ell-m+1)P_{\ell-1}^{m-1} + P_{\ell-1}^{m+1}]$$

3.
$$\sin\theta\frac{dP_\ell^m}{d\theta} = \frac{1}{2\ell+1}[\ell(\ell-m+1)P_{\ell+1}^m - (\ell+m)(\ell+1)P_{\ell-1}^m]$$

4.
$$\cos\theta P_\ell^m = \frac{1}{2\ell+1}[(\ell-m+1)P_{\ell+1}^m + (\ell+m)P_{\ell-1}^m]$$

5.
$$P_\ell^{m+1} = 2m\cot\theta P_\ell^m - (\ell+m)(\ell-m+1)P_\ell^{m-1}$$

6.
$$(\ell-m+1)P_{\ell+1}^m = (2\ell+1)\cos\theta P_\ell^m - (\ell+m)P_{\ell-1}^m$$

7.
$$(2\ell+1)\sin\theta P_\ell^m = P_{\ell+1}^{m+1} - P_{\ell-1}^{m+1}$$

8.
$$P_{\ell+1}^{m+1} = \cos\theta P_\ell^{m+1} + (\ell+m+1)\sin\theta P_\ell^m$$

9.
$$P_{\ell-1}^{m+1} = \cos\theta P_\ell^{m+1} - (\ell-m)\sin\theta P_\ell^m$$

10.
$$\frac{dP_\ell^m}{d\theta} = m\cot\theta P_\ell^m - P_\ell^{m+1}$$

11.
$$\frac{dP_\ell^m}{d\theta} = -m\cot\theta P_\ell^m + (\ell+m)(\ell-m+1)P_\ell^{m-1}$$

In the tables to follow, order is ℓ, degree is m, and $\chi = \cos\theta$. Polynomials are even or odd functions of χ as $(\ell+m)$ is even or odd; absence of a superscript indicates degree 0. Normalization is chosen to make $P_\ell(1) = 1$.

Table A.15.2 and A.15.3 contain values of associated Legendre polynomials. Useful recursion relationships for constructing the tables are:

$$P_\ell^m(\chi) = \frac{1}{\ell-m}\left[(2\ell-1)\chi P_{\ell-1}^m(\chi) - (\ell-1+m)P_{\ell-2}^m(\chi)\right]$$

$$P_\ell^m(\chi) = \left[2(m-1)\cot\theta P_\ell^{m-1}(\chi) - (\ell-1+m)(\ell+2-m)P_\ell^{m-2}(\chi)\right]$$

$$P_\ell^1(\chi) = -\frac{dP_\ell}{d\theta}$$

Table A.15.2. Table of Legendre polynomials, $P_\ell(\chi)$, $\chi = \cos\theta$.

ℓ	$m = 0$, Values for $P_\ell(\chi)$
0	1
1	χ
2	$\dfrac{1}{2}\left(3\chi^2 - 1\right)$
3	$\dfrac{\chi}{2}\left(5\chi^2 - 3\right)$
4	$\dfrac{1}{8}\left(35\chi^4 - 30\chi^2 + 3\right)$
5	$\dfrac{\chi}{8}\left(63\chi^4 - 70\chi^2 + 15\right)$
6	$\dfrac{1}{16}\left(231\chi^6 - 315\chi^4 + 105\chi^2 - 5\right)$
7	$\dfrac{\chi}{16}\left(429\chi^6 - 693\chi^4 + 315\chi^2 - 35\right)$

Table A.15.3. Table of associated Legendre Polynomials $P_\ell^{\,1}(\chi)$), $\chi = \cos\theta$.

ℓ	Values for $P_\ell^1(\chi)$
0	
1	$\sin\theta$
2	$3\chi \sin\theta$
3	$\dfrac{3}{2}\left(5\chi^2 - 1\right)\sin\theta$
4	$\dfrac{5\chi}{2}\left(7\chi^2 - 3\right)\sin\theta$
5	$\dfrac{15}{8}\left(21\chi^4 - 14\chi^2 + 1\right)\sin\theta$
6	$\dfrac{21\chi}{8}\left(33\chi^4 - 30\chi^2 + 5\right)\sin\theta$
7	$\dfrac{7}{16}\left(429\chi^6 - 495\chi^4 + 135\chi^2 - 5\right)\sin\theta$

The sum of all coefficients in the expansion for $P_\ell^m(\chi)$ is $(\ell+m)!/2^m(m)!(\ell-m)!$. To normalize, multiply each associated Legendre function by $[(2\ell+1)(\ell-m)!/2(\ell+m)!]^{1/2}$.

Table A.15.4. Table of spherical angular function, $\chi = \cos\theta$.

ℓ	$m = 1, \quad dP_\ell^1/d\theta = -\cot\theta P_\ell^1 + \ell(\ell+1)P_\ell$
1	χ
2	$3(2\chi^2 - 1)$
3	$\dfrac{3}{2}\chi(15\chi^2 - 11)$
4	$\dfrac{5}{2}(28\chi^4 - 27\chi^2 + 3)$
5	$\dfrac{15\chi}{8}(105\chi^4 - 126\chi^2 + 29)$
6	$\dfrac{21}{8}(198\chi^6 - 285\chi^4 + 100\chi^2 - 5)$
7	$\dfrac{7\chi}{16}(3003\chi^6 - 5049\chi^4 + 2385\chi^2 - 275)$

Table A.15.4 contains values of spherical angular function $dP_\ell^1/d\theta$, based upon the relationship:

$$dP_\ell^1/d\theta = -\cot\theta P_\ell^1 + \ell(\ell+1)P_\ell = \cot\theta P_\ell^1 - P_\ell^2$$

A.16 Integrals of Legendre Functions

Solutions of problems within this book require integrals of several functional combinations of Legendre polynomials. One is the integral:

$$I_3 = m\int_0^\pi \frac{d}{d\theta}\left(P_\ell^m P_n^m\right)d\theta = 0 \tag{A.16.1}$$

The equality follows since the integrand is a perfect differential and $P_\ell^m(\pm1) = 0$ for $m > 0$.

Consider the integral:

$$I_4 = \int_0^{\pi} \left\{ \frac{dP_\ell^m}{d\theta} \frac{dP_n^m}{d\theta} + \frac{m^2 P_\ell^m P_n^m}{\sin^2 \theta} \right\} \sin \theta d\theta \qquad (A.16.2)$$

The first term in the integrand may be written:

$$\frac{dP_\ell^m}{d\theta} \frac{dP_n^m}{d\theta} = \frac{1}{\sin \theta} \frac{d}{d\theta} \left[\sin \theta P_\ell^m \frac{dP_n^m}{d\theta} \right] - \frac{P_\ell^m}{\sin \theta} \frac{d}{d\theta} \left[\sin \theta \frac{dP_n^m}{d\theta} \right]$$

The second term in the integrand, after using the differential equation, may be written:

$$\frac{m^2 P_\ell^m P_n^m}{\sin^2 \theta} = \frac{P_\ell^m}{\sin \theta} \frac{d}{d\theta} \left[\sin \theta \frac{dP_n^m}{d\theta} \right] + n(n+1) P_n^m P_\ell^m$$

Combining:

$$I_4 = \int_0^{\pi} \sin \theta d\theta \left\{ \frac{1}{\sin \theta} \frac{d}{d\theta} \left[\sin \theta P_\ell^m \frac{dP_n^m}{d\theta} \right] + n(n+1) P_\ell^m P_n^m \right\} \qquad (A.16.3)$$

The first term is an exact differential that integrates to zero, leaving:

$$I_4 = n(n+1) \int_0^{\pi} \sin \theta d\theta P_\ell^m P_n^m = \frac{2\ell(\ell+1)(\ell+m)!}{(2\ell+1)(\ell-m)!} \delta(\ell,n) \qquad (A.16.4)$$

Consider the integral

$$I_5 = \int_0^{\pi} \left\{ \frac{dP_\ell^m}{d\theta} \frac{dP_n^m}{d\theta} + \frac{m^2 P_\ell^m P_n^m}{\sin^2 \theta} \right\} \cos \theta \sin \theta d\theta \qquad (A.16.5)$$

The procedure is similar to that for I_4. Replace the first term using the differential equation, then sum, and then partially integrate once:

$$I_5 = \int_0^\pi \left\{ \left[\sin\theta P_\ell^m \frac{dP_n^m}{d\theta} \right] + n(n+1)\cos\theta P_\ell^m P_n^m \right\} \sin\theta \, d\theta$$

Combining with recursion relationships Table A.15.1.3 and A.15.1.4:

$$I_5 = \int_0^\pi \left\{ \frac{n(n-m+1)(n+2)}{(2n+1)} P_{n+1}^m + \frac{(n+m)(n^2-1)}{(2n+1)} P_{n-1}^m \right\} P_\ell^m \sin\theta \, d\theta$$

Evaluation using Eq. A.16.11 gives:

$$I_5 = \left\{ \begin{array}{l} \dfrac{2n(n+2)(n+m+1)!}{(2n+1)(2n+3)(n-m)!} \delta(\ell, n+1) \\[12pt] + \dfrac{(n^2-1)(n+m)!}{(2n-1)(2n+1)(n-m-1)!} \delta(\ell, n-1) \end{array} \right\} \tag{A.16.6}$$

The next integral is:

$$I_6 = \int_0^\pi \frac{d}{d\theta} \left(P_\ell^m P_n^m \right) \sin^2\theta \, d\theta \tag{A.16.7}$$

Expanding the differential and using recursion relationship Table A.15.1.3 gives:

$$I_6 = \int_0^\pi \left\{ \begin{array}{l} P_\ell^m \left[\dfrac{n(n-m+1)}{(2n+1)} P_{n+1}^m - \dfrac{(n+m)(n-1)}{(2n+1)} P_{n-1}^m \right] \\[12pt] + P_n^m \left[\dfrac{(\ell-m+1)}{(2\ell+1)} P_{\ell+1}^m - \dfrac{(\ell+m)(\ell-1)}{(2\ell+1)} P_{\ell-1}^m \right] \end{array} \right\} \sin\theta \, d\theta$$

Term-by-term evaluation shows that:

$$I_6 = 0 \tag{A.16.8}$$

The next integral of interest is:

$$I_7 = \int_0^\pi \left\{ \frac{dP_\ell^m}{d\theta} \frac{dP_n^{m+1}}{d\theta} + \frac{m(m+1)}{\sin^2\theta} P_\ell^m P_n^{m+1} \right\} \sin\theta\, d\theta \qquad (A.16.9)$$

Substituting the differential equation into the first term, summing, and taking one partial integration gives:

$$\int_0^\pi \left\{ -\cos\theta \frac{dP_\ell^m}{d\theta} + \left[\ell(\ell+1)\sin\theta + \frac{m}{\sin\theta} \right] P_\ell^m \right\} \sin\theta\, d\theta$$

After using the recursion relationship of Table A.15.1.10 on the first term then summing the curly bracket becomes:

$$\left\{ \left[\ell(\ell+1) + m \right] \sin\theta P_\ell^m + \cos\theta P_\ell^{m+1} \right\}$$

With the use of the recursion relationships of Tables A.15.1.4 and A.15.1.7 the bracket becomes:

$$\left\{ \frac{\ell(\ell+2)}{(2\ell+1)} P_{\ell+1}^{m+1} - \frac{(\ell-1)(\ell+1)}{(2\ell+1)} P_{\ell-1}^{m+1} \right\}$$

Putting the bracket back under the integral sign and integrating gives:

$$\begin{aligned}
I_7 &= \frac{2\ell(\ell+2)(\ell+m+2)!}{(2\ell+1)(2\ell+3)(\ell-m)!} \delta(n, \ell+1) \\
&\quad - \frac{2(\ell-1)(\ell+1)(\ell+m)!}{(2\ell-1)(2\ell+1)(\ell-m-2)!} \delta(n, \ell-1)
\end{aligned} \qquad (A.16.10)$$

The next integral of interest is:

$$I_8 = \int_0^\pi \left\{ (m+1) P_\ell^{m+1} \frac{dP_n^m}{d\theta} + m P_n^m \frac{dP_\ell^{m+1}}{d\theta} \right\} \sin\theta\, d\theta \qquad (A.16.11)$$

This may be rewritten as:

$$\int_0^\pi \left\{ \sin\theta P_\ell^{m+1} \frac{dP_n^m}{d\theta} + m\sin\theta \frac{d}{d\theta}\left[P_n^m P_\ell^{m+1} \right] \right\} d\theta$$

Integrating the perfect differential by parts gives:

$$\int_0^\pi P_\ell^{m+1}\left(\frac{dP_n^m}{d\theta} - m\cot\theta P_n^m \right)\sin\theta\, d\theta = -\int_0^\pi P_\ell^{m+1} P_n^{m+1} \sin\theta\, d\theta$$

The second equality is in the proper form to use Table A.15.1.10. The result is:

$$I_8 = -\frac{2}{(2\ell+1)}\frac{(\ell+m+1)!}{(\ell-m-1)!}\delta(\ell,n) \tag{A.16.12}$$

Similarly, using Table A.15.1.11, integrals with $(m-1)$ replacing $(m+1)$ may be evaluated, and are listed in Table A.16.1.

Table A.16.1. Table of integrals of Legendre polynomials.

1. $$\int_0^\pi P_\ell^m P_n^m \sin\theta\, d\theta = \frac{2(\ell+m)!}{(2\ell+1)(\ell-m)!}\delta(\ell,n)$$

2. $$\int_0^\pi P_\ell^m P_\ell^{m+2s} \sin\theta\, d\theta = \frac{2(-1)^s}{(2\ell+1)}\frac{(\ell+m)!}{(\ell-m-2s)!}$$

3. $$m\int_0^\pi \frac{d}{d\theta}\left(P_\ell^m P_n^m \right) d\theta = 0$$

4. $$m\int_0^\pi \frac{d}{d\theta}\left(P_\ell^m P_n^m \right)\cos\theta\, d\theta = \frac{2m(\ell+m)!}{(2\ell+1)(\ell-m)!}\delta(\ell,n)$$

5. $$\int_0^\pi \frac{d}{d\theta}\left(P_\ell^m P_n^m \right)\sin^2\theta\, d\theta = 0$$

(*Continued*)

Table A.16.1. (*Continued*)

6.
$$\int_0^\pi \left[\frac{dP_\ell^m}{d\theta}\frac{dP_n^m}{d\theta} + \frac{m^2 P_\ell^m P_n^m}{\sin^2\theta} \right]\sin\theta\, d\theta = \frac{2\ell(\ell+1)(\ell+m)!}{(2\ell+1)(\ell-m)!}\delta(\ell,n)$$

7.
$$\int_0^\pi \left[\frac{dP_\ell^m}{d\theta}\frac{dP_n^m}{d\theta} + \frac{m^2 P_\ell^m P_n^m}{\sin^2\theta} \right]\cos\theta\sin\theta\, d\theta$$

$$= \left\{ \begin{array}{l} \dfrac{2\ell(\ell+2)(\ell+m+1)!}{(2\ell+1)(2\ell+3)(\ell-m)!}\delta(n,\ell+1) \\[1.5em] +\dfrac{2(\ell-1)(\ell+1)(\ell+m)!}{(2\ell-1)(2\ell+1)(\ell-m-1)!}\delta(n,\ell-1) \end{array} \right\}$$

8.
$$\int_0^\pi \left[\frac{dP_\ell^m}{d\theta}\frac{dP_n^{m+1}}{d\theta} + \frac{m(m+1) P_\ell^m P_n^{m+1}}{\sin^2\theta} \right]\sin^2\theta\, d\theta$$

$$= \left\{ \begin{array}{l} \dfrac{2\ell(\ell+2)(\ell+m+2)!}{(2\ell+1)(2\ell+3)(\ell-m)!}\delta(n,\ell+1) \\[1.5em] -\dfrac{2(\ell-1)(\ell+1)(\ell+m)!}{(2\ell-1)(2\ell+1)(\ell-m-2)!}\delta(n,\ell-1) \end{array} \right\}$$

9.
$$\int_0^\pi \left[\frac{dP_\ell^m}{d\theta}\frac{dP_n^{m-1}}{d\theta} + \frac{m(m-1) P_\ell^m P_n^{m-1}}{\sin^2\theta} \right]\sin^2\theta\, d\theta$$

$$= \left\{ \begin{array}{l} -\dfrac{2\ell(\ell+2)(\ell+m)!}{(2\ell+1)(2\ell+3)(\ell-m)!}\delta(\ell,n+1) \\[1.5em] +\dfrac{2(\ell-1)(\ell+1)(\ell+m)!}{(2\ell-1)(2\ell+1)(\ell-m)!}\delta(\ell,n-1) \end{array} \right\}$$

10.
$$\int_0^\pi P_\ell^m(\cos\theta)\sin^{\ell+1}\theta\, d\theta = (-1)^{(\ell-m)/2}\frac{2^{\ell+1}(\ell)!(\ell+m)!}{(2\ell+1)!}\delta(\ell+m,2q)$$

11.
$$\int_0^\pi \left((m+1)P_\ell^{m+1}\frac{dP_n^m}{d\theta} + mP_n^m\frac{dP_\ell^{m+1}}{d\theta} \right)\sin\theta\, d\theta$$

$$= -\frac{2}{(2\ell+1)}\frac{(\ell+m+1)!}{(\ell-m-1)!}\delta(\ell,n)$$

12.
$$\int_0^\pi \left\{ (m-1)P_\ell^{m-1}\frac{dP_n^m}{d\theta} + mP_n^m\frac{dP_\ell^{m-1}}{d\theta} \right\}\sin\theta\, d\theta = -\frac{2}{(2\ell+1)}\frac{(\ell+m)!}{(\ell-m)!}\delta(\ell,n)$$

A.17 Integrals of Fractional Order Legendre Functions

The Legendre differential equation of fractional order is:

$$\frac{1}{\sin\theta}\frac{d}{d\theta}\left(\sin\theta\frac{d\Theta_v^m(\cos\theta)}{d\theta}\right)+\left(v(v+1)-\frac{m^2}{\sin^2\theta}\right)\Theta_v^m=0 \qquad (A.17.1)$$

Functions $\Theta_v^m(\cos\theta)$ represents $L_v^m(\cos\theta)$, $M_v^m(\cos\theta)$, or any linear combination thereof. Useful boundary conditions are:

$$\qquad (A.17.2)$$

$$M_v^m(\cos\theta)\Big|_{\theta=\psi}=0 \quad \text{and} \quad \frac{dL_v^m(\cos\theta)}{d\theta}\Big|_{\theta=\psi}=0$$

This evaluation of integrals over noninteger order Legendre functions includes the boundary conditions of Eq. A.17.2.

Consider the integral

$$I_9=\int_\psi^{\pi-\psi}\left(\frac{dM_v^m}{d\theta}\frac{dP_n^m}{d\theta}+\frac{m^2M_v^mP_n^m}{\sin^2\theta}\right)\sin\theta d\theta \qquad (A.17.3)$$

The evaluation procedure is to use the differential equation and rewrite the first term as:

$$\frac{dM_v^m}{d\theta}\frac{dP_n^m}{d\theta}=\frac{1}{\sin\theta}\frac{d}{d\theta}\left(\sin\theta M_v^m\frac{dP_n^m}{d\theta}\right)-\frac{M_v^m}{\sin\theta}\frac{d}{d\theta}\left(\sin\theta\frac{dP_n^m}{d\theta}\right) \qquad (A.17.4)$$

The first term on the right side of Eq. A.17.4 forms a perfect differential and, after imposing Eq. A.17.2, the integral of that differential is equal to zero. With the differential equation substituted into the remaining term, the result is:

$$I_9=n(n+1)\int_\psi^{\pi-\psi}M_v^mP_n^m\sin\theta d\theta \qquad (A.17.5)$$

To evaluate Eq. A.17.5, since $M_v^m(\cos\theta)$ has odd parity and $P_v^m(\cos\theta)$ is even or odd as $(v+m)$ is even or odd, the integral vanishes if $(v+m)$ is even. If $(v+m)$ is odd, repeat the procedure used in Eq. A.16.1 through A.16.3. The result is:

$$I_9 = \int_{\psi}^{\pi-\psi} M_v^m P_n^m \sin\theta d\theta = \frac{\sin\theta\left[P_n^m \, dM_v^m/d\theta - M_v^m \, dP_n^m/d\theta\right]_{\psi}^{\pi-\psi}}{n(n+1)-v(v+1)}$$

After imposing the boundary condition of Eq. A.17.2:

$$\int_{\psi}^{\pi-\psi} M_v^m P_n^m \sin\theta d\theta = 2\sin\psi \, \frac{P_n^m(\cos\psi)\dfrac{dM_v^m(\cos\psi)}{d\theta}}{v(v+1)-n(n+1)} \delta(n,2q+1) \qquad \text{(A.17.6)}$$

The integer 'q' represents any positive integer, including zero.

The next integral to be evaluated is:

$$I_{10} = \int_{\psi}^{\pi-\psi} \sin\theta d\theta \left(\frac{dL_\mu^m}{d\theta}\frac{dP_n^m}{d\theta} + \frac{m^2 L_\mu^m P_n^m}{\sin^2\theta}\right) \qquad \text{(A.17.7)}$$

The same technique that was applied to Eq. A.17.2 applied to Eq. A.17.7 results in:

$$I_{10} = \mu(\mu+1)\int_{\psi}^{\pi-\psi} L_\mu^m P_n^m \sin\theta d\theta \qquad \text{(A.17.8)}$$

The same technique applied to Eq. A.17.4 results in:

$$I_{10} = \int_{\psi}^{\pi-\psi} L_\mu^m P_n^m \sin\theta d\theta = 2\sin\psi \, \frac{L_\mu^m(\cos\psi)\dfrac{dP_n^m(\cos\psi)}{d\theta}}{n(n+1)-\mu(\mu+1)} \delta(n,2q) \qquad \text{(A.17.9)}$$

It follows from the parity of the functions that the integral:

$$I_{11} = \int_{\psi}^{\pi-\psi} \sin\theta d\theta \left(\frac{dM_v^m}{d\theta}\frac{dL_\mu^m}{d\theta} + \frac{m^2 M_v^m L_\mu^m}{\sin^2\theta}\right) = 0 \qquad \text{(A.17.10)}$$

The next integral of interest is:

$$I_{12} = \int_{\psi}^{\pi-\psi} \sin\theta d\theta \left(\frac{dM_\nu^m}{d\theta}\frac{dM_\mu^m}{d\theta} + \frac{m^2 M_\nu^m M_\mu^m}{\sin^2\theta} \right) \tag{A.17.11}$$

The technique used to evaluate the previous integrals when applied to Eq. A.17.2 gives:

$$I_{12} = \nu(\nu+1) \int_{\psi}^{\pi-\psi} \sin\theta d\theta M_\nu^m M_\mu^m \tag{A.17.12}$$

Applying the technique to Eq. A.17.4 gives:

$$\int_{\psi}^{\pi-\psi} M_\nu^m M_\mu^m \sin\theta d\theta = \frac{\sin\theta \left[M_\mu^m \dfrac{dM_\nu^m}{d\theta} - M_\nu^m \dfrac{dM_\mu^m}{d\theta} \right]_{\psi}^{\pi-\psi}}{\mu(\mu+1)-(\nu+1)}$$

The boundary condition shows that the result is zero unless $\mu(\mu+1) = \nu(\nu+1)$. For that case, evaluating the indeterminate form gives:

$$I_{12} = \int_{\psi}^{\pi-\psi} M_\nu^m M_\mu^m \sin\theta d\theta = \frac{2\sin\psi}{2\nu+1}\left[\frac{\partial M_\nu^m(\cos\psi)}{\partial\nu} \frac{\partial M_\nu^m(\cos\psi)}{\partial\theta} \right]\delta(\nu,\mu) \tag{A.17.13}$$

The delta function indicates a Kronecker delta function with a noninteger argument.

Consider the integral

$$I_{13} = \int_{\psi}^{\pi-\psi} \sin\theta d\theta \left(\frac{dL_\nu^m}{d\theta}\frac{dL_\mu^m}{d\theta} + \frac{m^2 dL_\nu^m dL_\mu^m}{\sin^2\theta} \right) \tag{A.17.14}$$

In a way similar to the earlier integrals:

$$I_{13} = \mu(\mu+1) \int_{\psi}^{\pi-\psi} \sin\theta d\theta L_\nu^m L_\mu^m \tag{A.17.15}$$

$$I_{13} = \int_{\psi}^{\pi-\psi} L_v^m L_\mu^m \sin\theta d\theta = -\frac{2\sin\psi}{2\mu+1}\left[L_\mu^m(\cos\psi)\frac{\partial^2 L_\mu^m(\cos\psi)}{\partial v\partial\theta}\right]\delta(v,\mu)$$

$$(A.17.16)$$

Values are calculated and tabulated in Table A.17.1.

Table A.17.1. Table of integrals, noninteger order Legendre functions.

1

$$I_{vn} = \int_{-\psi}^{\psi} M_v^m P_n^m \sin\theta d\theta = -2\sin\psi\left\{ \frac{P_n^m(\cos\psi)\dfrac{dM_v^m(\cos\psi)}{d\theta}}{n(n+1)-v(v+1)}\right\}\delta(m+n,2q+1)$$

2

$$\int_{-\psi}^{\psi}\left(\frac{dM_v^m}{d\theta}\frac{dP_n^m}{d\theta}+\frac{m^2 M_v^m P_n^m}{\sin^2\theta}\right)\sin\theta d\theta = n(n+1)I_{vn}$$

3

$$\int_{\psi}^{\pi-\psi} L_\mu^m P_n^m \sin\theta d\theta = 2\sin\psi\left\{ \frac{L_\mu^m(\cos\psi)\dfrac{dP_n^m(\cos\psi)}{d\theta}}{n(n+1)-\mu(\mu+1)}\right\}\delta(n,2q)$$

4

$$\int_{\psi}^{\pi-\psi}\left(\frac{dL_\mu^m}{d\theta}\frac{dP_n^m}{d\theta}+\frac{m^2 L_\mu^m P_n^m}{\sin^2\theta}\right)\sin\theta d\theta = \mu(\mu+1)I_{\mu n}$$

5

$$\int_{\psi}^{\pi-\psi}\left(\frac{dM_v^m}{d\theta}\frac{dL_\mu^m}{d\theta}+\frac{m^2 M_v^m L_\mu^m}{\sin^2\theta}\right)\sin\theta d\theta = 0$$

6

$$\int_{\psi}^{\pi-\psi} M_v^m M_\mu^m \sin\theta d\theta = \frac{2\sin\psi}{2v+1}\left[\frac{\partial M_v^m(\cos\psi)}{\partial v}\frac{\partial M_v^m(\cos\psi)}{\partial\theta}\right]\delta(v,\mu)$$

7

$$\int_{\psi}^{\pi-\psi}\left(\frac{dM_v^m}{d\theta}\frac{dM_\mu^m}{d\theta}+\frac{m^2 M_v^m M_\mu^m}{\sin^2\theta}\right)\sin\theta d\theta = v(v+1)I_{vv}$$

8

$$\int_{\psi}^{\pi-\psi} L_v^m L_\mu^m \sin\theta d\theta = -\frac{2\sin\psi}{2\mu+1}\left[L_\mu^m(\cos\psi)\frac{\partial^2 L_\mu^m(\cos\psi)}{\partial v\partial\theta}\right]\delta(v,\mu)$$

9

$$\int_{\psi}^{\pi-\psi}\left(\frac{dL_v^m}{d\theta}\frac{dL_\mu^m}{d\theta}+\frac{m^2 L_v^m L_\mu^m}{\sin^2\theta}\right)\sin\theta d\theta = \mu(\mu+1)K_{\mu\mu}$$

A.18 Hankel Functions, The First Solution Form

The radial differential equation Eq. 1.11.7 does not contain separation parameter m and, therefore, neither do its solutions. The spherical Bessel differential equation with separation parameter v is:

$$\frac{1}{\sigma^2}\frac{d}{d\sigma}\left(\sigma^2\frac{dR}{d\sigma}\right)+\left(1-\frac{v(v+1)}{\sigma^2}\right)R=0 \tag{A.18.1}$$

Solutions valid for all points $r>0$ follow using a power series expansion. The series and its first two derivatives are:

$$R_v(\sigma)=\sum_{s=0}^{\infty}a_s\sigma^{s+p};\quad \frac{dR_v(\sigma)}{d\sigma}=\sum_{s=0}^{\infty}(s+p)a_s\sigma^{s+p-1}$$

$$\frac{d^2R_v(\sigma)}{d\sigma^2}=\sum_{s=0}^{\infty}(s+p)(s+p-1)a_s\sigma^{s+p-2} \tag{A.18.2}$$

Substituting Eq. A.18.2 into Eq. A.18.1 leads to:

$$\left[p(p+1)-v(v+1)\right]a_0\sigma^{p-2}+\left[(p+1)(p+2)-v(v+1)\right]a_1\sigma^{p-1}$$

$$+\sum_{s=0}^{\infty}\left\{\left[(s+p+2)(s+p+3)-v(v+1)\right]a_{s+2}+a_s\right\}\sigma^{s+p}=0$$

$$\tag{A.18.3}$$

Since the series Eq. A.18.3 is an identity in σ, the coefficient of each power of v is separately equal to zero. There are but two nontrivial ways that the coefficients of σ^{p-2} and σ^{p-1} can both equal zero. One is if a_0 is equal to zero and $(p+1)(p+2)=v(v+1)$; the other is if a_1 is equal to zero and $p(p+1)=v(v+1)$. Arbitrarily making the second choice, the condition the equality is met either of two ways: $p=v$ or $p=-(v+1)$; the choice determines the two independent solutions.

For the case $p=v$ the terms in the curly brackets sum to zero. This gives the recursion relationship:

$$\frac{a_{s+2}}{a_s}=-\frac{1}{(s+2)(2v+s+3)} \tag{A.18.4}$$

This relationship, after redefining the dummy index, leads to the functional form of the radial function $R_\nu(\sigma)$:

$$R_\nu(\sigma) = a_0 \sum_{s=0}^{\infty} \frac{(-1)^s (2\nu+1)!!}{2^s s!(2\nu+2s+1)!!} \sigma^{\nu+2s} \tag{A.18.5}$$

Making the definition that $a_0 = 1/(2\nu+1)!!$ the result is:

$$j_\nu(\sigma) = \sum_{s=0}^{\infty} \frac{(-1)^s}{2^s s!(2\nu+2s+1)!!} \sigma^{\nu+2s} \tag{A.18.6}$$

Functions $j_\nu(\sigma)$ are the spherical Bessel functions of order ν. The functional limit at small values of σ follows from Eq. A.18.6, and is equal to:

$$\lim_{\sigma \to 0}\left[j_\nu(\sigma) \right] = \frac{\sigma^\nu}{(2\nu+1)!!} \tag{A.18.7}$$

For the case $p = -(\nu+1)$, Eq. A.18.5 yields the recursion relationship:

$$\frac{a_{s+2}}{a_s} = \frac{1}{(s+2)(2\nu-s-1)} \tag{A.18.8}$$

This expression leads to the series solution:

$$y_\nu(\sigma) = \frac{a_0}{\sigma^{\nu+1}}\left\{ \begin{aligned} &1 + \frac{\sigma^2}{2(2\nu-1)} + \frac{\sigma^4}{2\cdot4(2\nu-1)(2\nu-3)} + ... + \frac{\sigma^{2p}}{(2p)!!(2\nu-1)...(2\nu-2p-1)} + ...\\ &+ \frac{\sigma^{2(\nu-1)}}{(2\nu-2)!!(2\nu-1)!!} - \frac{\sigma^{2\nu}}{1\cdot(2\nu)!!(2\nu-1)!!} + \frac{\sigma^{2(\nu+1)}}{3\cdot(2\nu+2)!!(2\nu-1)!!} - ... \end{aligned} \right\} \tag{A.18.9}$$

The series is monotone for s less than ν and oscillatory for s greater than ν. The combination is best described by separate sums over the monotone and oscillatory portions. After using the definition $a_0 = 1/(2\nu+1)!!$ and again redefining the dummy index:

$$y_v(\sigma) = -\sum_{s=0}^{[v]} \frac{(2v-2s-1)!!}{(2s)!!\sigma^{v+1-2s}} - \sum_{s=0}^{\infty} \frac{(-1)^s}{(2s-1)!!} \frac{\sigma^{v-1+2s}}{(2v+2s)!!} \qquad \text{(A.18.10)}$$

The symbol $[v]$ indicates the largest integer less than v. The first sum of Eq. A.18.10 describes a monotone power series with inverse powers of v, powers that range upward from $-(v+1)$ to $(v+1)$ and a second sum that represents an alternating series with positive powers of v. The sums are the spherical Neumann functions; the small argument value follows from Eq. A.18.10 by inspection. For integer orders, v equal integer ℓ and $\sigma \ll 1$, by Eq. A.18.6 and the first of Eq. A.18.10. The two solutions are:

$$\lim_{\sigma\to 0}\left[j_\ell(\sigma)\right] = \frac{\sigma^\ell}{(2\ell+1)!!}; \quad \lim_{\sigma\to 0}\left[y_\ell(\sigma)\right] = -\frac{(2\ell-1)!!}{\sigma^{\ell+1}} \qquad \text{(A.18.11)}$$

All orders of the Bessel function are regular at the origin. Quite differently, the Neumann functions have an $(\ell+1)$ order singularity. The first term of Eq. A.18.10 shows all terms with negative powers of σ have the same sign, from which we observe the magnitude remains large through at least through $\sigma = 1$.

For integer orders, v equal integer ℓ and $\sigma \gg 1$, Eq. A.18.6 and the second of Eq. A.18.10 determine the functions. The solutions are:

$$j_\ell(\sigma) = \sum_{s=0}^{\infty} \frac{(-1)^s}{(2s)!!(2\ell+2s+1)!!} \sigma^{\ell+2s}$$

$$y_\ell(\sigma) \cong -\sum_{s=0}^{\infty} \frac{(-1)^s}{(2s-1)!!(2\ell+2s)!!} \sigma^{\ell-1+2s} \qquad \text{(A.18.12)}$$

Term-by-term comparison of the series representation of expansions for the trigonometric functions and A.18.12 shows that:

$$\lim_{\sigma\to\infty} j_\ell(\sigma) = \frac{1}{\sigma}\cos\left[\sigma - \frac{\pi}{2}(\ell+1)\right]$$

$$\lim_{\sigma\to\infty} y_\ell(\sigma) = \frac{1}{\sigma}\sin\left[\sigma - \frac{\pi}{2}(\ell+1)\right] \qquad \text{(A.18.13)}$$

Using Eq. A.18.13, it follows that the Bessel and Neumann functions are related as:

$$\lim_{\sigma\to\infty}\left\{j_\ell(\sigma)=\frac{1}{\sigma}\frac{d}{d\sigma}\left[\sigma y_\ell(\sigma)\right]\right\}$$
$$\lim_{\sigma\to\infty}\left\{y_\ell(\sigma)=-\frac{1}{\sigma}\frac{d}{d\sigma}\left[\sigma j_\ell(\sigma)\right]\right\} \qquad\text{(A.18.14)}$$

A.19 Hankel Functions, The Second Solution Form

If the separation constant is an integer, the spherical Bessel differential equation, Eq. 1.11.7, is:

$$\frac{1}{\sigma^2}\frac{d}{d\sigma}\left(\sigma^2\frac{dR}{d\sigma}\right)+\left(1-\frac{\ell(\ell+1)}{\sigma^2}\right)R=0 \qquad\text{(A.19.1)}$$

This is a second order equation with two independent solutions. Because of a singularity at the origin, solutions are satisfied over the region $0<\sigma\le\infty$. For solutions at long range, $\sigma^2 >> \ell(\ell+1)$, the differential equation goes to:

$$\frac{1}{\sigma^2}\frac{d}{d\sigma}\left(\sigma^2\frac{dR(\sigma)}{d\sigma}\right)+R(\sigma)=0 \qquad\text{(A.19.2)}$$

If Eq. A.19.2 were an exact solution the result would be an exponential with constant coefficients. Although Eq. A.19.2 is not exact, it is helpful to write its solution in the form:

$$R_\ell(\sigma)=F_\ell(\sigma)e^{-i\sigma}+G_\ell(\sigma)e^{i\sigma} \qquad\text{(A.19.3)}$$

A requirement is that at large radii $F_\ell(\sigma)$ and $G_\ell(\sigma)$ vary much less rapidly with increasing radius than do the exponentials. Also since $F_\ell(\sigma)$ and $G_\ell(\sigma)$ are complex conjugates it is necessary to solve for only one of them.

A convenient method to find $F_\ell(\sigma)$ is by a power series expansion. The series and the first two derivatives are:

$$R_\ell(\sigma) = F_\ell(\sigma)e^{-i\sigma}; \quad \frac{dR_\ell(\sigma)}{d\sigma} = \left[\frac{dF_\ell(\sigma)}{d\sigma} - iF_\ell(\sigma)\right]e^{-i\sigma}$$

$$\frac{d^2R_\ell(\sigma)}{d\sigma^2} = \left[\frac{d^2F_\ell(\sigma)}{d\sigma^2} - 2i\frac{dF_\ell(\sigma)}{d\sigma} - F_\ell(\sigma)\right]e^{-i\sigma}$$

$$\text{(A.19.4)}$$

Substituting Eq. A.19.4 into Eq. A.19.1 results in the differential equation:

$$\frac{d^2F_\ell(\sigma)}{d\sigma^2} + 2\left(\frac{1}{\sigma} - i\right)\frac{dF_\ell(\sigma)}{d\sigma} - \left(\frac{2i}{\sigma} + \frac{\ell(\ell+1)}{\sigma^2}\right)F_\ell(\sigma) = 0 \qquad \text{(A.19.5)}$$

The most convenient method of solving Eq. A.19.5 is with a power series expansion. The series and the first two derivatives are:

$$F_\ell(\sigma) = \sum_{s=0}^{\infty} a_s\sigma^{s+p}$$

$$\frac{dF_\ell(\sigma)}{d\sigma} = \sum_{s=0}^{\infty}(s+p)a_s\sigma^{s+p-1} \qquad \text{(A.19.6)}$$

$$\frac{d^2F_\ell(\sigma)}{d\sigma^2} = \sum_{s=0}^{\infty}(s+p)(s+p-1)a_s\sigma^{s+p-2}$$

Inserting Eq. A.19.6 into Eq. A.19.5 and gathering similar powers of σ gives:

$$\left\{\frac{p(p+1)-\ell(\ell+1)}{\sigma^{p-2}} + \sum_{s=0}^{\infty}\sigma^{s+p+1} \right. $$
$$\left. \times\left\{a_{s+1}\left[(s+p+1)(s+p+2)-\ell(\ell+1)\right]-2ia_s(s+p+1)\right\}\right\} = 0 \quad \text{(A.19.7)}$$

Since the series is an identity, the coefficient of each power of σ is equal to zero. It follows from the σ^{p-2} term that either $p = \ell$ or $p = -(\ell+1)$ and it follows from the square brackets that:

$$\frac{a_{s+1}}{a_s} = \frac{2i(s+p+1)}{(s+p+1)(s+p+2)-\ell(\ell+1)} \tag{A.19.8}$$

With the option $p = \ell$, in the limit as σ becomes infinite Eq. A.19.8 goes to:

$$\lim_{s\to\infty}\left(\frac{a_{s+1}}{a_s}\right) = \frac{2i}{s} \tag{A.19.9}$$

Equation A.19.9 is also the limiting form for a series expansion of $\exp(2i\sigma)$. Therefore, since $F_\ell(\sigma)$ varies more slowly with σ than $\exp(-i\sigma)$, the recursion relationship of Eq. A.19.8 is not an acceptable solution. Returning to the option $p = -(\ell + 1)$, Eq. A.19.7 goes to:

$$\frac{a_{s+1}}{a_s} = \frac{2i(\ell-s)}{(s+1)(2\ell-s)} \tag{A.19.10}$$

Since the progression of Eq. A.19.10 terminates at $s = \ell$ the power series truncates as a polynomial of highest order ℓ. The general term is:

$$\frac{a_{s+1}}{a_s} = \frac{(2i)^s \ell!(2\ell-s)!}{s!(2\ell)!(\ell-s)!} \tag{A.19.11}$$

The series results in solution $R_\ell(\sigma)$ where:

$$R_\ell(\sigma) = \frac{e^{-i\sigma}}{\sigma}\sum_{s=0}^{\ell}\frac{(2i)^s \ell!(2\ell-s)!}{s!(2\ell)!(\ell-s)!}a_0 \tag{A.19.12}$$

To characterize the solution substitute $p = \ell - s$, rewrite Eq. A.19.12 as a sum over p, then change the dummy index back to s. The result is:

$$R_\ell(\sigma) = a_0\frac{e^{-i\sigma}}{\sigma}\frac{\ell!(2i)^\ell}{(2\ell)!}\sum_{s=0}^{\ell}\frac{(\ell+s)!}{s!(\ell-s)!}\left(\frac{1}{2i\sigma}\right)^s \tag{A.19.13}$$

With the definition $a_0 = i(2\ell-1)!!$, the full solution is the function $h_\ell(\sigma)$ where:

$$h_\ell(\sigma) = \frac{i^{\ell+1}e^{-i\sigma}}{\sigma} \sum_{s=0}^{\ell} \frac{(\ell+s)!}{s!(\ell-s)!}\left(\frac{1}{2i\sigma}\right)^s \tag{A.19.14}$$

Function $h_\ell(\sigma)$ is a spherical Hankel function of the second kind. The real part is a spherical Bessel function and the negative of the imaginary part is a spherical Neumann function. By definition:

$$h_\ell(\sigma) = j_\ell(\sigma) - iy_\ell(\sigma) \tag{A.19.15}$$

The complex conjugate of Eq. A.19.14 is the second independent solution of the equation. It is a spherical Hankel function of the first kind. In this work, we shall be concerned primarily with Hankel functions of the second kind.

For vanishingly small values of σ the dominant term in Eq. A.19.14 is:

$$\lim_{\sigma\to 0}\left[h_\ell(\sigma)\right] = \frac{i(2\ell-1)!!}{\sigma^{\ell+1}} \tag{A.19.16}$$

As the radius increases without limit, the dominant term is:

$$\lim_{\sigma\to\infty}\left[h_\ell(\sigma)\right] = \frac{i^{\ell+1}}{\sigma}e^{-i\sigma} \tag{A.19.17}$$

This shows the function $[\sigma h_\ell(\sigma)]$ does not converge even at limitlessly large values of range, σ.

A.20 Tables of Spherical Bessel Functions

To evaluate spherical Bessel, Neumann, and Hankel functions, it is helpful to factor each function into rational and transcendental parts. We define rational functions $A_\ell(\sigma)$ and $B_\ell(\sigma)$ and combine them with the functions of Eq. A.19.14 and A.19.15 as:

$$j_\ell(\sigma) = \frac{1}{\sigma}\left\{B_\ell(\sigma)\cos\sigma + A_\ell(\sigma)\sin\sigma\right\}$$

$$y_\ell(\sigma) = \frac{1}{\sigma}\left\{-A_\ell(\sigma)\cos\sigma + B_\ell(\sigma)\sin\sigma\right\} \qquad (A.20.1)$$

$$h_\ell(\sigma) = \frac{1}{\sigma}\left\{B_\ell(\sigma) + iA_\ell(\sigma)\right\}e^{-i\sigma}$$

Comparison of the equations shows, with q equal to any integer, that:

$$A_\ell(\sigma) = \sum_{s=0}^{\ell} \frac{(\ell+s)!}{s!(\ell-s)!}\left(\frac{1}{2\sigma}\right)^s (-1)^{(\ell-s)/2}\,\delta(\ell+s,2q)$$

$$\qquad (A.20.2)$$

$$B_\ell(\sigma) = \sum_{s=0}^{\ell} \frac{(\ell+s)!}{s!(\ell-s)!}\left(\frac{1}{2\sigma}\right)^s (-1)^{(\ell-s+1)/2}\,\delta(\ell+s,2q+1)$$

Comparison of the equations yields values of the letter functions. The primary recursion relationship used to develop the table follows from Eq. A.18.6:

$$j_{\ell+2}(\sigma) = \frac{2\ell+3}{\sigma}j_{\ell+1}(\sigma) - j_\ell(\sigma) \qquad (A.20.3)$$

Related functions are obtained by operating on the radial function to obtain the special function:

$$h_\ell^\bullet(\sigma) = \frac{1}{\sigma}\frac{d}{d\sigma}\left[\sigma h_\ell(\sigma)\right] \qquad (A.20.4)$$

Carrying out the operation of Eq. A.20.4 on Eq. A.19.14 gives:

$$h_\ell^\bullet(\sigma) = \frac{ie^{-i\sigma}}{\sigma}\sum_{s=0}^{\ell}\frac{(\ell+s)!}{s!(\ell-s)!}\left(1-\frac{is}{\sigma}\right)\left(\frac{1}{2i\sigma}\right)^s \qquad (A.20.5)$$

Similarly to Eq. A.20.1, the related functions factor into rational and transcendental parts:

$$j_\ell^{\cdot}(\sigma) = \frac{1}{\sigma}\left\{D_\ell(\sigma)\cos\sigma + C_\ell(\sigma)\sin\sigma\right\}$$

$$y_\ell^{\cdot}(\sigma) = \frac{1}{\sigma}\left\{-C_\ell(\sigma)\cos\sigma + D_\ell(\sigma)\sin\sigma\right\} \tag{A.20.6}$$

$$h_\ell^{\cdot}(\sigma) = \frac{1}{\sigma}\left\{D_\ell(\sigma) + iC_\ell(\sigma)\right\}e^{-i\sigma}$$

Term-by-term comparison shows that:

$$\frac{dA_\ell(\sigma)}{d\sigma} = B_\ell(\sigma) + C_\ell(\sigma); \quad \frac{dB_\ell(\sigma)}{d\sigma} = D_\ell(\sigma) - A_\ell(\sigma) \tag{A.20.7}$$

It follows upon combining Eq. A.20.4 with Eq. A.18.7 and A.18.12 that in the limit of a vanishingly small radius:

$$\lim_{\sigma \to 0} j_\ell^{\cdot}(\sigma) = \frac{(\ell+1)\sigma^{\ell-1}}{(2\ell+1)!!}; \quad \lim_{\sigma \to 0} y_\ell^{\cdot}(\sigma) = \frac{\ell(2\ell-1)!!}{\sigma^{\ell+2}} \tag{A.20.8}$$

It follows similarly upon combining Eq. A.20.4 with Eq. A.18.14 that in the limit of an infinitely large radius:

$$\lim_{\sigma \to \infty} j_\ell^{\cdot}(\sigma) = -\frac{1}{\sigma}\sin\left[\sigma - \frac{\pi}{2}(\ell+1)\right]; \quad \lim_{\sigma \to \infty} y_\ell^{\cdot}(\sigma) = \frac{1}{\sigma}\cos\left[\sigma - \frac{\pi}{2}(\ell+1)\right]$$

$$\lim_{\sigma \to \infty} h_\ell^{\cdot}(\sigma) = j_\ell^{\cdot}(\sigma) + \frac{i}{\sigma}\frac{d}{d\sigma}\left[\sigma j_\ell^{\cdot}(\sigma)\right]$$

$$\tag{A.20.9}$$

Tables of functional values follow.

Table A.20.1. Table of values of letter functions.

ℓ	$A_\ell(\sigma)$	$B_\ell(\sigma)$	$C_\ell(\sigma)$	$D_\ell(\sigma)$
0	1	0	0	1
1	$1/\sigma$	-1	$-1/\sigma^2+1$	$1/\sigma$
2	$3/\sigma^2-1$	$-3/\sigma$	$-6/\sigma^3+3/\sigma$	$6/\sigma^2-1$
3	$15\sigma^3-6/\sigma$	$-15/\sigma^2+1$	$-45/\sigma^4+21/\sigma^2-1$	$45/\sigma^3-6/\sigma$
4	$105/\sigma^4-45/\sigma^2+1$	$-105/\sigma^3+10/\sigma$	$-420/\sigma^5+195/\sigma^3-10/\sigma$	$420/\sigma^4-55/\sigma^2+1$

Table A.20.2. Radial function equalities.

1. $A_\ell D_\ell - B_\ell C_\ell = 1$

2. $A_{\ell-1} + A_{\ell+1} = \dfrac{2\ell+1}{\sigma} A_\ell; \quad B_{\ell-1} + B_{\ell+1} = \dfrac{2\ell+1}{\sigma} B_\ell$

3. $C_\ell = -\dfrac{\ell}{\sigma} A_\ell + A_{\ell-1} = \dfrac{\ell+1}{\sigma} A_\ell - A_{\ell+1}$

   $D_\ell = -\dfrac{\ell}{\sigma} B_\ell + B_{\ell-1} = \dfrac{\ell+1}{\sigma} B_\ell - B_{\ell+1}$

4. $A_\ell B_{\ell-1} - A_{\ell-1} B_\ell = 1; \quad C_\ell D_{\ell-1} - C_{\ell-1} D_\ell = -\left(1 + \dfrac{\ell^2}{\sigma^2}\right)$

   $A_{\ell+1} D_\ell - B_{\ell+1} C_\ell = \dfrac{\ell+1}{\sigma}; \quad A_\ell D_{\ell+1} - B_\ell C_{\ell+1} = \dfrac{\ell+1}{\sigma}$

5. $A_{\ell+2} B_\ell - A_\ell B_{\ell+2} = \dfrac{2\ell+3}{\sigma}$

   $C_{\ell+2} D_\ell - C_\ell D_{\ell+2} = \dfrac{\ell+3}{\sigma}\left(1 - \dfrac{(\ell+1)\ell+2)}{\sigma^2}\right)$

Table A.20.3. Radial differential equalities.

1. $dA_\ell/d\sigma = C_\ell + B_\ell; \quad dB_\ell/d\sigma = D_\ell - A_\ell$

2. $dA_\ell/d\sigma = -\dfrac{\ell}{\sigma} A_\ell + A_{\ell-1} + B_\ell; \quad dB_\ell/d\sigma = -\dfrac{\ell}{\sigma} B_\ell + B_{\ell-1} - A_\ell$

3. $dA_\ell/d\sigma = \dfrac{\ell+1}{\sigma} A_\ell - A_{\ell+1} + B_\ell; \quad dB_\ell/d\sigma = \dfrac{\ell+1}{\sigma} B_\ell - B_{\ell+1} - A_\ell$

4. $dC_\ell/d\sigma = -\left(1 - \dfrac{\ell(\ell+1)}{\sigma^2}\right) A_\ell + D_\ell$

   $dD_\ell/d\sigma = -\left(1 - \dfrac{\ell(\ell+1)}{\sigma^2}\right) B_\ell - C_\ell$

5. $\dfrac{d}{d\sigma}\left(A_\ell B_n - B_\ell A_n + C_\ell D_n - D_\ell C_n\right)$

   $= \dfrac{1}{\sigma^2}\left[\ell(\ell+1)\left(A_\ell D_n - B_\ell C_n\right) - n(n+1)\left(A_n D_\ell - B_n C_\ell\right)\right]$

(Continued)

Table A.20.3. (*Continued*)

6.
$$\frac{d}{d\sigma}\left(A_\ell C_n - C_\ell A_n + B_\ell D_n - D_\ell B_n\right)$$
$$=\frac{1}{\sigma^2}\left[n(n+1)-\ell(\ell+1)\right]\left(A_\ell A_n + B_\ell B_n\right)$$

7.
$$\frac{d}{d\sigma}\left(A_\ell C_n + C_\ell A_n - B_\ell D_n - D_\ell B_n\right)$$
$$=\left\{\begin{array}{l}\dfrac{1}{\sigma^2}\left[n(n+1)-\ell(\ell+1)\right]\left(A_\ell A_n + B_\ell B_n\right)\\[2mm] +2\left(A_\ell D_n + A_n D_\ell + B_\ell C_n + B_n C_\ell\right)\\[2mm] -2\left(A_\ell A_n - B_\ell B_n - C_\ell C_n + D_\ell D_n\right)\end{array}\right\}$$

8.
$$\frac{d}{d\sigma}\left(A_\ell B_n + B_\ell A_n - C_\ell D_n - D_\ell C_n\right)$$
$$=\left\{\begin{array}{l}\dfrac{1}{\sigma^2}\left[\ell(\ell+1)\left(A_\ell D_n + B_\ell C_n\right)+n(n+1)\left(A_n D_\ell + B_n C_\ell\right)\right]\\[2mm] +2\left(A_\ell D_n + A_n D_\ell + B_\ell C_n + B_n C_\ell\right)\\[2mm] -2\left(A_\ell A_n - B_\ell B_n - C_\ell C_n + D_\ell D_n\right)\end{array}\right\}$$

9.
$$\frac{d}{d\sigma}\left(\frac{A_\ell B_\ell}{A_\ell^2 - B_\ell^2}\right)=\frac{\left(A_\ell^2 + B_\ell^2\right)}{\left(A_\ell^2 - B_\ell^2\right)^2}\left[1-\left(A_\ell^2 + B_\ell^2\right)\right]$$

Table A.20.4. Radial integral equalities.

1.
$$2\ell(\ell+1)\int\frac{d\sigma}{\sigma^2}\left(A_\ell C_\ell + B_\ell D_\ell\right)=\left(A_\ell^2 + B_\ell^2 + C_\ell^2 + D_\ell^2\right)$$

2.
$$\int d\sigma\left\{\left(-A_\ell^2 - B_\ell^2 + C_\ell^2 + D_\ell^2\right)+\left(A_\ell^2 + B_\ell^2\right)\left(\frac{\ell(\ell+1)}{\sigma^2}\right)\right\}$$
$$=\left(A_\ell C_\ell + B_\ell D_\ell\right)$$

3.
$$\int d\sigma\left\{\left[\left(A_\ell - D_\ell\right)^2 +\left(B_\ell + C_\ell\right)^2\right]+\left(A_\ell D_\ell + B_\ell C_\ell\right)\frac{\ell(\ell+1)}{\sigma^2}\right\}$$
$$=-\left(A_\ell B_\ell - C_\ell D_\ell\right)$$

(*Continued*)

Table A.20.4. (*Continued*)

4.
$$2\int d\sigma\left\{2(A_\ell-D_\ell)(B_\ell+C_\ell)-(A_\ell C_\ell-B_\ell D_\ell)\frac{\ell(\ell+1)}{\sigma^2}\right\}$$
$$=\left(A_\ell^2-B_\ell^2-C_\ell^2+D_\ell^2\right)$$

5.
$$\int d\sigma\left\{\left[-(A_\ell-D_\ell)^2+(B_\ell+C_\ell)^2\right]+\left(A_\ell^2-B_\ell^2\right)\left(\frac{\ell(\ell+1)}{\sigma^2}\right)\right\}$$
$$=\left(A_\ell C_\ell-B_\ell D_\ell\right)$$

6.
$$\ell(\ell+1)\int\frac{d\sigma}{\sigma^2}\left[A_\ell(A_\ell-D_\ell)-B_\ell(B_\ell+C_\ell)\right]=-(A_\ell-D_\ell)(B_\ell+C_\ell)$$

7.
$$2\int d\sigma\left\{-(A_\ell-D_\ell)(B_\ell+C_\ell)+A_\ell B_\ell\left(\frac{\ell(\ell+1)}{\sigma^2}\right)\right\}$$
$$=\left[A_\ell D_\ell+B_\ell C_\ell-(-1)^\ell\right]$$

8.
$$\int\frac{d\sigma}{\sigma^2}\left\{\ell(\ell+1)\left(A_\ell C_n+B_\ell D_n\right)+n(n+1)\left(A_n C_\ell+B_n D_\ell\right)\right\}$$
$$=\left(A_\ell A_n+B_\ell B_n+C_\ell C_n+D_\ell D_n\right)$$

A.21 Sums Over Spherical Functions

An electromagnetic field may be expressed as the product of spherical Bessel, Neumann, Hankel functions of σ, or linear combinations thereof, times linear combinations of Legendre functions of θ, times linear combinations of trigonometric functions of azimuth angle ϕ.

To assist in the analysis of such functions we seek a table of sums over product functions and to obtain them we begin with a z-directed plane wave. For that case there is no dependence upon ϕ and hence only functions with $m=0$ are present, and since a plane wave is regular on the z-axis only spherical Bessel functions are present. The result, expressed using spherical coordinates, is a sum over the product form:

$$e^{-i\sigma\cos\theta}=\sum_{\ell=0}^\infty a_\ell j_\ell(\sigma)P_\ell(\cos\theta)\tag{A.21.1}$$

To evaluate each constant a_ℓ multiply both sides by $P_n(\cos\theta)$ and integrate over the full range of zenith angle. The result is:

$$\frac{2j_\ell(\sigma)}{(2\ell+1)}a_\ell = \int_0^\pi \sin\theta\, d\theta P_\ell(\cos\theta)e^{-i\sigma\cos\theta} \qquad (A.21.2)$$

Differentiating both sides of the equation ℓ times with respect to σ then going to limitlessly small radius, see Eq. A.20.8, gives:

$$\frac{2a_\ell i^\ell}{(2\ell+1)}\frac{\ell!}{(2\ell+1)!!} = \int_0^\pi \sin\theta\, d\theta P_\ell(\cos\theta)\cos^\ell\theta \qquad (A.21.3)$$

The integral is listed in Table A.20.1. Integrating and then solving for a_ℓ gives:

$$a_\ell = i^{-\ell}(2\ell+1) \qquad (A.21.4)$$

Combining Eq. A.21.1 with A.21.4 gives:

$$e^{-i\sigma\cos\theta} = \sum_{\ell=0}^\infty i^{-\ell}(2\ell+1)j_\ell(\sigma)P_\ell(\cos\theta) \qquad (A.21.5)$$

Differentiating with respect to θ gives:

$$\sigma\sin\theta e^{-i\sigma\cos\theta} = \sum_{\ell=1}^\infty i^{1-\ell}(2\ell+1)j_\ell P_\ell^1(\cos\theta) \qquad (A.21.6)$$

Operating on Eq. A.21.5 and A.21.6 to form sums over the associated Bessel function gives:

$$-\left[i\left(\cos\theta+\frac{1}{\sigma}\right)e^{-i\sigma\cos\theta}+\frac{\cos\sigma}{\sigma}\right]=\sum_{\ell=1}^{\infty}i^{-\ell}\left(2\ell+1\right)j_{\ell}^{\bullet}\left(\sigma\right)P_{\ell}\left(\cos\theta\right) \quad \text{(A.21.7)}$$

$$-i\sigma e^{-i\sigma\cos\theta}=\sum_{\ell=1}^{\infty}i^{-\ell}\left(2\ell+1\right)j_{\ell}\frac{P_{\ell}^{1}\left(\cos\theta\right)}{\sin\theta}$$

$$-i\sigma\left(\cos\theta+\frac{2i}{\sigma}\right)e^{-i\sigma\cos\theta}=\sum_{\ell=1}^{\infty}i^{-\ell}\left(2\ell+1\right)j_{\ell}^{\bullet}\frac{P_{\ell}^{1}\left(\cos\theta\right)}{\sin\theta} \quad \text{(A.21.8)}$$

We next seek to evaluate the four angular sums contained within Eq. 7.5.1. To do so, note the equality of Eq. 7.5.1 and 7.5.2 yields:

$$\cos\theta e^{-i\sigma\cos\theta}=\sum_{\ell=1}^{\infty}i^{-\ell}\frac{\left(2\ell+1\right)}{\ell\left(\ell+1\right)}\left[ij_{\ell}^{\bullet}\left(\sigma\right)\frac{dP_{\ell}^{1}}{d\theta}+j_{\ell}\left(\sigma\right)\frac{P_{\ell}^{1}}{\sin\theta}\right]$$

$$e^{-i\sigma\cos\theta}=\sum_{\ell=1}^{\infty}i^{-\ell}\frac{\left(2\ell+1\right)}{\ell\left(\ell+1\right)}\left[ij_{\ell}^{\bullet}\left(\sigma\right)\frac{P_{\ell}^{1}}{\sin\theta}+j_{\ell}\left(\sigma\right)\frac{dP_{\ell}^{1}}{d\theta}\right] \quad \text{(A.21.9)}$$

The solution of this equation is aided with identities from Table A.15.1:

$$\frac{d}{d\theta}P_{\ell}^{1}\left(\cos\theta\right)=\frac{1}{2}\left[\ell\left(\ell+1\right)P_{\ell}\left(\cos\theta\right)-P_{\ell}^{2}\left(\cos\theta\right)\right]$$

$$\frac{P_{\ell}^{1}\left(\cos\theta\right)}{\sin\theta}=\frac{1}{2\cos\theta}\left[\ell\left(\ell+1\right)P_{\ell}\left(\cos\theta\right)+P_{\ell}^{2}\left(\cos\theta\right)\right] \quad \text{(A.21.10)}$$

Inserting these identities in Eq. A.21.9 yields:

$$\cos\theta e^{-i\sigma\cos\theta}=\frac{1}{2}\left\{\begin{array}{l}\displaystyle\sum_{\ell=1}^{\infty}i^{-\ell}\left(2\ell+1\right)\left[j_{\ell}\left(\sigma\right)\frac{P_{\ell}\left(\cos\theta\right)}{\cos\theta}+ij_{\ell}^{\bullet}\left(\sigma\right)P_{\ell}\left(\cos\theta\right)\right]\\[12pt]\displaystyle+\sum_{\ell=1}^{\infty}i^{-\ell}\frac{\left(2\ell+1\right)}{\ell\left(\ell+1\right)}\left[j_{\ell}\left(\sigma\right)\frac{P_{\ell}^{2}\left(\cos\theta\right)}{\cos\theta}-ij_{\ell}^{\bullet}\left(\sigma\right)P_{\ell}^{2}\left(\cos\theta\right)\right]\end{array}\right\} \quad \text{(A.21.11)}$$

$$e^{-i\sigma\cos\theta} = \frac{1}{2}\left\{ \begin{array}{l} \displaystyle\sum_{\ell=1}^{\infty} i^{-\ell}(2\ell+1)\left[j_\ell(\sigma)P_\ell(\cos\theta)+ij_\ell^{\bullet}(\sigma)\frac{P_\ell(\cos\theta)}{\cos\theta}\right] \\[4mm] \displaystyle -\sum_{\ell=1}^{\infty} i^{-\ell}\frac{(2\ell+1)}{\ell(\ell+1)}\left[j_\ell(\sigma)P_\ell^2(\cos\theta)+ij_\ell^{\bullet}(\sigma)\frac{P_\ell^2(\cos\theta)}{\cos\theta}\right] \end{array}\right\}$$

The sums on the upper lines of the above equations are linear, algebraic equations that contain the sums on the bottom lines as unknowns. Writing 'x' and 'y' for unknowns the equations have the form:

$$f_1(\sigma,\cos\theta)=\frac{1}{2}\left(\frac{x}{\cos\theta}-iy\right) \quad f_2(\sigma,\cos\theta)=-\frac{1}{2}\left(x+\frac{iy}{\cos\theta}\right) \qquad \text{(A.21.12)}$$

Solving gives:

$$\sum_{\ell=1}^{\infty} i^{-\ell}\frac{(2\ell+1)}{\ell(\ell+1)} j_\ell(\sigma)P_\ell^2(\cos\theta)=$$

$$-\left(e^{-i\sigma\cos\theta}+\frac{\sin\sigma}{\sigma}\right)-\frac{2i}{\sigma\sin^2\theta}\left[e^{-i\sigma\cos\theta}\cos\theta-(\cos\sigma\cos\theta-i\sin\sigma)\right]$$

$$\text{(A.21.13)}$$

Multiply each of Eq. A.21.10 on the left by the operator:

$$O\Rightarrow \sum_{\ell=1}^{\infty} i^{-\ell}\frac{(2\ell+1)}{\ell(\ell+1)} j_\ell(\sigma) \qquad \text{(A.21.14)}$$

This gives the equality set:

$$\sum_{\ell=1}^{\infty} i^{-\ell}\frac{(2\ell+1)}{\ell(\ell+1)}\frac{d}{d\theta}P_\ell^1(\cos\theta)=\frac{1}{2}\left[\sum_{\ell=1}^{\infty} i^{-\ell}(2\ell+1)P_\ell(\cos\theta)-\sum_{\ell=1}^{\infty} i^{-\ell}\frac{(2\ell+1)}{\ell(\ell+1)}P_\ell^2(\cos\theta)\right]$$

$$\sum_{\ell=1}^{\infty} i^{-\ell}\frac{(2\ell+1)}{\ell(\ell+1)}\frac{P_\ell^1(\cos\theta)}{\sin\theta}=\frac{1}{2\cos\theta}\left[\sum_{\ell=1}^{\infty} i^{-\ell}(2\ell+1)P_\ell(\cos\theta)+\sum_{\ell=1}^{\infty} i^{-\ell}\frac{(2\ell+1)}{\ell(\ell+1)}P_\ell^2(\cos\theta)\right]$$

$$\text{(A.21.15)}$$

We next define two new functions, $U(\sigma,\theta)$ and $V(\sigma,\theta)$ to be:

$$U(\sigma,\theta) = \frac{1}{\sigma \sin^2 \theta}\left[e^{-i\sigma\cos\theta} - (\cos\sigma - i\sin\sigma\cos\theta)\right]$$

$$V(\sigma,\theta) = \frac{1}{\sigma \sin^2 \theta}\left[e^{-i\sigma\cos\theta}\cos\theta - (\cos\sigma\cos\theta - i\sin\sigma)\right] \tag{A.21.16}$$

The sums on the right side of Eq. A.21.15 follow from Eq. A.21.6 and A.21.13 and give exact values for the angular sums:

$$\sum_{\ell=1}^{\infty} i^{-\ell}\frac{(2\ell+1)}{\ell(\ell+1)}j_\ell(\sigma)\frac{P_\ell^1}{\sin\theta} = -iU(\sigma,\theta)$$

$$\sum_{\ell=1}^{\infty} i^{-\ell}\frac{(2\ell+1)}{\ell(\ell+1)}j_\ell(\sigma)\frac{dP_\ell^1}{d\theta} = e^{-i\sigma\cos\theta} + iV(\sigma,\theta) \tag{A.21.17}$$

Operating on Eq. A.21.17 yields sums over the associated function:

$$\sum_{\ell=1}^{\infty} i^{-\ell}\frac{(2\ell+1)}{\ell(\ell+1)}j_\ell^\bullet(\sigma)\frac{P_\ell^1}{\sin\theta} = -iV(\sigma,\theta)$$

$$\sum_{\ell=1}^{\infty} i^{-\ell}\frac{(2\ell+1)}{\ell(\ell+1)}j_\ell^\bullet(\sigma)\frac{dP_\ell^1}{d\theta} = e^{-i\sigma\cos\theta}\cos\theta + iU(\sigma,\theta) \tag{A.21.18}$$

The expressions for sums over spherical Bessel functions are listed in Table A.21.1 and Table 7.6.3.

Table A.21.1. Solutions for sums of spherical Bessel functions.

$$S_{11}(\sigma,\theta) = \sigma\sin\theta e^{-i\sigma\cos\theta}$$

$$S_{21}(\sigma,\theta) = e^{-i\sigma\cos\theta} + iV \qquad S_{31}(\sigma,\theta) = -iU$$

$$S_{21}(\sigma,\theta) = \cos\theta e^{-i\sigma\cos\theta} + iU \quad S_{31}(\sigma,\theta) = -iV$$

Table A.21.2 contains a list of sums over several functions of spherical Bessel functions.

Table A.21.2. Exact sums with spherical Bessel and related functions.

1.
$$\sigma \sin\theta\, e^{-i\sigma\cos\theta} = \sum_{\ell=1}^{\infty} i^{1-\ell}\left(2\ell+1\right) j_\ell(\sigma) P_\ell^1(\cos\theta)$$

2.
$$-iU = \sum_{\ell=1}^{\infty} i^{-\ell}\,\frac{\left(2\ell+1\right)}{\ell\left(\ell+1\right)}\, j_\ell(\sigma)\,\frac{P_\ell^1(\cos\theta)}{\sin\theta}$$

3.
$$-iV = \sum_{\ell=1}^{\infty} i^{1-\ell}\,\frac{\left(2\ell+1\right)}{\ell\left(\ell+1\right)}\, j_\ell^{\bullet}(\sigma)\,\frac{P_\ell^1(\cos\theta)}{\sin\theta}$$

4.
$$e^{-i\sigma\cos\theta} + iV = \sum_{\ell=1}^{\infty} i^{-\ell}\,\frac{\left(2\ell+1\right)}{\ell\left(\ell+1\right)}\, j_\ell(\sigma)\,\frac{dP_\ell^1}{d\theta}$$

5.
$$e^{-i\sigma\cos\theta}\cos\theta + iU = \sum_{\ell=1}^{\infty} i^{1-\ell}\,\frac{\left(2\ell+1\right)}{\ell\left(\ell+1\right)}\, j_\ell^{\bullet}(\sigma)\,\frac{dP_\ell^1}{d\theta}$$

The relationship between spherical Bessel and Hankel functions in the asymptotic limit as σ becomes large is given in Eq. 7.5.5. Using it to operate on the values of Table A.21.1 gives the asymptotic sums over spherical Hankel functions. There is a problem, however: Hankel functions supports an $(\ell+1)$ order singularity and this requires an upper limit to possible orders; hence the summation is truncated at maximum value L. With σ_0 the largest range to be considered equality, Eq. 7.5.5 remains valid only if $L(L+1) \gg \sigma_0^2$ and therefore the sums are valid only at large but not limitlessly large range. With that limitation Table A.21.2 is transformed to that Table A.21.3:

Table A.21.3. Sums over spherical Hankel and related functions,
valid only at large ranges.

1.
$$\sigma\sin\theta\left(\left(1+\cos\theta\right)+\frac{2i}{\sigma}\right)e^{-i\sigma\cos\theta}$$
$$= \lim_{\sigma\to\infty}\sum_{\ell=1}^{L} i^{1-\ell}\left(2\ell+1\right) h_\ell(\sigma) P_\ell^1(\cos\theta)$$

2.
$$-i\left(U+V\right) = \lim_{\sigma\to\infty}\sum_{\ell=1}^{L} i^{-\ell}\,\frac{\left(2\ell+1\right)}{\ell\left(\ell+1\right)}\, h_\ell(\sigma)\,\frac{P_\ell^1}{\sin\theta}$$

(Continued)

Table A.21.3. (*Continued*)

3.
$$\frac{i}{\sigma}e^{-i\sigma\cos\theta}-i(U+V)=\lim_{\sigma\to\infty}\sum_{\ell=1}^{L}i^{1-\ell}\frac{(2\ell+1)}{\ell(\ell+1)}h_\ell^\bullet(\sigma)\frac{P_\ell^1}{\sin\theta}$$

4.
$$\left[i(U+V)+(1+\cos\theta)e^{-i\sigma\cos\theta}\right]$$
$$=\lim_{\sigma\to\infty}\sum_{\ell=1}^{L}i^{-\ell}\frac{(2\ell+1)}{\ell(\ell+1)}h_\ell(\sigma)\frac{dP_\ell^1}{d\theta}$$

5.
$$\left[i(U+V)+\cos\theta\left((1+\cos\theta)+\frac{i}{\sigma}\right)e^{-i\sigma\cos\theta}\right]$$
$$=\lim_{\sigma\to\infty}\sum_{\ell=1}^{L}i^{1-\ell}\frac{(2\ell+1)}{\ell(\ell+1)}h_\ell^\bullet(\sigma)\frac{dP_\ell^1}{d\theta}$$

Values of product sums on the coordinate axes follow from the above. What follows is an example using spherical Bessel functions. Taking derivates with respect to σ evaluates sums over technique extends to the associated Bessel function. Operating on both functions gives the asymptotic, long range values of spherical and associated Neumann functions and, of course, the sums provide long-range values over spherical Hankel functions.

Taking the product of Eq. A.21.5 times its own complex conjugate, integrating, and then collecting terms gives:

$$1-\frac{\sin^2\sigma}{\sigma^2}=\sum_{\ell=1}^{\infty}(2\ell+1)\left[j_\ell(\sigma)\right]^2 \tag{A.21.19}$$

The fraction on the left side comes after changing the lower limit on the sum. Differentiation of Eq. A.21.5 by σ gives the equalities:

$$-i\cos\theta e^{-i\sigma\cos\theta}=\sum_{\ell=0}^{\infty}i^{-\ell}(2\ell+1)\frac{dj_\ell(\sigma)}{d\sigma}P_\ell(\cos\theta)$$

$$-\cos^2\theta e^{-i\sigma\cos\theta}=\sum_{\ell=0}^{\infty}i^{-\ell}(2\ell+1)\frac{d^2j_\ell(\sigma)}{d\sigma^2}P_\ell(\cos\theta)$$

(A.21.20)

Multiplying each by its own complex conjugate, integrating, and then collecting terms gives:

$$\frac{1}{3}-\left(\frac{\cos\sigma}{\sigma}-\frac{\sin\sigma}{\sigma^2}\right)^2=\sum_{\ell=1}^{\infty}(2\ell+1)\left[\frac{dj_\ell(\sigma)}{d\sigma}\right]^2$$

$$\frac{1}{5}-\left(2\frac{\cos\sigma}{\sigma^2}+\frac{\sin\sigma}{\sigma}\left\langle1-\frac{2}{\sigma^2}\right\rangle\right)^2=\sum_{\ell=1}^{\infty}(2\ell+1)\left[\frac{d^2j_\ell(\sigma)}{d\sigma^2}\right]^2 \tag{A.21.21}$$

The terms in parentheses on the left side arise by changing the lower limit of the sums from zero to one.

Still other related sums follow from Eq. A.21.5 by differentiating with respect to θ and using Table A.21.1:

$$\sigma e^{-i\sigma}=\sum_{\ell=1}^{\infty}i^{1-\ell}(2\ell+1)j_\ell(\sigma)\frac{P_\ell^1(\cos\theta)}{\sin\theta} \tag{A.21.22}$$

Evaluating Eq. A.21.5 on the positive *z*-axis gives the three series:

$$e^{-i\sigma}=\sum_{\ell=0}^{\infty}i^{-\ell}(2\ell+1)j_\ell(\sigma)$$

$$\sin\sigma=\sum_{\ell o;1}^{\infty}(-1)^{(\ell-1)/2}(2\ell+1)j_\ell(\sigma) \tag{A.21.23}$$

$$\cos\sigma=\sum_{\ell e;0}^{\infty}(-1)^{\ell/2}(2\ell+1)j_\ell(\sigma)$$

Subscripts "o;1" and "e;0" indicate respectively odd integers beginning with one and even integer beginning with zero.

Evaluation of Eq. A.21.5 and A.21.6 at $\theta=\pi/2$, see Table A.20.1, gives:

$$1=\sum_{\ell e;0}^{\infty}(2\ell+1)\frac{(\ell-1)!!}{(\ell!!)^2}j_\ell(\sigma)$$

$$\sigma=\sum_{\ell o;1}^{\infty}(2\ell+1)\frac{(\ell)!!}{(\ell-1)!!^2}j_\ell(\sigma) \tag{A.21.24}$$

Application of Eq. A.20.4 to Eq. A.21.8 gives:

$$\frac{1}{\sigma} = \sum_{\ell e;0}^{\infty} (2\ell+1) \frac{\ell!!}{(\ell!!)^2} \overset{\bullet}{j_\ell}(\sigma)$$

$$1 = \sum_{\ell o;1}^{\infty} \frac{(2\ell+1)}{2} \frac{\ell!!}{(\ell-1)!!^2} \overset{\bullet}{j_\ell}(\sigma)$$

$$(A.21.25)$$

Integrating Eq. A.21.21 over θ gives:

$$\left[\frac{e^{-i\sigma\cos\theta}}{\sigma} \right]_{\theta_1}^{\theta_2} = \sum_{\ell=0}^{\infty} i^{-\ell-1} j_\ell(\sigma) \left[P_{\ell+1}(\cos\theta) - P_{\ell-1}(\cos\theta) \right]_{\theta_1}^{\theta_2} \tag{A.21.26}$$

Evaluation of Eq. A.21.26 between limits $\theta = \pi$ and $\pi/2$ gives:

$$\left(\frac{1-e^{i\sigma}}{\sigma} \right) = \sum_{\ell=0}^{\infty} i^{-\ell-1} j_\ell(\sigma) \left[P_{\ell+1}(0) - P_{\ell-1}(0) \right] - i j_0(\sigma) \tag{A.21.27}$$

Evaluation of Eq. A.21.26 between limits $\theta = \pi/2$ and 0 gives:

$$\left(\frac{e^{-i\sigma}-1}{\sigma} \right) = -i j_0(\sigma) - \sum_{\ell=0}^{\infty} i^{-\ell-1} j_\ell(\sigma) \left[P_{\ell+1}(0) - P_{\ell-1}(0) \right] \tag{A.21.28}$$

Subtracting Eq. A.21.25 from Eq. A.21.26 gives:

$$\frac{\cos\sigma-1}{\sigma} = \sum_{\ell=0}^{\infty} (2\ell+1) \frac{(\ell-1)!}{(\ell-1)!!(\ell+1)!!} j_\ell(\sigma) \tag{A.21.29}$$

Operating on Eq. A.21.29 using Eq. A.20.4 gives:

$$\frac{\sin\sigma}{\sigma} = \sum_{\ell=0}^{\infty} (2\ell+1) \frac{(\ell-1)!}{(\ell-1)!!(\ell+1)!!} \overset{\bullet}{j_\ell}(\sigma) \tag{A.21.30}$$

Evaluation on the positive z-axis gives:

$$\sigma e^{-i\sigma} = \frac{1}{2} \sum_{\ell=1}^{\infty} i^{1-\ell} (2\ell+1)\ell(\ell+1) j_\ell(\sigma) \qquad (A.21.31)$$

The exponential contains the two equations:

$$\sigma \cos\sigma = \frac{1}{2} \sum_{\ell o;1}^{\infty} (-1)^{(\ell-1)/2} (2\ell+1)\ell(\ell+1) j_\ell(\sigma)$$

$$\sigma \sin\sigma = \frac{1}{2} \sum_{\ell e;2}^{\infty} (-1)^{\ell/2} (2\ell+1)\ell(\ell+1) j_\ell(\sigma)$$

$$(A.21.32)$$

Table A.21.4 contains a selected list of twelve axial sums:

Table A.21.4. Table of sums over spherical Bessel functions.

1.
$$\frac{1-\cos\sigma}{\sigma} = \sum_{\ell o;1}^{\infty} (2\ell+1)\frac{(\ell-2)!!}{(\ell+1)!!} j_\ell(\sigma)$$

2.
$$\frac{\sin\sigma}{\sigma} = \sum_{\ell=o;1}^{\infty} (2\ell+1)\frac{(\ell-2)!!}{(\ell+1)!!} \dot{j}_\ell(\sigma)$$

3.
$$1 = \sum_{\ell e;0}^{\infty} (2\ell+1)\frac{(\ell-1)!!}{\ell!!} j_\ell(\sigma)$$

4.
$$\frac{1}{\sigma} = \sum_{\ell e;0}^{\infty} (2\ell+1)\frac{(\ell-1)!!}{(\ell)!!} \dot{j}_\ell(\sigma)$$

5.
$$\sigma = \sum_{\ell o;1}^{\infty} (2\ell+1)\frac{(\ell)!!}{(\ell-1)!!} j_\ell(\sigma)$$

6.
$$1 = \sum_{\ell o;1}^{\infty} \frac{(2\ell+1)}{2}\frac{(\ell)!!}{(\ell-1)!!} \dot{j}_\ell(\sigma)$$

7.
$$\sigma e^{-i\sigma} = \frac{1}{2} \sum_{\ell=1}^{\infty} i^{1-\ell} (2\ell+1)\ell(\ell+1) j_\ell(\sigma)$$

8.
$$(2-i\sigma)e^{-i\sigma} = \frac{1}{2} \sum_{\ell=1}^{\infty} i^{1-\ell} (2\ell+1)(\ell+1) \dot{j}_\ell(\sigma)$$

(Continued)

Table A.21.4. (*Continued*)

9.
$$1 = \sum_{\ell=0}^{\infty} (2\ell+1)\left[j_\ell(\sigma) \right]^2$$

10.
$$\left[1/\sigma^2 + 1/3 \right] = \sum_{\ell=0}^{\infty} (2\ell+1)\left[j_\ell^{\bullet}(\sigma) \right]^2$$

11.
$$-1/3 = \sum_{\ell=0}^{\infty} (2\ell+1)\left[j_\ell(\sigma) y_\ell^{\bullet}(\sigma) \right]$$

12.
$$\frac{1}{\sigma} = \sum_{\ell=0}^{\infty} (2\ell+1) j_\ell(\sigma) j_\ell^{\bullet}(\sigma)$$

A.22 Static Scalar Potentials

To examine physical sources of TM modal coefficients $F(\ell,m)$, we start with the static scalar potential field, $\Phi(r)$. By Eq. 1.10.9 the differential equation that governs the static scalar potential is:

$$\nabla^2 \Phi = \rho/\varepsilon \qquad\qquad\qquad (A.22.1)$$

This equation shows the scalar potential has non-zero spatial curvature only where electric charges exist and a field function exists only if the curvature is other than zero. Therefore static scalar potentials arise only from electric charges.

To characterize such potentials we establish an origin near or in a region that contains static electric charges. Based upon that origin, the coordinates within the charged region are $r(x',y',z')$ and the coordinates at the point where the potential is to be evaluated are $r(x,y,z)$; the field point may be either interior or exterior to the charge-containing region. The distance from the source to the field point is R where, by definition:

$$R(r,r') = \left[(x-x')^2 + (y-y')^2 + (z-z')^2 \right]^{1/2} \qquad (A.22.2)$$

The potential at an arbitrary field point was previously calculated, Eq. 1.5.9, and is given by:

$$\Phi(r) = \frac{1}{4\pi\varepsilon} \int \frac{\rho(r')}{R} dV'$$

(A.22.3)

It is convenient to work with distance from the origin to field point r, but the function in the denominator of Eq. A.22.3 is the distance from the differential source point to field point R. To replace $1/R$ by a function of $1/r$, use the Taylor series expansion:

$$\frac{1}{R} = \left\{ \begin{array}{l} \frac{1}{r} + x_i \frac{\partial}{\partial x_i}\left(\frac{1}{R}\right)_r + \frac{1}{2}x_i x_j \frac{\partial}{\partial x_i}\frac{\partial}{\partial x_j}\left(\frac{1}{R}\right)_r + \frac{1}{6}x_i x_j x_k \frac{\partial}{\partial x_i}\frac{\partial}{\partial x_j}\frac{\partial}{\partial x_k}\left(\frac{1}{R}\right)_r \\ + \frac{1}{24}x_i x_j x_k x_m \frac{\partial}{\partial x_i}\frac{\partial}{\partial x_j}\frac{\partial}{\partial x_k}\frac{\partial}{\partial x_m}\left(\frac{1}{R}\right)_r + \dots \end{array} \right\}$$

(A.22.4)

Placing the expansion of Eq. A.22.4 into Eq. A.22.3 results in the desired form:

$$\Phi(r) = \frac{1}{4\pi\varepsilon} \left\{ \begin{array}{l} \left(\frac{1}{r}\right)\int \rho(r')dV + \frac{\partial}{\partial x_i}\left(\frac{1}{R}\right)_r \int x_i' \rho(r')dV' \\ + \frac{1}{2}\frac{\partial}{\partial x_i}\frac{\partial}{\partial x_j}\left(\frac{1}{R}\right)_r \int x_i' x_j' \rho(r')dV' \\ + \frac{1}{6}\frac{\partial}{\partial x_i}\frac{\partial}{\partial x_j}\frac{\partial}{\partial x_k}\left(\frac{1}{R}\right)_r \int x_i' x_j' x_k' \rho(r')dV' + \dots \end{array} \right\}$$

(A.22.5)

The first term of Eq. A.22.5 has a first order singularity at the origin:

$$\Phi_0(r,\theta,\phi) = \frac{1}{4\pi\varepsilon}\frac{q}{r} \quad \text{where} \quad q = \int \rho(r')dV'$$

(A.22.6)

Succeeding terms have successively higher order singularities. Fields associated with succeeding singularities are generated by equal values of positive and negative charge, spaced incremental distances apart. The total potential is the sum of that from each charge and, in the limit as the inter-charge spacing goes to zero, the mathematical affect on the potential is the same as obtained by differentiating with respect to x_i. For example, for the special case of charge separation z_0 all differentials in Eq. A.22.5 are z-directed.

Consider the multipolar electric moment of a charge distribution with ℓ displacements. The moment is of order ℓ and degree m:

$$p_\ell^m = \int \rho(\mathbf{r}') x_i' x_j' \cdots x_k' \, dV' \qquad (A.22.7)$$

The number of charges involved designates the moment: moment p_ℓ^m, where ℓ is the order, is formed by 2^ℓ charges. Charges are arrayed according to the coefficients of a binomial expansion of the same order; $(\ell\!-\!m)$ have z-directed displacements, (s) have x-directed displacements, and $(m\!-\!s)$ have y-directed displacements. Easily verified equalities satisfied by the Legendre polynomials are:

$$P_\ell(\cos\theta) = \frac{1}{\ell!}\frac{\partial^\ell}{\partial z^\ell}\left(\frac{1}{R}\right)_r \; ; \; P_\ell^1(\cos\theta)\sin\phi = \frac{1}{(\ell-1)!}\frac{\partial^\ell}{\partial y \partial z^{\ell-1}}\left(\frac{1}{R}\right)_r \qquad (A.22.8)$$

Consider, as examples, structures that generate the lowest order multipolar moments. A dipole consists of two discrete charges: charge q at $z_0/2$ and charge $-q$ at $-z_0/2$; the volume integral, Eq. A.22.7, over order one is qz_0 and over any even order is zero. A linear quadrupole consists of four discrete charges: charges q at $+z_0$ and $-z_0$ and charge $-2q$ at the origin; the volume integral over order two is $2qz_0^2$ and over any odd order is zero. A linear octupole consists of eight discrete charges: charge q at $3z_0/2$, $-3q$ at $z_0/2$, $3q$ at $-z_0/2$, $-q$ at $-3z_0/2$. For a source of order three, the volume integral is $qz_0^3/4$. The same integral over any even order is zero. The volume integral over order one is zero but the volume integral for odd orders greater than three is not.

In all cases, the volume integral of Eq. A.22.7 is zero if the charge distribution and the displacements have opposite parity. It is also equal to zero if there are fewer charges than the number of displacements. In all other cases, the integral is non-zero. Tables A.22.1 and A.22.2 show some basic features of common multipolar electric moments. In each table: column one shows the order of the source, column two shows the discrete charge distribution that generates that order of singularity, column three shows the volume integral of Eq. A.22.7, and column four shows the lowest non-vanishing order of the potential. With the aid of the static portion of Eq. A.22.6, column five shows the radial components of the generated electric field intensity. Table A.22.1 is for only z-directed displacements and Table A.22.2 is for one y- and $(\ell-1)$ z-directed displacements. Table A.22.1 uses scalar charge q as the unit cell and Table A.22.2 uses a y-directed dipole as the unit cell. For the two cases the multipolar moments are:

$$p_\ell = \ell!qz_0^{\ \ell} \quad \text{and} \quad p_\ell^1 = (\ell-1)!qy_0z_0^{\ \ell-1}$$

The scalar potential of an arbitrary charge distribution is the simple sum of values obtained from each moment:

$$\Phi(r,\theta,\phi) = \frac{1}{4\pi\varepsilon}\sum_{\ell=0}^{\infty}\sum_{m=-\ell}^{\ell}\frac{p_\ell^m}{r^{\ell+1}}P_\ell^m(\cos\theta)e^{-jm\phi} \tag{A.22.9}$$

Table A.22.1. Electrostatic potentials of a linear array of ℓ spaced charges.

ℓ	p_ℓ	$4\pi\varepsilon\Phi_\ell(r,\theta)$	$4\pi\varepsilon E_{r\ell}(r,\theta)$
0	$+q$ at 0	q	$\dfrac{q}{r^2}$
1	$-q$ at $-z_0/2$ $+q$ at $z_0/2$	$p_1 = qz_0$	$\dfrac{2p_1}{r^3}P_1(\cos\theta)$
2	$+q$ at $-z_0$ $-2q$ at 0 $+q$ at z_0	$p_2 = 2p_1z_0$ $= 2qz_0^2$	$\dfrac{3p_2}{r^4}P_2(\cos\theta)$
3	$-q$ at $-3z_0/2$; $+3q$ at $-z_0/2$; $-3q$ at $z_0/2$ $+q$ at $3z_0/2$	$p_3 = 3p_2z_0$ $= 6qz_0^3$	$\dfrac{4p_3}{r^5}P_3(\cos\theta)$
4	$+q$ at $-2z_0$ $-4q$ at $-z_0$ $+6q$ at 0 $-4q$ at z_0; $+q$ at $2z_0$	$p_4 = 4p_3z_0$ $= 24qz_0^4$	$\dfrac{5p_4}{r^6}P_4(\cos\theta)$

Table A.22.2. Electrostatic potentials of a linear array of $(\ell-1)$ spaced charges and one transverse charge.

ℓ	Charge and Sites	p_ℓ^1	$\dfrac{4\pi\varepsilon}{\sin\phi}\Phi_\ell^1(r,\theta,\phi)$	$\dfrac{4\pi\varepsilon}{\sin\phi}\mathrm{E}_{r\ell}(r,\theta,\phi)$
1	$+q$ at $y_0/2$ $-q$ at $-y_0/2$	$p_1^1 = qy_0$	$\dfrac{p_1^1}{r^2}P_1^1(\cos\theta)$	$\dfrac{2p_1^1}{r^3}P_1^1(\cos\theta)$
2	$+q$ at $(y_0+z_0)/2$ $-q$ at $(-y_0+z_0)/2$ $+q$ at $-(y_0+z_0)/2$ $-q$ at $(y_0-z_0)/2$	$p_2^1 = p_1^1 z_0$	$\dfrac{p_2^1}{r^3}P_2^1(\cos\theta)$	$\dfrac{3p_2^1}{r^4}P_2^1(\cos\theta)$
3	$+q$ at $(y_0\pm2z_0)/2$ $-2q$ at $y_0/2$ $+2q$ at $-y_0/2$ $-q$ at $(-y_0\pm2z_0)/2$	$p_3^1 = 2p_2^1 z_0$	$\dfrac{p_3^1}{r^4}P_3^1(\cos\theta)$	$\dfrac{4p_3^1}{r^5}P_3^1(\cos\theta)$
4	$+q$ at $\pm(y_0+3z_0)/2$ $-3q$ at $\pm(y_0+z_0)/2$ $-q$ at $\pm(y_0-3z_0)/2$ $3q$ at $\pm(y_0-z_0)/2$	$p_4^1 = 3p_3^1 z_0$	$\dfrac{p_4^1}{r^5}P_4^1(\cos\theta)$	$\dfrac{5p_4^1}{r^6}P_4^1(\cos\theta)$

This is the static scalar potential at an arbitrary, exterior field point due to the charge distribution. The radial component of the electric field intensity follows from Eq. A.22.9 and is equal to:

$$E_r(r,\theta,\phi) = \frac{1}{4\pi\varepsilon}\sum_{\ell=0}^{\infty}\sum_{m=-\ell}^{\ell}\frac{(\ell+1)p_\ell^m}{r^{\ell+2}}P_\ell^m(\cos\theta)e^{-jm\phi} \qquad (A.22.10)$$

This is the radial field on a circumscribing sphere if the radius of that sphere is vanishingly small. Comparison of the coefficients of Eq. 1.13.9 and A.22.10 in the limit as the frequency goes to zero gives:

$$E_r(r,\theta,\phi) = -\sum_{\ell=0}^{\infty}\sum_{m=0}^{\ell}i^{-\ell}F(\ell,m)\ell(\ell+1)\frac{(2\ell-1)!!}{\sigma^{\ell+2}}P_\ell^m(\cos\theta)e^{-jm\phi}$$

$$(A.22.11)$$

Comparison of Eq. A.22.10 and A.22.11 gives:

$$F(\ell,m) = -\frac{1}{4\pi\varepsilon}\frac{i^\ell k^{\ell+2}p_\ell^m}{\ell(2\ell-1)!!}$$

(A.22.12)

For the special case of a z-directed electric dipole of moment p_1, it follows from the third row of Table A.22.1 that:

$$\Phi_1 = \frac{p_1}{4\pi\varepsilon r^2}\cos\theta \quad \text{and} \quad E_r = \frac{2p_1}{4\pi\varepsilon r^3}\cos\theta$$

(A.22.13)

A.23 The Blind Men and the Elephant

John Godfrey Saxe's Famous Indian Legend, c.1873

It was six men of Indostan
To learning much inclined,
Who went to see the Elephant ~ (Though all of them were blind),
That each by observation ~ Might satisfy his mind.

The First approached the Elephant,
And happening to fall
Against his broad and sturdy side, ~ At once began to bawl:
"God bless me! but the Elephant ~ Is very like a wall!"

The Second, feeling of the tusk,
Cried, "Ho! what have we here?
So very round and smooth and sharp? ~ To me 'tis mighty clear
This wonder of an Elephant ~ Is very like a spear!"

The Third approached the animal,
And happening to take
The squirming trunk within his hands, ~ Thus boldly up and spake:
"I see," quoth he, "the Elephant ~ Is very like a snake!"

The Fourth reached out an eager hand,
And felt about the knee.
"What most this wondrous beast is like ~ Is mighty plain," quoth he;
"'Tis clear enough the Elephant ~ Is very like a tree!"

The Fifth who chanced to touch the ear,
Said: "E'en the blindest man
Can tell what this resembles most; ~ Deny the fact who can,
This marvel of an Elephant ~ Is very like a fan!"

The Sixth no sooner had begun
About the beast to grope,
Than, seizing on the swinging tail ~ That fell within his scope, "I see,"
quoth he, "the Elephant ~ Is very like a rope!"

And so these men of Indostan
Disputed loud and long,
Each in his own opinion ~ Exceeding stiff and strong,
Though each was partly in the right ~ And all were in the wrong!

power density distribution, 280
power in fields, 61
power-frequency proportionality, 307
Poynting vector, 19
Poynting's theorem, 25, 30, 52, 75
projectiles, 310
push or pull on a scatterer, 299

Q of time varying systems, 57
Q-bandwidth relationship, 59
quantized energy exchanges, 174
quantized radiation, 307
quantum mechanics, 147
quantum operators, 161
quantum theory, 147, 309

radar cross section, 78
radial fields, 309
radiated angular momentum, 231
radiated energy, 230
radiated momentum, 230
radiating electric dipole, 210
radiation expansion force, 224
radiation fields, 43
radiation Q, xvii, 43, 209, 213, 229
radiation reactance, 61
radiation reaction braking force,, 215
radiation reaction force, 215, 248
radiation reaction forces, 43
radiation resistance, 61
Rayleigh region, 84
Rayleigh-Jeans formula, 177
reactance, 63
reaction force, on a radiating electric
    dipole, 220
reactive power, 43, 61, 209
reactive power density, 65
reactive radiation reaction force, 43,
    216
real and reactive radiation reaction
    forces, 217
real power, 65
receiving biconical antenna, 116
regenerative feedback, 261
regenerative field buildup, 279
regeneratively, 307

retarded time, 12, 61
Ritz power-frequency combination law,
    256
rotational symmetry, 61

scattering by a sphere, 73
scattering coefficients, 82
scattering cross section, 77, 145
scattering spheres, specific examples,
    81
Schelkunoff, 147, 305, 315
Schrödinger, xvi, 148, 306, 312, 315
Schrödinger's time-independent
    differential equation, 155
selected historical developments, 309
self-consistent fields, 235
single wave function, 302
solenoidal currents, 46
source modes resulting from an electric
    dipole, 264
speed of light, 3, 28
spherical Bessel function, 74
spherical Bessel functions, 36, 387
spherical coordinates, 34, 71, 225
spherical Hankel function, 74
spherical Hankel functions, 36, 376,
    387
spherical Laplacian operator, 170
spherical Neumann functions, 36, 387
spherical scatterer, 74
spherical shell dipole, 339
spontaneous emission, 255, 277, 308
standing energy density, 212
standing energy field, 211
standing wave pattern, 301
standing waves, 309
statistical nature of radiation onset, 314
stimulated photon emission, 309
stress in a dipole radiation field, 221
stress tensor, 222
sums over spherical functions, 387
superheterodyne radio receiver, 314
superimposed fields, 309
superregenerative feedback, 280, 315
superregenerative modal buildup, 256
superregenerative response, 303